Global Catastrophes and Trends

Also by Vaclav Smil

China's Energy

Energy in the Developing World (editor, with W.E. Knowland)

Energy Analysis in Agriculture (with P. Nachman and T.V. Long II)

Biomass Energies

The Bad Earth

Carbon—Nitrogen—Sulfur

Energy Food Environment

Energy in China's Modernization

General Energetics

China's Environmental Crisis

Global Ecology

Energy in World History

Cycles of Life

Energies

Feeding the World

Enriching the Earth

The Earth's Biosphere

Energy at the Crossroads

Creating the 20th Century

Transforming the 20th Century

Energy: A Beginner's Guide

Energy in Nature and Society

Oil: A Beginner's Guide

Global Catastrophes and Trends

The Next 50 Years

Vaclav Smil

The MIT Press
Cambridge, Massachusetts
London, England

First MIT Press paperback edition, 2012

For information about special quantity discounts, please email special_sales@mitpress.mit.edu

This book was set in Sabon by SNP Best-set Typesetter Ltd., Hong Kong.
Printed on recycled paper and bound in the United States of America.

Library of Congress Cataloging-in-Publication Data

Smil, Vaclav.
Global catastrophes and trends: the next fifty years / Vaclav Smil.
p. cm.
Includes bibliographical references and index.
ISBN 978-0-262-19586-7 (hc : alk. paper)—978-0-262-51822-2 (pb.)
1. Natural disasters. 2. Environmental risk assessment. I. Title.
GB5014.S58 2008
363.34—dc22

2007046675

10 9 8 7 6

Contents

Preface

What to Expect

Some lifelong endeavors, many old (and later resurrected) skills, and a great deal of new work have gone into this book. As a scientist, I have been always interested in global environmental change, and in natural catastrophes and anthropogenic risks (particularly in the failures of modern techniques) and the quantification of their probabilities. My study of unfolding national trends has been made easier by my personal experiences and fondness for languages. As a European who emigrated first to the United States and then to Canada and who has frequently visited Asia, I have decades of direct experience with most of the societies whose fortunes will shape the global future of the twenty-first century.

Although my dominant research interests have shifted during the past 40 years, I have always followed European, Russian, and Middle Eastern affairs. For two decades I have studied China's energy use and environment, with frequent visits to the country, usually combined with stays in Japan. During my undergraduate days at the Faculty of Natural Sciences at the Carolinum University in Prague in the early 1960s, I developed a distaste for rigid compartmentalization of knowledge. Ever since that time I have tried to understand complex environmental and engineered systems as they interact with social and economic forces; hence my keen interest in history, demography, and economics. Many of my publications could be assigned to these categories. My interest in risk assessment and patterns of technical innovation began shortly after emigration from Europe to the United States in 1969; Robert Ayres and Chauncey Starr were my intellectual guides.

Given this background, my intent is to present as wide-ranging and interdisciplinary a perspective on the next 50 years as practicable in a book that amounts to less than 100,000 words. The book's principal aims need more than a single sentence to summarize. Above all, this is not a book of forecasts: I do not make a single claim that by a certain date a particular event will take place or a given trend will peak or end. Nor is this a volume of scenarios: I do not offer imaginative fables describing alternative worlds of 2050. This book is simply a multifaceted attempt to identify major factors that will shape the global future and to evaluate their probabilities and potential impacts.

This work is based on recognizing a simple dichotomy—fundamental shifts in human affairs come mostly in two guises, as low-probability events that could (in an instant) "change everything," and as persistent, gradually unfolding trends that have no less far-reaching impacts in the long term. A close, critical, interdisciplinary look at both these factors can be beneficial in reminding us—as individuals and as polities—to pay adequate attention to the consequences of unpredictable (or poorly predictable) catastrophic events and to the clearly discernible outcomes of worrisome long-term trends.

Better understanding and heightered awareness should help us lessen the impact of unpredictable events, even prevent some whose timing could not be known but whose coming might have been anticipated. (9/11, the September 11, 2001, destruction of the World Trade Center's Twin Towers by terrorists, which came after the World Trade Center bombing on February 26, 1993, and after the publication of al-Qaeda's training manuals during the trial of Umar Abdul Rahman in 1995, is an obvious instance.) They should also improve our efforts at moderating or reversing deleterious trends at a stage when changes are tolerable and sacrifices reasonable, before such trends bring unavoidable economic collapse, protracted social turmoil, heightened risks of widespread violent conflicts, or a global environment altered to a greater degree than at any time since the emergence of our species.

Consequently, in chapter 2, I begin by identifying key fatal discontinuities—sudden catastrophic events that can change the course of world history. These events include rare but recurrent natural phenomena, such as the Earth's encounters with extraterrestrial bodies, volcanic mega-eruptions, and viral pandemics, as well as destructive human actions, such as major wars and terrorist attacks. I evaluate these phenomena in order to provide the best current understanding and, where possible, to quantify the probabilities of their occurrence during the first half of the twenty-first century.

Chapter 3, devoted to principal trends of global importance, examines key resource, demographic, economic, political, strategic, and social shifts. First is a fundamental universal trend that will affect the global history of the next two generations: a complex energy transition from a world powered largely by the combustion of fossil fuels to an as-yet-uncertain mix of new resources and conversions. Few other factors will be as important in determining the economic and social fortunes of both affluent and poor countries as the tempo and eventual success or failure of this unfolding energy transition.

Second, I look at other gradual shifts by focusing on the principal actors on the world stage today: Europe, Japan, Russia, China, the United States, and the Muslim world. Global civilization has a relatively small number of leading actors (equivalent to keystone species in ecosystems) whose aspirations, commitments (or lack thereof), internal changes, and external postures disproportionately affect the future and fortunes of all.

Three examples illustrate the point of disproportionate influence. (1) although the demographic trends in Hungary and Japan appear to be similarly bleak, Japan's rapidly aging population is a matter of global consequence because the country is still the world's third largest economy and a principal technical innovator. (2) Continuation of the chronic and legendary mismanagement of the Italian economy will have only a marginal effect on global investment and trade, but the very foundations of the world's economy could be entirely remade if the United States does not soon end its economic excesses. (3) During the past generation Hindu extremists and Serbian nationalists have instigated acts of violence that have caused many casualties, but the global import of their violence and hate speech is minimal compared to the rise of the unyieldingly militant, terrorizing version of Islam whose threats extend to all inhabited continents.

The assessments of states and the Muslim world in chapter 3 consider factors ranging from demographic trends and immigration to technical innovation and macroeconomic performance. For each of these specific surveys I provide historical background (often contradictory) evidence regarding the strength and durability of the unfolding trends, and the likelihood of particular future developments (these trends, unlike recurrent natural catastrophes, are not subject to meaningful quantification because they are contingent on so many events).

The third part of chapter 3 addresses two aspects of who is on top. The first is a strategic, collectivist matter of ever-shifting global primacy (a more accurate term than dominance), a multifaceted and hard-to-evaluate quest for power, influence, and advantage. The second concerns individual fortunes in life, a worrisome and apparently global trend of growing economic and social inequalities that results, to a large extent, from vigorous (and seemingly interminable) globalization of resource use, production, and consumption.

Although most of the events that will mold the future can be categorized either as sudden catastrophic events or as unfolding trends, environmental change warrants separate treatment because it is such an inimitable amalgam of shocking discontinuities (especially given that sudden environmental change is measured on a different time scale) and gradual trends, and because these two classes of phenomena are intertwined in multiple (and still poorly understood) feedbacks. In chapter 4, I review the best available evidence regarding the magnitude and tempo of environmental changes that have the potential to affect the course of planetary civilization seriously during the coming two generations. This assessment includes not only the still insufficiently appreciated complexities of global warming but also brief looks at other profound environmental changes, such as a multifaceted assault on the global water cycle, a massive human alteration of the global nitrogen cycle, and a trend of increasing resistance of common pathogenic bacteria to antibiotics.

I close the book by offering in chapter 5 a rational framework for assessing potential risks and evaluating unfolding trends. Quantification of risks offers a useful basis for rational perception and effective preparation for threats ranging from recurrent natural catastrophes to technical failures and terrorist attacks. Our understanding of unfolding trends and any attempts to change them in desirable directions benefit from setting them in appropriate historical context, not mistaking short-lived phenomena for long-term processes, and stressing the unpredictable nature of complex, interwoven social, economic, political, strategic, and environmental developments. These realities preclude meaningful long-range forecasting, but they do not prevent us from acting as responsible risk minimizers.

In sum, do not expect any grand forecasts or prescriptions, any deliberate support for euphoric or catastrophic views of the future, any sermons or ideologically slanted arguments. Instead, expect eclectic inquiries, reliance on long-term historical perspectives, reminders that limited understanding and inherent uncertainties are our constant companions in appraising the risks of globally fatal discontinuities and the strength and ultimate outcomes of unfolding trends.

Complex realities often produce contradictory evidence and seemingly incompatible arguments. For example, the assessment of the future of the United States is more pessimistic in the chapter on national trends than in the book's concluding discussion. This is understandable. While it is hard to escape a rather gloomy feeling after a systematic, cumulative look at a series of trends (economic, demographic, social, strategic) whose only common denominator appears to be their wrong direction, the overall assessment of the country's prospects brightens considerably when its recent failings and misfortunes are seen alongside its great residual strengths and historically tested capacity for reinvention and restructuring, and are then compared with the weaknesses, handicaps and rigidities of other major actors: only the youngest readers of this book will be able to judge the eventual outcomes.

My intent is to illuminate, not to prescribe; to question and to convince readers of the fundamental openness of contingent futures. The framework chosen to accomplish this is a wide-ranging, historically based interdisciplinary appraisal of sudden discontinuities and unfolding trends, of the contest for global primacy, and of underlying energy needs and worrisome environmental changes. All of this is neither soothing nor grimly satisfying, but I believe that such a realistic, searching, amalgamative, mosaic-building approach is superior to grand prescriptions, and it offers the best way to power our imagination, to mobilize our creativity, and to deploy our considerable capacity for adapting to new, unforeseen and unforeseeable circumstances.

Finally, two technical notes and a paragraph of thanks. Being able to get insights unfiltered by translations has been a very useful asset for me in understanding the histories and appraising the fortunes of different societies. Besides reading in all principal European languages (Russian and Italian are my favorites), I have studied both *putonghua* (Chinese) and *nihongo* (Japanese), and I also spent five years working on literary Arabic and the Egyptian dialect. That is why I prefer to use consistent and linguistically accurate transcriptions in this book. For readers' convenience, exceptions were made for terms that are now commonly used in English-language publications: al-Qaeda (*al-qāʿida*) and the Koran (*al-qur'ān*). And, as in all of my books, all statistics are in metric units used with appropriate SI prefixes, listed in appendix A.

My thanks, above all, to Paul Demeny for asking me for an unorthodox contribution to his journal and hence unwittingly launching this book: the two papers about the next 50 years published in *Population and Development Review* (Smil 2005a; 2005b) became its core. Thanks also to Clay Morgan for giving me the latitude to do my seventh MIT Press book; to John Katzenberger, Granger Morgan, Peter Nolan, Simon Upton, Daniel Vining, and an anonymous reviewer for reading the entire text or parts of the typescript and offering their criticism and suggestions; and, once again, to Douglas Fast for creating a fine set of illustrations.

1

How (Not) to Look Ahead

Inusitatis atque incognitis rebus magis confidamus vehementiusque exterreamur.
(The unusual and the unknown make us either overconfident or overly fearful.)
Gaius Julius Caesar, *Commentarii de Bello Civili*, II. 4

Any one of us may indulge in speculations about global futures tailored to particular moods or biases, from Francis Fukuyama's (1992) ahistorical "end of history," forseeing the universal triumph of liberal democracy, to the Ehrlichs' (2004) lament that the fate of liberal democracy will be similar to Nineveh's. Fukuyama rightly complains that he has been misunderstood, that he did not suggest events' coming to an end. Rather, he maintains, no matter how large and grave any future events will be, history itself ("as a single, coherent, evolutionary process") is over because nothing else awaits but an eventual triumph of liberal democracy. This claim irritates because of its combination of wishful thinking and poverty of imagination. If we were to believe it, then 9/11, fundamentalist Islam, terrorism, nuclear blackmail, globalization of the labor force, and the resurgence of China are inconsequential because "all of the really big questions had been settled."

As for our Nineveh-like fate, I am far from convinced, despite the enormous challenges we face, that our civilization will be soon transmuted into a defunct heap. Even if that were the case, we would still not be one with Nineveh: myriads of our artifacts made of steel, other metals, glass, and plastics that we leave behind will be better preserved than the Assyrian Empire's short-lived capital of clay that was so thoroughly destroyed by invading Babylonians. But these are just asides provoked by Fukuyama's and the Ehrlichs' claims, which were introduced in order to illustrate something that such grand forecasts have in common: their outcomes are preconceived, and their arguments are predetermined by strongly held visions, whether of inexorable progress or unavoidable collapse.

Visions of unavoidable collapse have been in the ascendant. Diamond's *Collapse* (2004), a derivative, unpersuasive, and simplistically deterministic book, gained a cult following with its tales of failed societies prefiguring our approaching demise. Martin Rees, a Cambridge don and the Astronomer Royal, tipped his hand with a very unforgiving title, *Our Final Hour* (2003) followed by a bleak subtitle listing terror, error, and environmental disasters as the greatest threats to humankind's future. Kunstler's (2005) book is another notable contribution to the literature of catastrophes, and Lovelock (2006) sees the Earth goddess Gaia taking revenge on her human abusers. Only Posner (2004) kept his usual analytical cool while looking at catastrophic risks and our response to them.

And then there is the burgeoning field of specific point forecasts that quantify numerous attributes of populations, environments, techniques, or economies. The Internet has made it a matter of seconds to find the requisite data for particular years: total number of females in Yemen in 2040, CO_2 concentrations in the atmosphere in 2030, the aggregate U.S. national debt in 2010, and so on. For all those curious but unwilling to search, here are the forecasts: a medium variant of the UN's latest population forecasts (United Nations 2005) lists 25 million Yemeni females in 2040 (10 million in 2005); according to scenarios published by the Intergovernmental Panel on Climatic Change (IPCC 2001; 2007), the average global atmospheric CO_2 level should be ~450 ppm by 2030 (~380 ppm in 2006); and the U.S. federal debt was expected to approach $11 trillion in 2010, ($7.9 trillion in 2005) (OMB 2006).

Given prevailing life expectancies, most male readers in their early 40s and female readers who have just turned 50 will still get the chance to check the 2030 outcome and find how badly mistaken the original forecast was. This conclusion (the only reliable forecast being our inability to forecast) rests on a voluminous, increasing amount of evidence: the only sensible way to appraise the reliability of modern forecasts is to look back and see how well their counterparts foretold yesterday's and today's realities. Such backward-looking exercises are particularly valid because during the past generation most of these specific point forecasts relied on the same suite of intellectual approaches and techniques as do today's prognoses that look 5–50 years ahead.

Retrospectives reveal that most of the truly long-range quantitative forecasts (spanning at least one generation, or 20–25 years) turned out to be useless within years or even months of their publication. I have demonstrated these failures by a detailed examination of more than a century's worth of every possible category of long-range energy forecasts (Smil 2003). Trend forecasts fail so rapidly because they tend to be unrealistically static. But trends are finite: they weaken or deepen suddenly; they can be reversed abruptly.

Population forecasts provide pertinent examples of these failed anticipations. A comparison of the revision for 2004 (United Nations 2005) with the 1990–2025 global population forecast (United Nations 1991) shows a difference of about 600 million people, a reduction about 10% greater than today's entire population of Latin America. Thus even forecasts that deal with given biophysical realities (most of the females that will give birth during the next 20 years are already alive) and that are issued only a dozen years apart can differ by continent-sized margins.

When looking at the global prospects for the next 50 years I have no desire to add to this almost instantly irrelevant mountain of specific point forecasts. Nor do I want to become an inventive fabulist and proffer assorted scenarios, a practice that has been embraced by individual forecasters (e.g., Hammond 1998), international institutions (e.g., WBCSD 1997; WEF 2006), major corporations (e.g., Shell Group 2006), and government agencies. An excellent example of this genre on the global scale (limited to only four visions of the world in 2020) is an effort by the National Intelligence Council (NIC 2004): it offers a Pax Americana (continuing U.S. predominance), a Davos World (robust economic growth led by China and India), a Cycle of Fear (proliferation of weapons leads to large-scale intrusive security measures in an Orwellian world), and a New Caliphate ("a global movement fueled by radical religious identity politics [that] could constitute a challenge to Western norms and values as the foundation of the global system").

The principal reason that even the cleverest and the most elaborate scenarios are ultimately so disappointing is that they may get some future realities approximately right, but they will inevitably miss other components whose dynamic interaction will create profoundly altered wholes. Suppose that in 1975 (years before the adoption of the one-child policy in China) a group of scenario writers correctly predicted the decrease in China's total fertility rate (and hence the country's much reduced population total). Would they—would anybody in 1975 (during the last phase of the Maoist Cultural Revolution and a year before Mao's death)—have set that number amidst a more than quadrupled quasi-capitalist economy absorbing annually tens of billions of dollars of direct foreigninvestment and serving as the leading workshop for the world (fig. 1.1)? What expert group gathered in 1985 to rank relative standings of major powers in 1995 would have forecast the collapse of the Soviet Union, Japan's economic retreat, the first Gulf War, and the resurgent U.S. economy against the background of surging globalization and the emergence of the Internet?

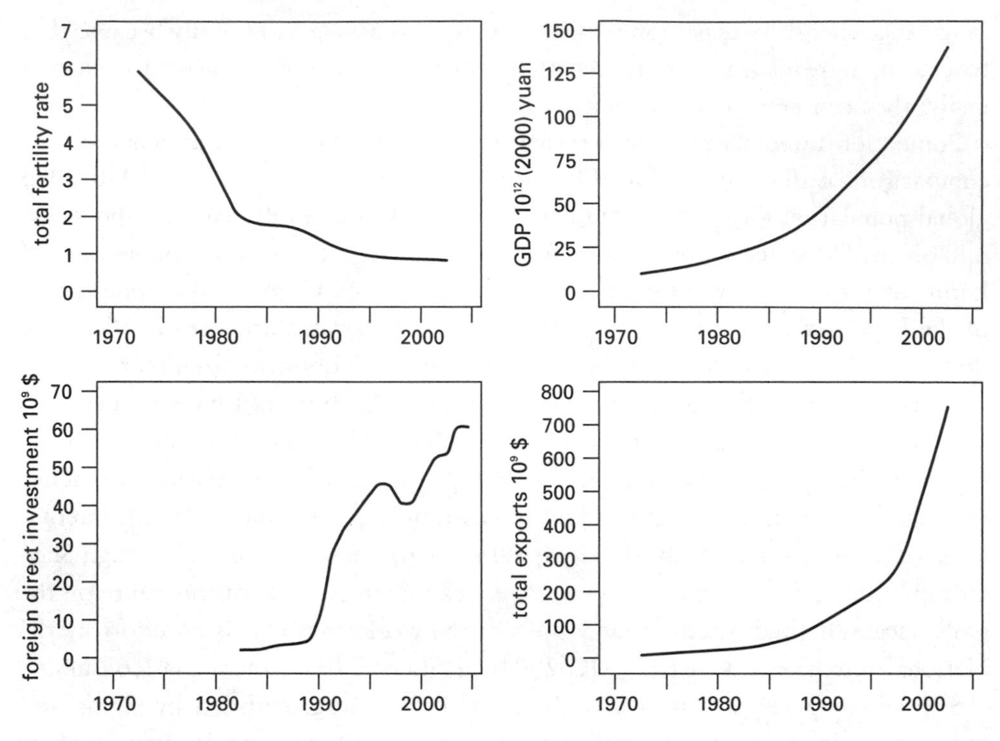

Fig. 1.1
China's unpredictable changes, 1975–2005: fertility, GDP, foreign investment, exports. Plotted from data in United Nations (2005) and NBS (2006).

As I have mentioned, I offer no quantitative point forecasts and no alternative scenarios. My intent is to explore those key variables whose impact is likely to be large enough to shape the course of world history during the first half of the twenty-first century. My firm belief is that looking far ahead is done most profitably by looking far back and that this approach works both for natural catastrophes and socioeconomic trends. Naturally, there are no specifics to be learned from such an exercise, yet those extended retrospectives impress with one key lesson: history advances as much by saltations—sudden discontinuities—as it does by gradual unfolding of long-lasting trends.

In this respect, history mirrors, in a much contracted fashion, the record of life's evolution on Earth, which is marked both by very slow (Darwinian) transformations and by relatively sudden (saltationary) changes (Simpson 1983; Eldredge and Gould 1972). Gradual, but cumulatively astonishing, evolutionary advances are much more widely appreciated than are several remarkable saltations embedded in the fossil record. None was more stunning than the great Cambrian explosion of highly organized and highly diversified terrestrial life. This great evolutionary saltation began about 533 million years ago and it produced within a geologically short spell of just 5–10 million years, or less than 0.3% of the entire evolutionary span, virtually all of the animal lineages that are known today (McMenamin and McMenamin 1990). And modern science also came to appreciate the role of rare catastrophic episodes in shaping the life's evolution (Albritton 1989; Ager 1993).

The increasingly frequent attempts at long-range forecasting (mostly dynamic modeling and scenario building) are of a gradualistic variety, resting largely on following a number of critical trends. I turn to these gradual processes in chapters 3 and 4, which look at the new population realities (differential growth, regional redistributions, aging, migration), socioeconomic trends with capacity for long-lasting global impacts (marginalization of Japan, Islam's role, Russia's reemergence as a major power, China's rise and its checks), the perils of nuclear proliferation, changing global leadership, and worrisome environmental trends.

But I start by focusing in chapter 2 on those unpredictable saltations whose consequences, in terms of lives lost and disrupted, economies destroyed and transformed, and outlooks dashed and altered, could change humanity's collective fortunes during the next 50 years.

Before I do so, a few paragraphs on the meaning of *global*, certainly one of the most overused adjectives of the new century. This seemingly straightforward term actually has a number of contextual meanings. It is often used as a synonym for *worldwide* even if the phenomenon does not encompass the entire planet. There are natural processes operating on truly global scales: atmospheric circulation is a fundamental example of a unified, planetwide, climate-shaping flux that is powered by a single source (solar radiation). Plate tectonics is another example of a planetwide process that determines the basic physical features of every continent and ocean.

Other natural phenomena are global in different sense: their extent is limited either to land or to the ocean, but they are widespread within these confines. Soil erosion and ocean currents belong to this category. Other processes, natural or anthropogenic, are ubiquitous but spatially discontinuous, found in numerous locations on all continents; in this sense there are definitely global problems with invasive species, losses of agricultural nitrogen, increasing income disparities, or governmental pension liabilities. Economic, political, and military uses of *global* have their analogs of natural "global" categories. Trade is now truly global because

Fig. 1.2
Hurricane Katrina landfall, August 29, 2005. Satellite image at <http//goes.gsfc.nasa.gov/pub/goes/050829.katrina.jpg>.

no country can be economically autarkic, and affluent nations could not support their high quality of life without intensive selling and buying of goods and services.

International finance is global: money in modest savings accounts is commingled with the legal but excessive profits of multinational companies and with the illegal and even more excessive profits of cocaine and marijuana wholesalers. So is international telecommunication. The U.S. military reach is global because its vessels cruise all oceans, and its strategic lift and amphibious capabilities can put forces on land wherever there is a suitable runway or a beach. And *global* is now applied also to individual events that make a distinct worldwide difference. Henisz et al. (2005) asked if hurricane Katrina (fig. 1.2) was a global event and answered yes, based on three considerations: disruption of oil and gas production in the Gulf of Mexico, which helped drive up the world price of oil; worldwide insurance and reinsurance implications of this major loss (more than $40 billion); and a tarnished image of the United States as billions of people saw televised images of distress and devastation with a tardy and limited response from government.

In this book I focus on truly global phenomena that can directly affect the entire planet, either as instant catastrophes or as gradually unfolding trends. Yet some events and processes that are much more restricted can change the course of world history; their eventual consequences are undeniably global. The terrorist attacks of 9/11 are a perfect example of this kind. No individual, no expert group can be prescient enough to separate the matters that will be truly consequential from those that appear important but will eventually make little difference. Inevitably, this book shares that fundamental shortcoming; some of its hoped-for hits will surely turn out to be misses.

2

Fatal Discontinuities

Mors ultima linea rerum est.
(Death is everything's final limit.)
Quintus Horatius Flaccus (Horace)

Bostrom (2002) classified existential risks—those that could annihilate intelligent life or permanently or drastically curtail its potential, in contrast to such "endurable" risks as moderate global warming or economic recessions—into bangs (extinctions due to sudden disasters), crunches (events that thwart future developments), shrieks (events resulting in very limited advances), and whimpers (changes that lead to the eventual demise of humanity). I divide them, less dramatically, into (1) known catastrophic risks, whose probabilities can be assessed owing to their recurrence; (2) plausible catastrophic risks, which have never taken place and whose probabilities of occurrence are thus much more difficult to quantify satisfactorily; and (3) entirely speculative risks, which may or may not materialize.

Known catastrophic risks encompass discontinuities whose probabilities of recurrence can be meaningfully appraised because of reasonably well-understood natural realities and historical precedents. Their probabilities of near- or long-term recurrence can be quantified with a degree of accuracy that is useful for assessing relative risks and allocating resources for preventive actions or eventual mitigation. This category includes natural catastrophes such as the Earth's encounters with extraterrestrial bodies, volcanic mega-eruptions, and virulent pandemics as well as transformational wars and terrorist attacks.

Although *plausible catastrophic risks* have never yet occurred, their potentially enormous impacts require that they not be excluded from any comprehensive assessment of future fatal discontinuities. Some of these catastrophes have been widely anticipated for decades. The fear of accidental nuclear war has been with us since

November 1951, when the Soviets deployed their first deliverable fission bomb (RDS-1 Tatyana), although a more appropriate dating might be 1955, when both superpowers acquired their first nuclear-tipped long-range missiles, Matador and R-5M Pobeda (Johnston 2005). Unlike the strategic bombers (the first jet-powered plane, B-47, flew in 1947), the launched ballistic missiles could not be recalled, and there was no way to intercept them during the decades of the Cold War. Despite enormous expenditures initiated during the first term of the Reagan presidency, there is still no reliable antimissile defense in place.

Other events in this category have been matters of occasional speculation (e.g., a pandemic caused by a previously unknown pathogen), but overall the likelihood of occurrence and extent of impact elude any meaningful quantification.

Entirely speculative risks include both the fanciful—for instance, Joy's (2000) vision of new omnivorous "bacteria" capable of reducing the biosphere to dust in a matter of days—and the completely unknown. Clearly, no one can give examples of the latter, but the likelihood of such unknowable surprises increases as the time span under consideration lengthens. Still, it is worthwhile to comment on key speculative unquantifiable risks and assign them to two basic categories of more and less worrisome events. This division can be based on the best relative ranking of (guess)timated probabilities, the most likely overall impact of such developments, or both.

Many critics would argue that discontinuities whose very occurrence remains speculative belong in the realm of science fiction. The rationale for addressing these matters here is captured in Tom Wolfe's (1968) description of U.S. business leaders' reaction to the quasi-prophetic statements of Marshall McLuhan: What if he is right?

Several of these speculative concerns were popularized by Joy's (2000) paper about the dangers for humanity of three powerful twenty-first-century techniques: robotics, genetic engineering, and nanotechnology.

The robotics part of Joy's publication was largely a derivative effort based on the work of two artificial intelligence enthusiasts, Hans Moravec (1999) and Ray Kurzweil (1999), who maintain that robotic intelligence will soon rival human capability (fig. 2.1). Kurzweil (2005) placed the arrival of "singularity"—when computer power will reach 10^{23} floating operations per second, vastly surpassing the power and intelligence of the human brain—quite precisely in 2045.

We have been promised superintelligent, omnipotent robots for several generations (Čapek 1921; Hatfield 1928). There are no such machines today; even the

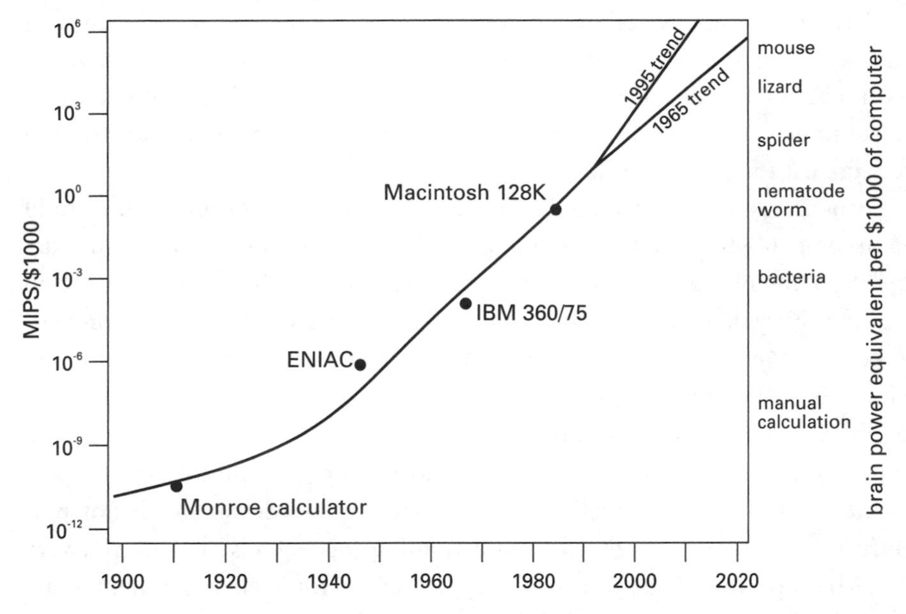

Fig. 2.1
Faster-than-exponential evolution of computing power since 1900 (graphed here as millions of instructions per second per thousand 1997 dollars) led Hans Moravec to conclude that humanlike robots should be possible before the middle of the twenty-first century. Adapted from Moravec (1999).

"intelligent" software installed in IBM's Deep Blue II in order to play chess against the world champion Garry Kasparov in 1998 did not show the coming triumph of machines but merely that "world-class chess-playing can be done in ways completely alien to the way in which human grandmasters do it" (Casti 2004, 680). And while computers have been used for many years to write software and to assemble other computers and machines, such deployments do not indicate any imminent self-reproductive capability. All those processes require human actions to initiate them, raw materials to build the hardware, and above all, energy to run them. I find it hard to visualize how those machines would (particularly in less than a generation) launch, integrate, and sustain an entirely independent exploration, extraction, conversion, and delivery of the requisite energies.

Joy's (2000) most sensational claim concerned the aforementioned omnivorous "bacteria" that could swiftly reduce the entire biosphere to dust. This claim might have been modified had Joy acknowledged some fundamental ecological realities and considered the necessary resource and interspecific competition checks on such a runaway scenario. Microorganisms have been around for some 3.5 billion years, and evolutionary biologists have difficulty envisaging a new one that could do away almost instantaneously with all other organisms that have survived, adapted, and prospered against such cosmic odds.

If the biosphere were prone to rapid takeover by a single microorganism, it could not have become differentiated into millions of species, thousands of them interdependent within complex food webs of rich ecosystems and all of them connected through global biogeochemical cycles. Symbiosis rather than interspecific competition has been the most fundamental driver of life's evolution and survival (Sapp 1994; Margulis 1998; Smil 2002).

There are even more speculative, ostensibly science-based suggestions regarding civilization's demise, including the idea that we are living in a simulation of a past human society run by a superintelligent entity that can choose to shut it down at any time (Bostrom 2002). Clearly, the mind running this exercise has been a very patient one because the simulation has been going on for nearly 4 billion years (unless one dismisses the evidence of the Earth's evolution and our emergence as one of its results).

In any case, there is little we can do about the frightening (or liberating: no human worries anymore) aspects of such scenarios. If the emergence of superior machines or all-devouring gooey nanospecies is only a matter of time, then all we can do is wait passively to be eliminated. If such developments are possible, we have no rational way to assess the risk. Is there a 75% or a 0.75% chance of self-replicating robots' taking over the Earth by 2025 or nanobots' being in charge by 2050? And if such "threats" are nothing more than pretentious, upscale science fiction, then they have a massive amount of lower-grade company in print, film, and television and are good for little else than producing an intellectual frisson.

In this chapter, I look in some detail only at those natural catastrophes that take place rapidly, in a matter of minutes to months. Global climate change, a natural event that has commonly been posited as the most worrisome environmental crisis, can take place rapidly only when measured on an evolutionary time scale. Consequently, its assessment belongs to chapter 4, which deals with unfolding environmental trends.

And I consider only those catastrophes that do not have a vanishingly low probability of occurring during the next 50 years, that is, those that recur at intervals no longer than 10^5–10^6 years and that could change the course of global history and perhaps even eliminate the modern civilization. This is why I do not give a closer attention to such very rare events as the Earth's exposure to supernova explosions or periods of enormous lava flows such as those that created Deccan and Siberian Traps.

Supernovae are rare, taking place only about once every 100 years in a spiral galaxy like the Milky Way (Wheeler 2000). The solar system is within 10 parsecs (3×10^{17} m) of a supernova only once every 2 billion years (2 Ga) and the explosion (typically yielding 10 billion times more energy than the Sun) would flood the top of the atmosphere with X-ray and very short UV flux about 10,000 times higher than does the incoming solar radiation. The Earth would receive in just a few hours a dose of ionizing radiation of 500 roentgens that would be fatal to most unprotected vertebrates. Their 50% effective lethal dose is mostly 200–700 roentgens, but many would survive given the differences in exposure and specific resistance. Invertebrates and microbes would remain largely unaffected. Terry and Tucker (1968) calculated that the Earth has received at least this dose ten times since the Precambrian, or roughly once every 50 million years (50 Ma), an interval that yields a negligibly low probability of occurrence during the next 50 years.

Similarly, the periods of massive and prolonged effusions of basaltic lavas accumulating in thick layers are uncommon even when measured on a geological time scale. The oldest identified episode of this kind (508–505 Ma ago) produced more than 190,000 km^3 of Australia's Kalkarindji basalts and was the most likely cause of the first major animal extinction (Glass and Phillips 2006). The past 250 Ma have seen only eight giant plumes of magma penetrating the Earth's crust and forming massive basalt deposits. India's Deccan Traps, containing more than 500,000 km^3 of basalt, were formed over a period of 5 Ma beginning 65 Ma ago, and these effusions, rather than an impact of an extraterrestrial body, may have killed the dinosaurs or at least greatly contributed to their demise (fig. 2.2). And the Siberian Traps, covering some 2.5 million km^2 with perhaps as much as 3 million km^3 of lavas, were formed about 250 Ma ago (Renne and Basu 1991).

Natural Catastrophes

Natural catastrophes range from relatively common events such as cyclones, floods, and landslides to less frequent violent releases of energies associated with geotectonic processes (earthquakes and volcanic eruptions, both capable of generating

Fig. 2.2
Exposed layers of Deccan flood basalt, more than 1 km thick, at Mahabaleshwar, Maharashtra, India. Photo courtesy of Hetu Sheth, Indian Institute of Technology, Mumbai.

tsunamis) to uncommon encounters of the Earth with large extraterrestrial bodies. Older data on the frequency and death tolls of natural catastrophes are incomplete, but recent statistics capture all major events and have fairly accurate fatality counts. Annual global compilations by the Swiss Reinsurance Company (Swiss Re 2006a) show that floods and storms are by far the most frequent events; during the first years of the twenty-first century they accounted for 70%–75% of all natural catastrophes. These are followed by earthquakes, tsunamis, and the effects of extreme temperatures, including droughts, fires, heat waves, blizzards, and frost. However, in terms of worldwide victims, earthquakes were the worst natural catastrophes between 1970 and 2005, when they killed nearly 900,000 people, compared to about 550,000 deaths from floods and cyclones (fig. 2.3).

These compilations also show the expected highly skewed frequency distribution of fatalities as a single event dominates the annual death toll. Most of the time this event is a major earthquake (including an earthquake-generated tsunami), and this dominance has been particularly pronounced during the recent past. In 2003, Iran's Bam earthquake was responsible for 80% of that year's fatalities caused by all

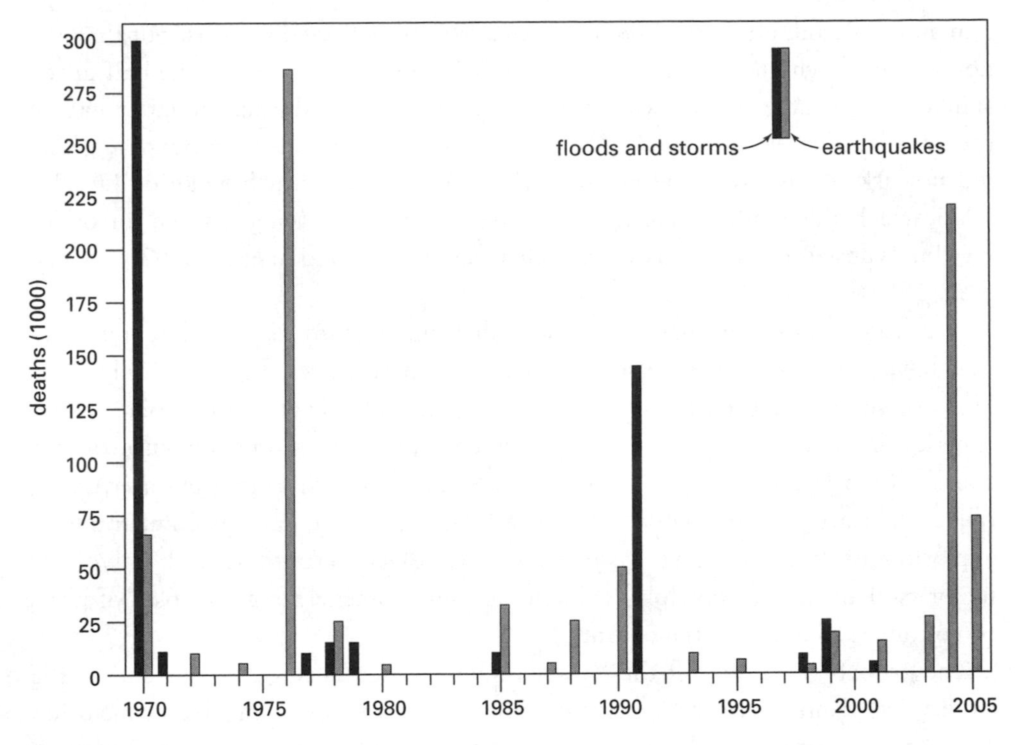

Fig. 2.3
Death tolls from major natural disasters (at least 4,000 deaths per event), 1970–2005. Plotted from data in Swiss Re (2006b).

natural disasters; in 2004 the Sumatra-Andaman earthquake and tsunami accounted for 95% of the total; and in 2005 the Kashmir earthquake's accounted for nearly 85% of the total (Swiss Re 2004; 2005; 2006a). Relatively frequent events with localized impacts often cause tens or hundreds, and less commonly thousands, of fatalities, but the most damaging catastrophes claim hundreds of thousands, even millions, of lives. The most disastrous cyclone of the twentieth century, Bangladesh's Bhola on November 13, 1970, killed at least 300,000 people; the most deadly earthquake, in northern China's Shaanxi on January 23, 1556, claimed 830,000 lives; and the Huanghe flood of 1931 claimed at least 850,000.

But the mostly deadly natural disaster of the first years of the twenty-first century, the Indian Ocean earthquake and tsunami on December 26, 2004 (Lay et al. 2005; Titov et al. 2005), illustrated that even these massive natural catastrophes do not alter the course of world history. They generate worldwide headlines, elicit humanitarian aid, and have long-term effects on the affected nations, but they are not among epoch-making events on the global scale. Indeed, one of the half dozen similarly devastating natural catastrophes that took place during the latter half of the twentieth century remained an entirely internal affair because xenophobic China did not ask for international aid following the Tangshan earthquake of July 28, 1976, which killed (officially) 242,219 people in that coal-mining city and surroundings but whose toll was estimated as high as 655,000 (Huixian et al. 2002; Y. Chen et al. 1988).

In contrast to frequent natural disasters that kill as many as 10^5–10^6 people and that have severe local and regional economic consequences, there are only three kinds of sudden, unpredictable, but recurrent natural events whose global, hemispheric, or large-scale regional impacts could have a profound influence on the course of world history. They are the Earth's collision with nearby extraterrestrial objects that are large enough to cause death and economic damage comparable to explosions of strategic nuclear weapons; massive volcanic eruptions (with or without major tsunamis); and (possibly) voluminous tsunami-generating collapses of parts of volcanoes sliding into the ocean.

The probability of any of these events' taking place during the first half of the twenty-first century is very low (well below 1%), but this comforting conclusion must be counterbalanced by the fact that if any one of them were to take place, it would be an event without counterpart in recorded history. The near-instant death toll would involve 10^6–10^9 people, 1–4 orders of magnitude (OM) greater than for frequent localized natural catastrophes. Moreover, if these events were to affect the densely populated core areas of the world's largest economies, their global impact would be considerable even if the spatial extent of destruction amounted to only a tiny fraction of the Earth's surface.

Encounters with Extraterrestrial Objects

The Earth constantly passes through a widely dispersed (but in aggregate quite massive) amount of universal debris (McSween 1999). Common sizes of these meteoroids range from microscopic particles to bodies with diameters <10 m. As a result, the planet is constantly showered with microscopic dust, and even the bits with diameter 1 mm, large enough to leave behind a light path as they self-destruct in the atmosphere (meteors), come every 30 s. This constant infall (about 5 t per day) poses virtually no risk to the evolution of life or to the functioning of modern civilization because these objects disintegrate during their passage through the

Fig. 2.4
Closeups of large asteroids. The Earth's collision with asteroids of this size would almost certainly destroy civilization. *Left*, composite image of Ida (~52 km long); *right*, Gaspra (illuminated portion ~18 km long). Galileo spacecraft images (1993 and 1991). From NASA (2006).

atmosphere, and only dust or small fragments reach the ground. But the planet's orbit is also repeatedly crossed by much larger objects, above all by stony asteroids with diameters >10 m and as large as tens of kilometers across (fig. 2.4), and by comets.

The risk of encounters with extraterrestrial bodies was first recognized during the 1940s. It began to receive greater attention during the 1980s, but until the early 1990s no systematic effort was made to comprehensively identify such objects, assess the frequencies of their encounters with the Earth, and devise possible defensive measures. Known Earth-crossing asteroids numbered 236 at the beginning of 1992 (compared to 20 in 1900), the year in which NASA proposed the Spaceguard Survey (Morrison 1992), whose goal is to identify 90% of all near-Earth asteroids (NEAs) by the year 2008. NASA funded and coordinated monitoring began in 1995, and ten years later the U.S. House of Representatives approved the Near-Earth Object Survey Act, which directs NASA to expand its detection and tracking program. These actions have been accompanied by publications assessing the threat (Chapman and Morrison 1994; Gehrels 1994; J. S. Lewis 1995; 2000; Atkinson, Tickell, and Williams 2000).

The progress in discovering new near-Earth objects (NEOs) has been rapid (NASA 2007). By the end of 1995 the total number of known objects was 386; by

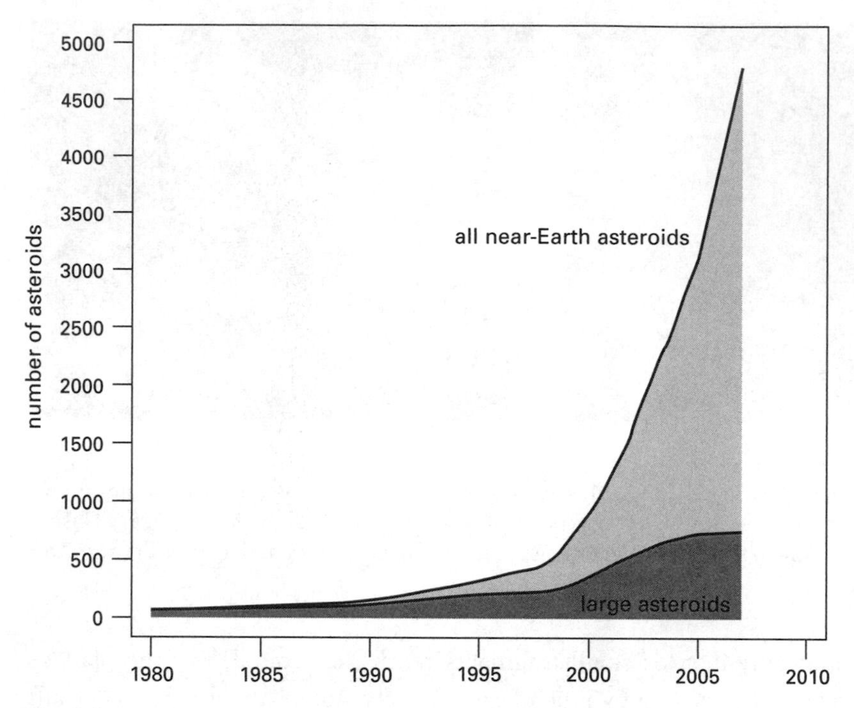

Fig. 2.5
Cumulative discoveries of near-Earth asteroids, 1980–2007. From NASA (2007).

the end of 2000, 1,254; and by June 2007, more than 4,100, of which nearly 880 were bodies with diameters ≥1 km (fig. 2.5). As the findings accumulate, there has been an expected decline in annual discoveries of NEAs with diameters >1 km, and the search has been asymptotically approaching the total number of such NEAs. Consequently, we are now much better able to assess the size-dependent impact frequencies and to quantify the probabilities of encounters whose consequences range from local damage through regional devastation to a global catastrophe.

There are perhaps as many as 10^9 asteroids orbiting the sun in a broad and constantly replenished belt between Mars and Jupiter as well as a similar number of comets moving in more distant orbits within the Öpik-Oort cloud beyond Pluto. Gravitational attraction of nearby planets constantly displaces a small portion of these bodies (remnant debris from the time of the solar system's formation 4.6 Ga ago) into elliptical orbits that move them toward the inner solar system and into the vicinity of the Earth. Several million near-Earth objects cross the Earth's orbit, and at least 1,000 of them have diameters ≥1 km. Because of their high impact

Fig. 2.6
Oblique aerial view of Meteor Crater in Arizona. USGS photo by David J. Roddy.

velocities, even small NEOs have kinetic energy equivalent to that of a small nuclear bomb; larger bodies can bring regional devastation, and the largest can cause a global catastrophe.

Craters provide the most obvious evidence of major past impacts (fig. 2.6) (Grieve 1987; Pilkington and Grieve 1992). More than 150 of these structures have been identified so far, but it must be kept in mind that most impacts have been lost in the ocean, and the evidence of most of the older terrestrial impacts has been erased by tectonic and geomorphic processes. The largest known crater, the now buried Chicxulub structure in Yucatan with diameter 300 km (Sharpton et al. 1993), was created 65 Ma ago by an asteroid whose impact has been credited with the great extinction at the Cretaceous-Tertiary (K-T) boundary (Alvarez et al. 1980). The most recent impact of an NEO with diameter >1 km took place less than 1 million years ago in Kazakhstan (NRCanada 2007). Asteroids and short-period comets make up about 90% of NEOs; the remaining risk is posed by intermediate and long-period comets that cross the planet's orbit only once in several decades. The frequency of NEO impacts declines exponentially with the increasing size of the impacting objects, and their kinetic energy determines the extent of damage (fig. 2.7).

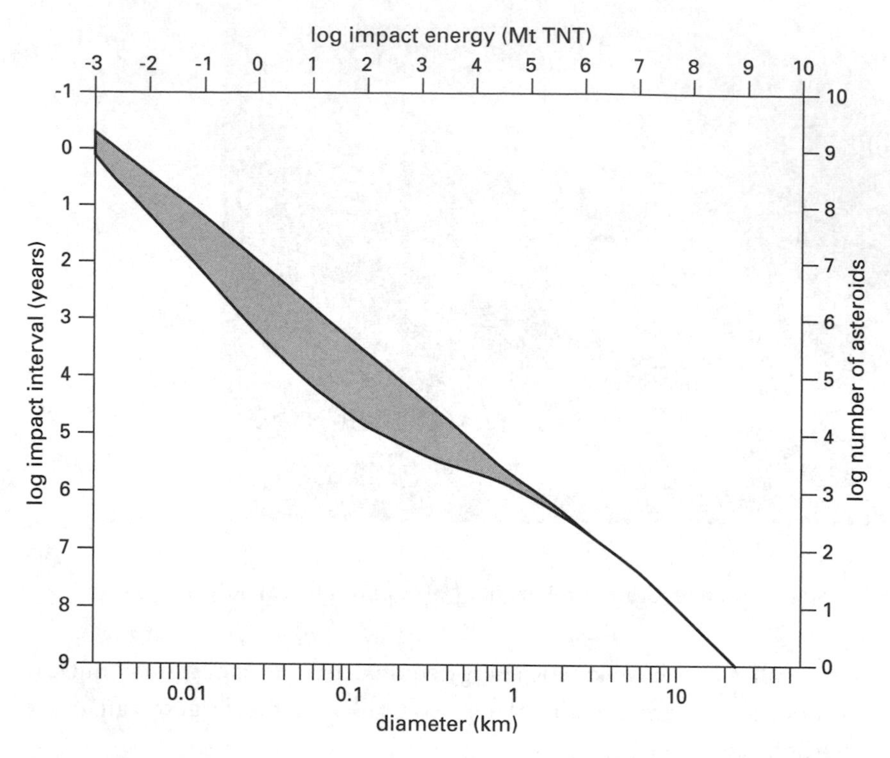

Fig. 2.7
Size, impact frequency, and impact energy of near-Earth asteroids. All four axes are logarithmic; the band indicates the range of uncertainty regarding the numbers and impact intervals of objects with diameter <1 km. Based on NASA (2003), Bland and Artemieva (2003), and Chapman (2004).

Roughly once a year the Earth encounters an extraterrestrial body whose size is 5 m across and whose air burst releases nearly 21 TJ, equivalent to 5 kt TNT (explosive power of 1 t TNT is equal to 4.18 GJ). This makes it about one-third as powerful as the Hiroshima bomb; there is no definite number for the explosive yield of that bomb, but the most authoritative source (Malik 1985) puts it at 15 (±3) kt TNT. Only if this body's center of disintegration were right above the U.S. Capitol during the President's State of the Union speech would the effect be felt globally. But the probability of such an encounter is vanishingly small, at least 8 OM smaller than that of a similar object's disintegrating at any time above any densely populated area.

Stony objects with diameters 10 m are intercepted by the Earth's atmosphere every decade, and their entry (at speeds ~20 km/s) discharges energy equivalent to about 100 kt TNT, roughly seven times the energy released by the Hiroshima bomb. As these bolides disintegrate during atmospheric deceleration, a fireball and a shock wave are the only phenomena felt on the ground within a radius of 10^2 km around their entry point. Brown et al. (2002) used data from a satellite designed to detect nuclear explosions in order to identify light records of bolide detonations (objects 1–10 m in size) in the atmosphere. From these observations they concluded that on average the Earth is struck by an object with diameter 50 m (equivalent to 10 Mt TNT) every 1,000 years.

The probability of such an impact is thus about 5% (uncertainty band of about 3%–12%) during the next 50 years, and its effects would be similar to those caused by the famous Tunguska meteor of June 30, 1908. Atmospheric disintegration of that stony object released energy equivalent to 12–20 Mt TNT, produced a shock wave that flattened trees over an area of about 2150 km^2 but killed nobody in that unpopulated region of central Siberia (Dolgov 1984). If a similar object were to disintegrate over a densely populated urban area, it could cause great damage. Its explosion about 15 km above the ground would release energy equivalent to at least 800 Hiroshima bombs and result in 10^5 casualties and $\$10^{11}$ of material damage. But the chances of such an event are roughly 2 OM smaller than the probability of hitting an unpopulated or thinly inhabited region because densely populated areas cover only about 1% of the planet's surface.

As was clearly demonstrated by the contrast of casualties in Hiroshima and Nagasaki, the actual destruction would depend on the physical configuration of the affected area. Hiroshima, with a bowl-like setting that acted as a natural concentrator of the blast, had about 40% more fatalities and more destruction from a 15-kt blast than did Nagasaki from a 21-kt explosion (CCM 1981). Another complicating factor is that a Tunguska-like blast may not be a point-source event (similar to a nuclear bomb) but rather a plume-forming event (similar to a line of explosive charges) and hence could be caused by much less powerful objects (NASA 2003).

Asteroids with diameters ≥100 m reach the atmosphere once every 2,000–3,000 years, and their energy (equivalent to >60 Mt TNT) is as large as the yield of the largest tested thermonuclear devices. Hills and Goda (1993) calculated that stony objects with diameters up to ~150 m will release most of their energy in the atmosphere and will not hit the surface and create impact craters (however, heavier metallic objects of that diameter might penetrate). Stony objects with diameter

>150 m hit the Earth once every 5,000 years, and their terrestrial impacts create only local effects, small craters with adjacent areas covered by ejecta. Using as a reference point a stony body that produces only air blast to 220 m diameter, Bland and Artemieva (2003) estimated that bodies with a larger diameter would hit the Earth once in 170,000 years. There is broad consensus that the threshold size for an impact producing a global effect is a body with diameter at least 1 km and possibly closer to 2 km.

Toon et al. (1997) concluded that only bodies with kinetic energies equivalent to at least 100 Gt TNT (diameters >1.8 km) would cause global damage beyond the historic experience, and objects with diameters between 850 m and 1.4 km (energy equivalents of 10–100 Gt TNT) would cause globally significant atmospheric water vapor injection and ozone loss but would not inject enough submicrometer particulates into the stratosphere to have major, longer-term climatic effects. A 1-km body (density 2.5–3.3 g/cm^3, velocity 20–22 km/s) colliding with the Earth would release energy equivalent to about 62–105 Gt TNT, almost 1 OM more than the energy that would have been expended by an all-out thermonuclear war between the two superpowers in 1980 (Sakharov 1983). A 3-km asteroid would liberate energy equivalent to about 2 Tt TNT, possibly enough to terminate modern civilization regardless of where the asteroid hit (fig. 2.8).

The consequences of a collision with a 1-km body would depend greatly on the impact site. Odds are roughly 7 : 3 that the object would hit the ocean and damage the land indirectly by generating tsunamis, but a terrestrial impact would create a crater with diameter 10–15 times the object's size and pose an unprecedented threat to the survival of civilization. Such a collision would vaporize and fragment both the projectile and the impacted area, and enormous masses of dust would reach the stratosphere. While the larger dust fractions would rapidly settle, submicrometer-sized particles would remain in the atmosphere for weeks to months.

Simulations using the global circulation model show that ocean heat storage would prevent a global freeze even if the impact were equivalent to the K-T event (with kinetic energy perhaps as high as 1 Pt TNT) but that surface land temperatures would drop by more than 10°C and still be some 6°C lower a year later (Covey et al. 1994; Toon et al. 1997). In addition, hot ejecta would produce significant amounts of nitrogen oxides, whose presence in the stratosphere would degrade (and in extreme cases, largely destroy) the ozone shield that protects the Earth against UV radiation. A 1-km object would have much less effect because it would not

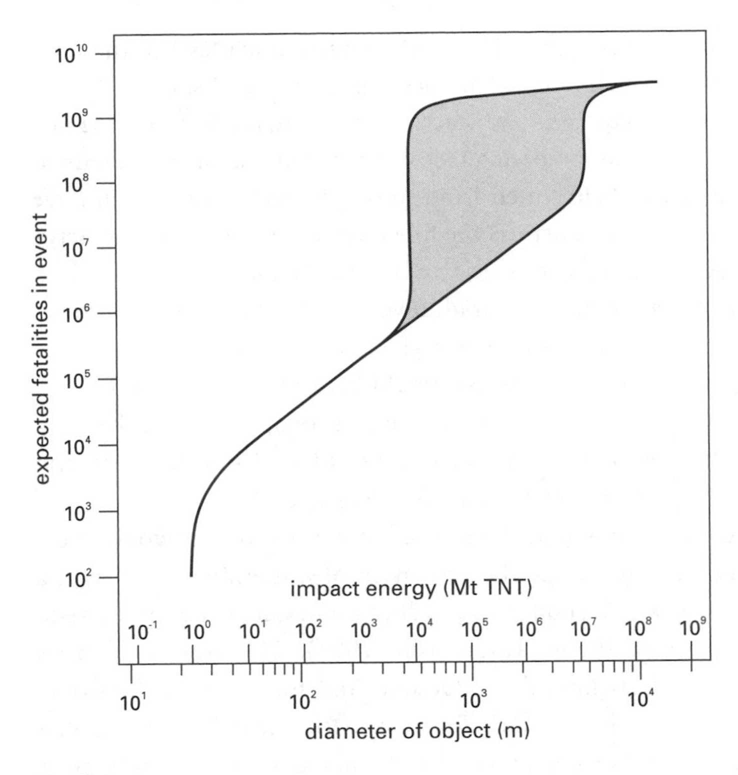

Fig. 2.8
Expected fatalities from impacts of near-Earth objects. From Morrison (1992).

generate enough dust to cause temporary planetwide darkness and shut down photosynthesis.

At least 10 Gt of submicrometer-sized dust would be required to make the minimum amount of light unavailable for photosynthesis (Toon et al. 1997), but using the analogy of a ground-level nuclear explosion—which produces about 25 t of submicrometer-sized dust per kt of yield (Turco et al. 1983)—means that a 1-km body would produce only about 1.5 Gt of fine dust, 4 OM less than a K-T-sized object (25 Tt). Moreover, Pope (2002) questioned the assumptions regarding the fine dust fraction in the ejecta produced by the K-T impact. Pope's calculations, coupled with observations of the deposited coarse fraction, indicated that a minor share was laid down as submicrometer-sized dust and that little debris diffused to high southern latitudes. These conclusions invalidate the original attribution of K-T extinction to the shutdown of photosynthesis by submicrometer-sized dust. Pope calculated that the impact released only 0.1% (and perhaps much less) of the total amount as fine dust (but his conclusions were questioned as unrealistic).

In any case, it is impossible to quantify satisfactorily the actual effect because fine dust would not be the only climate-modifying factor. Soot from massive fires ignited by hot ejecta and sulfate aerosols liberated from impacted rocks could each have as much cooling effect on the atmosphere as the fine dust. However, lingering aerosols would also increase the intercept of the outgoing terrestrial radiation and contribute to tropospheric warming. A rapid reversal of ground temperatures could follow once the debris settled, and water vapor and CO_2 injected into the stratosphere (from impacted carbonate rocks) would greatly enhance the natural greenhouse gas effect. With positive feedbacks (higher temperatures enhancing evaporation as well as plant respiration and the release of CO_2 from the ocean and soils), this bout of global warming could persist for decades.

The only defensible conclusion is that the impact of a 1-km object would most likely not have consequences resembling the aftermath of a thermonuclear war: a drop in ground temperature severe enough to produce a nuclear winter and a temporary cessation of all photosynthesis (Turco et al. 1991). The overall effect on photosynthesis, biodiversity, agricultural production, and human survival would depend critically on the mass of ejecta and their atmospheric perseverance. Specifics are impossible to enumerate, but extensive forest and grassland destruction by fires, a temporary but substantial reduction of precipitation due to the disrupted global water cycle, sharp declines in food production, and extensive interference in industrial, commercial, and transport activities are all easy to imagine. The impact would not bring an abrupt end to modern civilization, but it could be an enormously costly setback.

Earlier estimates put the number of NEOs with diameter >1 km at about 2,000, but Rabinowitz et al. (2000) used improved detection techniques to conclude that there were only about 1,000 such objects, and Stuart (2001) put the total number of kilometer-sized NEAs at just over 1,200 (he also found them less likely to collide with the Earth than previously assumed). If 1,100 were the actual total, then 80% of them had been discovered by June 2007. Certainly the most notable outcome of this effort is the good news that the likelihood of near-term impacts has been decreasing. On a 10-point Torino scale, measuring the severity of the collision threat (Binzel 2000), 0 indicates no hazard (white zone) with effectively no likelihood of collision, and 1 (normal, green zone) indicates an object whose close path near the Earth poses no unusual danger and which will very likely be reassigned to level 0 after additional observations. Levels 3 and 4 indicate close encounters with 1% or greater chance of collision capable of localized or regional destruction; and significant threats of close encounters causing a global catastrophe begin only with level 6.

As of 2007, only two objects, 2004 VD17 and 2004 MN4, were rated 2 and all other NEAs scored 0 on the Torino scale during the twenty-first century. The first of these objects is about 580 m across; the other is 320 m across, and it became the subject of short-lived concern when initial calculations indicated its collision with the Earth on April 13, 2029. That is not going to happen, but there is still a distant possibility of an encounter with MN4 between 2036 and 2056, and VD17 may come close by 2102 (Yeomans, Chesley, and Choclas 2004). By far the highest known probability of an NEO's colliding with the Earth is nearly a millennium away, on March 16, 2880. Analysis by Giorgini et al. (2002) suggests a very close approach by asteroid 29075, an asymmetrical spheroid with mean diameter 1.1 km that was discovered in 1950 (as 1950 DA), lost from view after 17 days, and rediscovered in 2000 (fig. 2.9). The impact probability was put as high as 0.33%, but because of the unknown direction of the asteroid's spin pole, the range of the actual risk may be closer to 0.

While it is very likely that we have already discovered all existing NEAs with diameter >2 km, we can never be quite sure that we know of every large NEA that

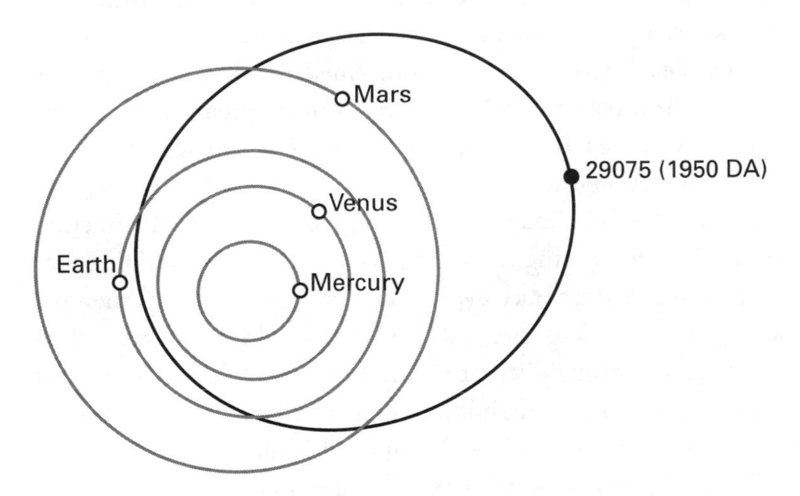

Fig. 2.9
A collision that is not going to happen: the orbits of four planets and asteroid 29075 (1950 DA). Based on NASA (2007).

is already on an Earth-crossing orbit and we will not be able to identify promptly every new addition to this dynamic collection of extraterrestrial objects. Consequently, assessing the risks of collision will always require assumptions regarding the impact frequency of various-sized objects. The general size frequency distribution of NEOs is now fairly well known (see fig. 2.7), but there are different assumptions about the most likely frequency of impacts; the estimates differ by up to 1 OM. For example, Ward and Asphaug (2000) assume that an object with diameter 400 m hits the Earth once every 10,000 years, and with diameter 1 km, once every 100,000 years. By contrast, Brown et al. (2002) would expect a body with diameter 400 m to hit once every 100,000 years, and with diameter 1 km, once every 2 million years; Chapman (2004) would expect an object with diameter 400 m to hit once every 1 million years; and Jewitt (2000), an object with diameter 400 m, once every 400,000 years.

Another important consideration enters at this point: even impacts of bodies with diameter <1 km could have global consequences if they destroyed a core area of a major nation. For example, an object 500 m across would devastate an area of about 70,000 km^2; Tokyo and its surrounding prefectures cover less than half that area (~30,000 km^2) and are inhabited by about 30 million people. Alternatively, calculations by Ward and Asphaug (2000) show that if an asteroid with diameter 400 m were to hit a 1-km-deep ocean site at 20 km/s, the maximum amplitude of a tsunami generated by this impact would be more than 200 m at a distunce of 100 km and 20 m at a distance of 1000 km. A near-shore impact off eastern Honshū or in the North Sea between London and Amsterdam would instantly obliterate core regions of the world's two leading economies, Japan and the EU, and unlike with tsunami generated by a distant earthquake, there would not be sufficient time for mass population evacuation.

Naturally, the probability for such a site-specific impact (P_S) is a small fraction of that for an unspecified location on the Earth (P_E): $P_S = P_E(A_E/A_S)$. Assuming that an object with diameter 400 m arrives once every 100,000 years ($P_E = 1^{-5}$) then the probability of its destroying the Tokyo area ($A_S = 3^{10}$ m^2) would be no more than 6^{-10} ($A_E = 5.1^{14}$ m^2), an annual probability of about 1 in 1.66 billion. Ward and Asphaug (2000) calculated specific probabilities of a 5-m-high tsunami wave's hitting Tokyo and New York at, respectively, 4.2% and 2.1% during the next 1,000 years, or roughly 0.2% and 0.1% during the next 50 years. In contrast, Bland and Artemieva (2003) estimated the frequencies of bolides that would most likely cause hazardous tsunami at only about one-fiftieth of the rate reported by Ward and

Asphaug (2000). Chesley and Ward (2006) calculated the overall long-term casualties that would be caused by impacts of objects with diameters 200–400 m at fewer than 200 deaths per year (or fewer than 10,000 total during the next half century).

The highest risk of collision-related fatalities comes from the land impact of smaller, and hence more common, bodies, with a more than 1% chance that such an impact will kill about 100,000 people during the twenty-first century, whereas somewhat larger objects (diameter 150–600 m) will pose the greatest tsunami hazards (Chapman 2004). By contrast, probabilities of encounters with NEOs with diameter ≥1 km are orders of magnitude smaller. If the average recurrence interval for a 1-km asteroid were 400,000 years, then the probability of impact during the next 50 years would be 0.0125%; bracketing this by uncertainties of 100,000 years to 2 million years gives a range of 0.0025%–0.05%. The minimum size of an asteroid whose impact would have severe global consequences depends not only on its diameter but also on its density and speed (an asteroid traveling at 30 km/s would have 2.25 times more kinetic energy than an equally massive counterpart moving at 20 km/s) and on the impacted area.

If a large asteroid were to enter the ocean, such an impact would generate tsunamis that would hit even distant shores with high-amplitude waves: the impact's location would determine the extent of global fatalities and economic damage. For example, 2.15 Ma ago the Eltanin asteroid, whose diameter may have been as much as 4 km, entered a deep (about 5-km) spot in the Pacific Ocean off southern Chile without forming a seabed crater and without resulting in a mass extinction (Mader 1998). The resulting tsunami (total energy of 200 EJ) would have completely destroyed the South Pacific islands, but the wave height along the coasts of North America and East Asia would have been less than 15 m.

But even if we were to discover all NEAs and determine that none of them is on a collision course with the Earth, we would still face an inherently much more difficult challenge of identifying cometary impactors. These bodies, made of rocky material and volatile ice, account for no more than 10% of all NEOs, but because they have higher encounter velocities (as much as 60 km/s compared to 15–25 km/s for asteroids) their kinetic energy is much higher, and they have been responsible for some 25% of all craters with diameters >20 km (Brandt and Chapman 2004). Fortunately, these more powerful impacts are rarer than the encounters with similar-sized asteroids. The closest approaches by historic comets missed the Earth by 3.7 lunar distance (LD = 384,000 km) in 1491, 5.9 LD in 1770, and 8.9 LD in 1366;

all other misses were >10 LD (NASA 2006). Consequently, probabilities of the Earth's catastrophic encounter with a comet are likely less than 0.001% during the next 50 years, a chance approaching the level of 1 out of 1 million.

Volcanic Mega-eruptions and Collapses

About a half billion people live within a 100-km radius of a volcano that has been active during the historical era, but the number of fatalities and the extent of material damage caused by volcanic eruptions have been highly variable (fig. 2.10). Fortunately, even with eruptions as large as the largest historic events, the potential for immediate fatalities is relatively limited. Hot lava usually spreads only over several square kilometers, ballistic projectiles fall on areas ≤10 km^2, and tephra deposits affect areas of 10^2–10^6 km^2. But tsunamis generated by large eruptions can cross an ocean, and volcanic dust from a major eruption can be transported worldwide. Loss of life and property depends on the prevailing form of energy release: slow-flowing, glowing Hawaiian lavas give plenty of time to evacuate houses, whereas the pyroclastic flows, such as these that swept down Vesuvius in 79 C.E.

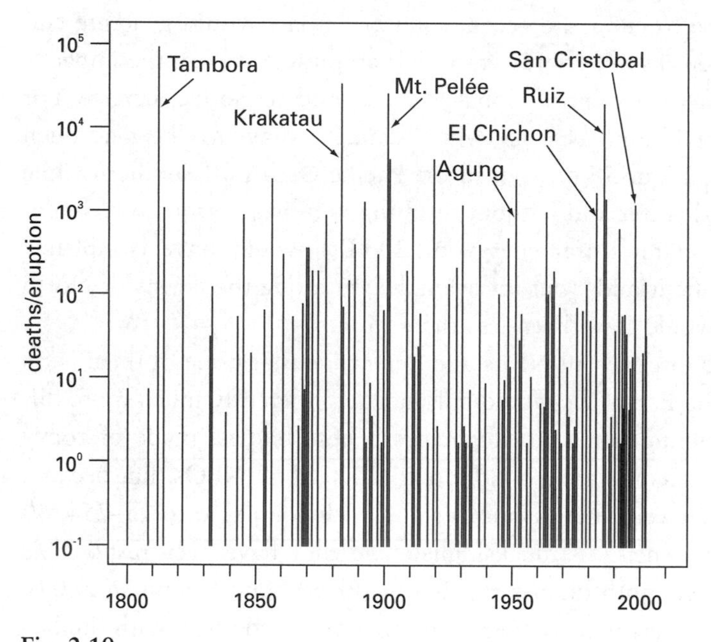

Fig. 2.10
Volcanic eruptions and fatalities, 1800–2000. Plotted from data at <http://www.volcanolive.com/>.

and buried Pompei and Herculaneum, or the flows from Mount Pelée in 1902, which killed all but 2 of 28,000 residents of St. Pierre on Martinique, produce instant mass burials (Sigurdsson et al. 1985; Heilprin 1903).

Because of larger populations the frequency of eruptions involving fatalities rose from fewer than 40 per century before 1700 to more than 200 during the twentieth century (Simkin 1993; Simkin, Siebert, and Blong 2001). Nearly 30% of the ~275,000 fatalities between 1500 and 2000 were due to pyroclastic flows and 20% to tsunamis. The four greatest disasters in terms of fatalities were Tambora (92,000), Krakatau (36,000), Mount Pelée (28,000), and Colombia's Nevado de la Ruiz in 1985 (23,000). As for the frequency of eruptions, it rose from fewer than 20 per year before 1800 to more than 60 per year by the late twentieth century, largely because of improved reporting. Ammann and Naveau (2003) analyzed sulfate spikes in polar ice and discovered a strong 76-year cycle of tropical explosive volcanism during the last six centuries.

The most common way to measure the magnitude of eruptions is the volcanic explosivity index (VEI), devised by Newhall and Self (1982). This logarithmic scale combines the volume of ejecta and the height of ash column. VEI values less than 4 include eruptions that take place somewhere on the Earth daily or weekly and that produce less than 1 km^3 of tephra (airborne fragments ranging from fairly large blocks to very fine dust) with maximum plume heights below 25 km. Mount St. Helens (1980) had VEI 5 (paroxysmal eruption, the same magnitude as Vesuvius in 79 C.E.) and produced just 1 km^3 of ejecta (Lipman and Mullineaux 1981).

Krakatau (1883) had VEI 6 (colossal eruption), and Tambora (1815), had VEI 7 (supercolossal eruption). The Bronze Age Minoan eruption in the Aegean Sea, about 3,650 years ago, was the largest release of volcanic energy (100 EJ) during the historic period, and it created the great Santorini caldera (surrounded by islands Thera, Therasia, and Aspronisi). But its total volume of ejecta, about 70 km^3, was less than half of Tambora's volume (Friedrich 2000), a comparison that exemplifies the lack of clear correlation between spectacular ash plumes and the total energy released by a volcanic eruption.

Historic eruptions are dwarfed by VEI 8 events variously labeled as gigantic, mega- or supereruptions (Sparks et al. 2005; Mason, Pyle, and Oppenheimer 2004). The two most recent ones created the Taupo caldera in New Zealand (26,500 years ago, VEI 8.1, 530 km^3) and the giant Toba caldera in northern Sumatra (fig. 2.11) (74,000 years ago, VEI 8.8, 30 km × 100 km), an oval now filled by a lake (Rose and Chesner 1990). The Toba event produced about 2800 km^3 of ejecta, and La

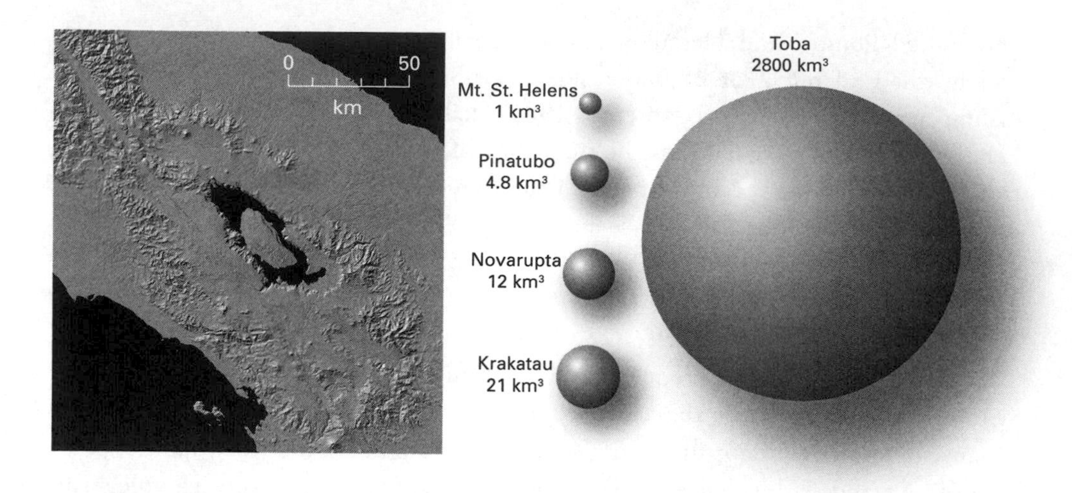

Fig. 2.11
Toba caldera, and comparison of Toba's volcanic ash volume with the largest nineteenth- and twentieth-century eruptions. Plotted from data in Mason, Pyle, and Oppenheimer (2004) and USGS (2005).

Garita (27.8 Ma ago, VEI 9.1), the largest supereruption identified so far, which produced the Fish Canyon Tuff in Colorado, ejected about 4500 km^3.

Toba's eruption sent trillions of tonnes of volcanic ash thousands of kilometers downwind and dispersed in both westerly and easterly directions, a pattern suggesting that the eruption happened during the summer monsoon (Bühring and Sarnthein 2000). Ash fall covered most of Southeast Asia, reached as far west as the northeastern Arabian Sea, and deposited several centimeters over the South China Sea and parts of southern China (Pattan et al. 2001; Ambrose 2003). Its greatest terrestrial impact was on the Indian subcontinent, where cores show layers of 40–80 cm in central India and very thick (2–5 m) deposits, possibly reworked by redeposition, close to the eastern coast (Acharya and Basu 1992). Rose and Chesner (1990) estimated that 1% of the Earth's surface was covered with more than 10 cm of Toba's ash.

Toba's impact must have been quite severe. It is perhaps the best explanation for a genetically well-documented late Pleistocene population bottleneck, when small and scattered groups of humans were reduced to a global total of fewer than 10,000 individuals and our species came very close to ending its evolution (Rampino and

Self 1992; Ambrose 1998). This explanation relies on studies of mitochondrial DNA that indicate severe population shrinkage between 80,000 and 70,000 years before the present (Harpending et al. 1993); as with any reconstructions of this kind, it has been criticized and defended (Ambrose 2003).

Quantifying the probabilities of future supereruptions is a highly uncertain undertaking. Their frequency cannot be extrapolated from the power law relation, which is based on much better records of size and frequency of smaller events. Such an extrapolation would suggest a recurrence interval on the order of 1,000 years, whereas the most recent event (Taupo) was 26,500 years ago. Our enumeration of supereruptions is certainly incomplete, but the best available account lists 42 events with VEI 8 or above during the past 36 million years, with two distinct pulses. The first one peaked about 29 million years ago, the other one began about 13.5 million years; analysis indicates that an eruption with VEI 8 or above could be expected to take place at least once every 715,000 years and that there is a 1% probability of such an eruption during the next 460–7,200 years (Mason, Pyle, and Oppenheimer 2004). This would translate to a 0.007%–0.1% probability during the next 50 years.

Impacts of supereruptions must be deduced from the effects of smaller events described in historic records and studied by modern volcanology. These extrapolations are also subject to many uncertainties, as is the use of modern global climate models to simulate the effect of high loads of stratospheric ash. Besides the highly variable gas composition (some eruptions produce relatively little or almost no SO_2, the precursor of sulfates) and different magma/ash ratios, the severity of regional impact and the overall extent of climatic effects would also be determined by the location of the event. Supereruptions close to densely populated areas would cause many more instant casualties and more physical destruction. Of the 14 supereruptions younger than 10 million years, 6 took place in the western United States close to or in California and upwind from prime agricultural regions (fig. 2.12).

Virtually instant fatalities would be caused by pyroclastic flows, heavy ash fall, and inhalation of highly acidic gases and aerosols, all effects well documented from Vesuvius and a number of modern eruptions. Severe damage to plants and acute respiratory effects would be limited to areas of relatively high concentrations of acidic and halogen gases. By far the most important global impact of supereruptions would be their short- to medium-term climatic consequences: dust injected all the way into the stratosphere produces hemispheric or even global temperature changes during the subsequent months or years (Angell and Korshover 1985; Robock 1999;

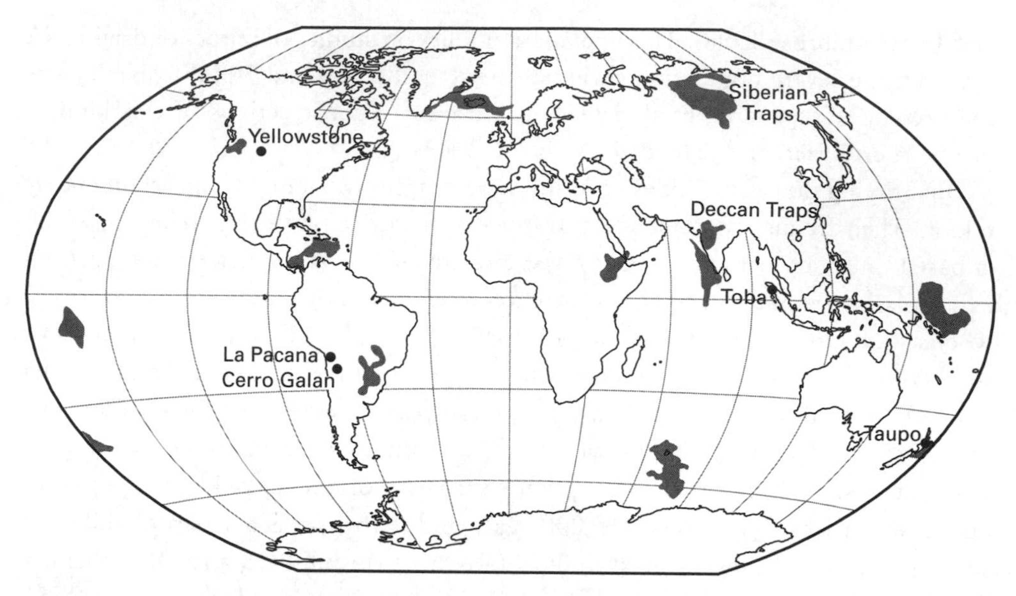

Fig. 2.12
Largest flood basalt provinces created during the past 250 million years, and locations of volcanic supereruptions during the last 5 million years. Based on Coffin and Eldholm (1994), Sparks et al. (2005), and Mason, Pyle, and Oppenheimer (2004).

Robock and Oppenhemier 2003). Sulfate aerosols, formed from the released SO_2, have a dual effect on the atmosphere: they reflect the incoming solar radiation, thereby cooling the troposphere, but they also absorb both the short-wave solar radiation and the outgoing long-wave terrestrial radiation and warm the stratosphere above the tropics.

Moreover, the stratospheric sulfates take part, together with emitted HCl, in complex reactions that destroy ozone. The outcomes are not easily predictable. Statistical analyses by Angell and Korshover (1985) indicated that out of 96 studied eruption events only 27 were followed by significant temperature declines. Because the maximum distance to the tropopause is 15–16 km in the tropics (compared to 9–11 km near the poles), tropical eruptions have to be more powerful in order to inject their plumes all the way into the stratosphere (Halmer and Schmincke 2003). But tropical eruption plumes that inject the ash into the atmosphere within ±30° of the equator produce a worldwide cooling effect more readily (as the general atmospheric circulation spreads the aerosols fairly rapidly over the both hemispheres) than those taking place in high latitudes.

Fig. 2.13
Enormous eruption plume of Mount Pinatubo, June 15, 1991. USGS photo by Dave Harlow.

This effect was closely observed and successfully modeled after the eruption of Mount Pinatubo (VEI 5–6) on the Philippines island of Luzon on June 15, 1991 (fig. 2.13) (Soden et al. 2002). This eruption was the twentieth century's largest injection of SO_2 into the stratosphere: about 20 Mt in a matter of days, and reaching as high as 45 km (McCormick, Thomason, and Trepte 1995; Newhall and Punongbayan 1996). Satellite monitoring showed that the resulting sulfate aerosols cooled the lower troposphere globally by about 0.5°C. Reduced global water vapor concentrations closely tracked this temperature decrease. But some regions experienced pronounced seasonal warming against the global background of cooling; during winter of 1991–1992 parts of North America and Western Europe were up to 4°C warmer than normal (Robock 2002).

Examination of tree ring densities indicates that the strongest Northern Hemisphere summer anomaly of the past 600 years was –0.8°C in 1601, most likely as the consequence of Peru's Huaynaputina eruption in 1600 (Briffa et al. 1998). Ash from Tambora's Plinian eruptions in April 1815 reached up to 43 km (Sigurdsson and Carey 1989), and during 1816 its global acid fallout of some 150 Mt caused average temperature deviation of –0.7°C in the Northern Hemisphere, producing the famous year without the summer in 1816, with reduced harvests, spikes in food prices, and localized famines (Stothers 1984). The Toba supereruption is credited with temperature decreases of up to 15°C below normal in latitudes between 30° N and 70° N, and with hemispheric cooling of as much as 3°C–5°C that may have persisted for several years, intense and long enough to trigger a "volcanic winter," a worldwide phenomenon akin to the effects of a nuclear winter, which was hypothesized to be (next to the radiation hazard) the most debilitating consequence of a thermonuclear war (Rampino and Self 1992; Turco et al. 1991).

Today a Toba-sized eruption in a similar location would, besides killing tens of millions of people throughout Southeast Asia, destroy at least one or two seasons of crops needed to feed some 2 billion people in one of the world's most densely populated regions. This alone would be a catastrophe unprecedented in history, and it could be compounded by much reduced harvests around the world. Compared to these food-related impacts, the damage to machinery, or the necessity to suspend commercial flights until the concentrations of ash in the upper troposphere returned to tolerable levels, would be a minor consideration.

But the VEI >8 supereruptions are not the only events with global impacts large enough to affect the course of the modern world. A moderate VEI 7 eruption would almost certainly have a global effect if it were located within ±30° of the equator and if it ejected at least 100 km^3 of magma, that is, 250–300 km^3 of ash. Sparks et al. (2005) estimated its frequency once every 3,000 (1,700–10,000) years. Its probability would then be 1.7% (0.5%–2.9%) during the next 50 years. VEI 7 eruptions releasing at least 300 km^3 of magma (750 km^3 of ash) could take place as often as once every 10,000 years, and their probability range for the next 50 years would be between 0.005%–0.5%. Even with all these uncertainties it is clear that with the (rounded) probabilities of 0.01%–0.1% for a supereruption (VEI >8) and 0.01%–3% for smaller events (VEI 7), globally significant volcanic eruptions are at least 1 or 2 OM more frequent than impacts of extraterrestrial bodies releasing comparable amounts of energy (Mason, Pyle, and Oppenheimer 2004).

For North America the most likely threat is presented by recurrent eruptions of the Yellowstone hotspot (Smith and Braile 1994). Past eruptions of this supervolcano left behind nine massive calderas during the last 15 million years. The last three eruptions took place 2.1 million, 1.3 million, and 640,000 years ago, and the last one produced about 1000 km^3 of volcanic ash (USGS 2005). There are three ways to interpret this sequence. First, it has too few members to allow for any conclusions. Second, the interval between the Yellowstone hotspot eruptions has actually decreased from about 800,000 years to 660,000 years; a repeat of the last interval leaves only 20,000 years before the next event is due. Third, the three events had an average interval of 730,000 years, and hence there are still some 90,000 years to go before the most likely repeat.

In either case, another such event has a very low probability of occurring (~0.007%) during the next 50 years. The overall impact of a Yellowstone eruption would depend on the prevailing winds. Dominant wind direction in the Yellowstone region is northwesterly flow; based on the effect of previous eruptions (Fisher, Heiken, and Hulen 1997), the area most affected by ash fall would encompass Wyoming, Colorado, Nebraska, Kansas, Oklahoma, and parts of South Dakota, Texas, New Mexico, and Utah. If a new eruption were to produce as much ash as the last one and affect approximately 2 million km^2, then all of the leading wheat-producing states would be buried under half a meter of ash. This calculation assumes an even downwind distribution; the actual ash cover would range from several meters in central Wyoming to a few centimeters in Texas.

But the past eruptions show that ash fall could affect all states west of the Mississippi (fig. 2.14). Thinner layers of volcanic ash could be incorporated into soils by plowing (and might actually improve productivity in years ahead), but even the most powerful tractors could not handle deposits of 0.5–1 m, and an inevitable consequence would be at least temporary abandonment of cropping on large areas of the Great Plains. Moreover, unstable ash layers would be easily dislodged by heavy rains and spring snow melt, creating enormous flooding and stream silting hazards. The economic costs of such an event could fully assessed only generations later.

There are two sets of circumstances when even a volcanic event of less than supereruption magnitude could have enormous socioeconomic consequences. The first would be if the eruption were to produce huge volumes of acidic gases upwind from a major, densely populated region whose economy would be severely damaged by the effects of sulfate aerosols. The absorption and scattering of visible light would create atmospheric haze, temporarily cool the troposphere and reduce photosynthesis, and cause damage to human and animal health. The second would be if an eruption were to cause a massive collapse of volcanic flanks into a nearby ocean and hence generate an extraordinarily large tsunami.

By far the greatest risk of the first event is presented by a repeat eruption of the Laki fissure (Skaftár Fires) in Iceland. The last episode, in 1783–1784, produced nearly 15 km^3 of lava and released about 122 Mt of SO_2 in eight months (for

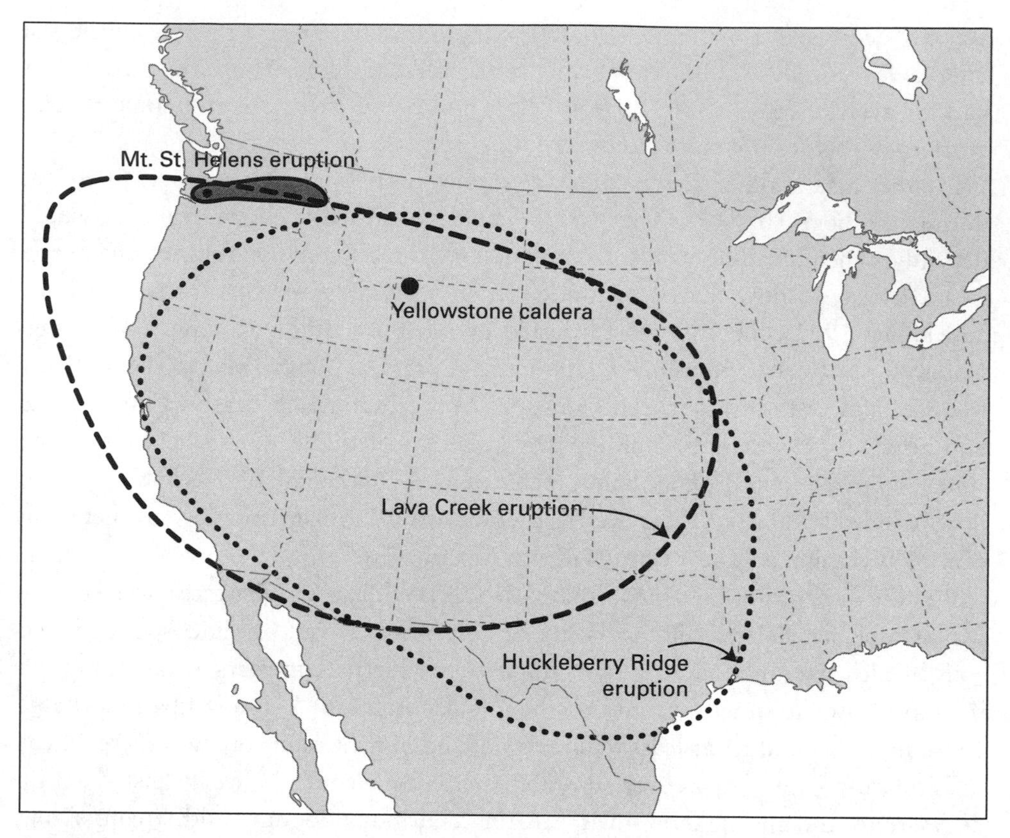

Fig. 2.14
Approximate volcanic ash fall zones from the two most recent Yellowstone eruptions, Lava Creek (640,000 years ago) and Huckleberry Ridge (2.1 million years ago), and for comparison, the area of heavy Mount St. Helens ash fall in 1980. Based on USGS (2005).

comparison, the global emissions of the gas from the combustion of fossil fuels were about 150 Mt SO_2/year during the early 2000s) as well as about 7 Mt of hydrochloric acid and 15 Mt of hydrofluoric acid; these emissions were carried by eruption columns to altitudes of 6–13 km (Thordarson et al. 1996; Thordarson and Self 2003). The emissions were then dispersed eastward across the Atlantic by the prevailing westerlies. The oxidation of SO_2 eventually produced some 200 Mt of H_2SO_4 aerosols, and nearly 90% of these particulates were removed as acid precipitation, resulting in heavy and extensive atmospheric haze (dry fog) locally and downwind in Atlantic Europe as well as subsequent severe winter and reduced temperatures (by as much as –1.3°C) for the next two or three years.

Nearly a quarter of Iceland's population (about 9,000 people) died because of the haze-induced famine, and deposited hydrofluoric acid contaminated the island's food and water. Volcanic fog over parts of Europe increased local mortality, caused respiratory complications, and damaged vegetation (Stothers 1996; Durand and Grattan 1999). The health impact was particularly pronounced in France and England. In France the excess deaths added 25% to the expected mortality between August 1973 and May 1784—more than the 16,000 premature deaths from the extreme heat wave of 2003.

Examination of English parish records by Witham and Oppenheimer (2005) concluded that there were nearly 20,000 excess deaths in the Laki eruption's aftermath. The odds of another eruption like it during the next 50 years are very low (<0.05%), but there is a high probability that a similar event will take place again during this millennium. Today its most immediate effects (besides the fatalities in Iceland) would be a temporary shutdown and extended disruption of the world's busiest intercontinental air routes between North America and Europe. Eastbound flights take advantage of the shifting jet streams, whose course often puts them close to Iceland; westbound flights approximate the great circle route, which puts them close to or right above Iceland. Given the chronically precarious financial situation of most airlines operating these routes, such an event would precipitate a number of major bankruptcies.

Huge volcano-triggered landslides, recurring roughly once every 100,000 years and creating waves in excess of 100 m, were first documented in the early 1960s as massive hummocks of debris on the sea floor surrounding the Hawaiian Islands (Moore, Normark, and Holcomb 1994). Subsequent research uncovered catastrophic landslides associated with eruptions off the Canary Islands, the Reunion, and Tahiti-Nui. Total volume removed by several episodes (870,000–850,000 years ago) of the north-directed Tahiti-Nui landslides is 400–450 km^3; the south-directed slides amount to about 300 km^3 (Hildenbrand, Gillot, and Bonneville 2006). Perhaps the greatest risk of this kind is now posed by a future eruption of Cumbre Vieja volcano, which created La Palma in the Canary Islands about 3 million years ago (Ward and Day 2001). During the 1949 eruption, the seventh known since 1470, the volcano's western half moved some 4 m oceanward; another eruption could cause a catastrophic failure of the entire western flank.

The resulting landslide of up to 500 km^3 of volcanic rock would generate a rapidly moving (up to 350 km/h) mega-tsunami, which would, after crossing the Atlantic, hit the eastern coast of North America with repeated walls of water up to 25 m high, assuring the destruction of Miami and severe damage to all coastal U.S. cities, including New York and Boston (fig. 2.15). Trombley and Toutain (2001) used all available historic and current eruption data and seismic, deformation, and thermal analyses to predict a >50% probability of another eruption by 2027, a near-certainty (>95% probability) by 2214. But we do not know which future eruption will cause a catastrophic landslide. The best evidence regarding the frequency of Canary Islands landslides would indicate intervals of about 70,000 years and hence a probability no higher than 0.07% during the next 50 years.

A 500-km^3 landslide is the worst-case scenario, which also assumes that the slide would take place instantaneously during a single event and enter the sea at high velocity. Halving the sliding mass would reduce the maximum waves to 5–10 m, and tsunamis from a landslide of 150 km^3 would reach only 3–8 m along the eastern cost of the United States. Wynn and Masson (2004) argued, on the basis of their studies of offshore deposits, that if each landslide were to be composed of multiple stages of gradual failure, then the average collapsed mass could be as low as 10–25 km^3 and the resulting tsunami would not inflict severe damage on the eastern coast of North America.

Influenza Pandemics

Modern hygiene, nationwide and worldwide inoculation, constant monitoring of infectious outbreaks, and emergency vaccinations have either completely eliminated or drastically reduced a number of previously lethal, deeply injurious, or widely discomforting epidemic diseases, including cholera, diphtheria, pertussis, polio, smallpox, tuberculosis, and typhoid. I hasten to add that these have been battles with no assured permanence. Pertussis (whooping cough) is coming back among children too young to be vaccinated (Tozzi et al. 2005). More than 10 million people worldwide still contract tuberculosis every year. The number of multidrug-resistant cases is increasing, and four decades have passed since the introduction of the last new effective anti-TB drugs, rifampicin in 1965 and ethambutol in 1968 (Glickman et al. 2006; Murray 2006).

The eradication goal has been particularly elusive in the case of polio. The number of cases dropped worldwide from 350,000 in 1988 to only about 500 in 2001. The next year the number of cases rebounded to some 2,000 a year and after another drop returned to nearly 2,000 in 2005 because of the suspension of vaccination in northern Nigeria, the virus's persistence in the slums of India, and a sudden increase of infections in Yemen, Somalia, Indonesia (Roberts 2006). In 2005 active transmis-

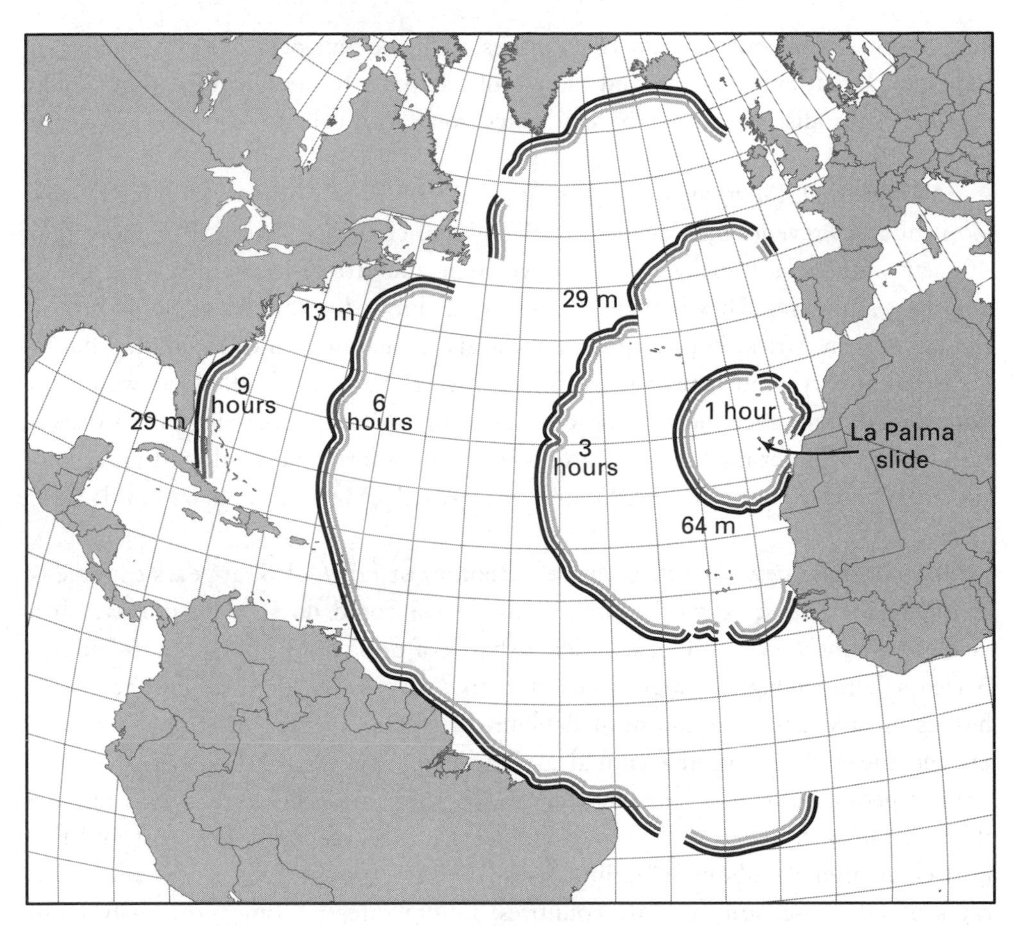

Fig. 2.15
Massive collapse of western flank of Cumbre Vieja volcano on La Palma, Canary Islands, would generate a tsunami that would hit the eastern coast of North America with a sequence of waves 10–25 m high. Tsunami progress across the Atlantic would allow for ample warning, and a staged collapse of a volcanic flank would produce much smaller trans-Atlantic waves. Based on Ward and Day (2001).

sion of polio virus took place in 16 countries, with endemic presence in Afghanistan, Pakistan, India, and Nigeria, and many polio experts have concluded that, unlike smallpox, the disease cannot be eradicated, only controlled (Arita, Nakane, and Fenner 2006).

A new journal, *Emerging Infectious Diseases*, published by the Centers for Disease Control and Prevention, is devoted to this global challenge. Since 1975 more than 40 new pathogens (mostly viruses) have been added to the ever-growing list of contagious diseases. They include such scary but limited outbreaks as Ebola hemorrhagic fever in Africa, Nipah virus in Malaysia, Singapore, and Bengal, and hantavirus pulmonary syndrome in the U.S. Southwest (Yates et al. 2002) as well more widespread and hence more worrisome cases of variant Creutzfeldt-Jakob disease (the human form of mad cow disease, bovine spongiform encephalopathy), cryptosporidiosis, cyclosporiasis, SARS and HIV/AIDS (Morens, Folkers, and Fauci 2004).

None of these new threats, with the exception of HIV/AIDS, appears capable of changing the course of world history, and AIDS could do so only if new, more virulent strains were to afflict significant shares of populations outside sub-Saharan Africa, where the highest rates of infection now surpass 20% and where the disease has its most widespread and most debilitating (social, mental, economic) impacts. During the early 2000s the annual global death toll from AIDS was about 2.8 million people, less than mortality due to diarrhea and tuberculosis, two diseases that we know perfectly well how to eradicate at an acceptably low cost and that now claim annually about 3.4 million lives (WHO 2006). Moreover, low and steady rates of HIV infection in many countries, falling rates in some previously badly affected nations (particularly Uganda and Thailand), new multiple drug regimens that extend productive lives (and the hopes for eventual vaccination) show that the disease can be managed.

As far as unpredictable discontinuities are concerned, only one somatic threat trumps all of this: we remain highly vulnerable to another episode of viral pandemic. High-frequency natural catastrophes have their somatic counterpart in recurrent epidemics of influenza, an acute infection of the respiratory tract caused by serotype A and B viruses belonging to the family *Orthomyxoviridae*. Influenza epidemics sweep the world annually, mostly during the winter months, but with different intensities. In the United States there are between 250,000–500,000 new cases every year, about 100,000 people are hospitalized, and 20,000 people die (less than 0.01% of the U.S. population).

Infection rates are by far the highest in young children (10%–30% annually) and in people over 65 years of age (Harper et al. 2000). Influenza pandemics occur when one of the 16 subtypes (H1–H16) of serotype A viruses, different from strains already present in humans, suddenly emerges, rapidly diffuses around the world (usually within six months), and afflicts between 30%–50% of people. The illness, with its characteristic symptoms of fever, myalgia, headache, cough, coryza, debilitation, and discomfort, spreads rapidly (latent period is just 1–4 days), and it is often complicated by bacterial or viral pneumonia. The former can be treated by antibiotics, but because there is no treatment for the latter, it becomes a common cause of death during influenza epidemics.

The first fairly well-documented influenza pandemic occurred in 1580, and there have been six known episodes during the last two centuries (Gust, Hampson, and Lavanchy 2001). In 1830–1833 an unknown subtype originated in Russia; in 1836–1837 another unknown subtype originated possibly in Russia; in 1889–1890 subtypes H2 and H3, originated possibly in Russia; in 1918–1919 subtype H1 (despite its common name "Spanish flu") originated most likely in the United States; in 1957–1958 subtype H2N2 originated in southern China, with total global excess mortality of more than 2 million people; and in 1968–1969 subtype H3N2 originated in Hong Kong, with excess worldwide mortality of about 1 million people. This low death rate was attributable to protection conferred on many people by the 1957 infection. None of the post-1969 epidemics reached virulent pandemic status (Kilbourne 2006).

All of the nineteenth-century pandemics as well as the 1957 and 1968 events were relatively mild and hence did not make any noticeable upticks in the secular trend of declining mortality. By contrast, the 1918–1919 pandemic was by far the largest sudden infectious burden in modern times (fig. 2.16). A common assumption is that its first, moderately virulent wave began in early March 1918 with the first infections at the U.S. Army Camp Funston in Kansas, but Langford (2005) proposed an origin in China. By May the virus had spread throughout most of the United States, Western Europe, north Africa, Japan, and the eastern coast of China; by August it was in Australia, Latin America, and India (Patterson and Pyle 1991; Davies 1999; Kolata 1999; Phillips and Killingray 2003; Barry 2004). The second wave, between September and December 1918, was responsible for most of the pandemic's deaths, with mortality as high as 2.5% (fig. 2.17); the third wave (February to April 1919) was less virulent.

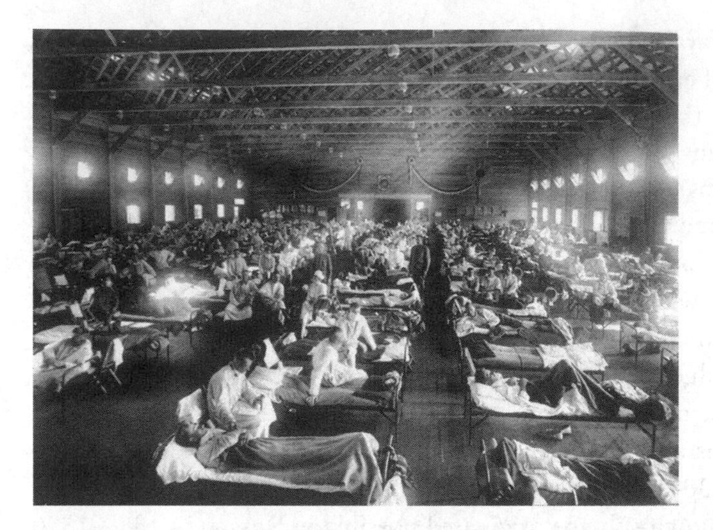

Fig. 2.16
Emergency hospital during the 1918 influenza pandemic at Camp Funston, Kansas. Image 1603, National Museum of Health and Medicine, Washington, D.C.

Scientific advances of the 1980s (polymerase chain reaction, permitting replication of genetic material) made it possible to identify the virus, which was initially retrieved from formalin-fixed, paraffin-embedded lung tissue samples and used to sequence first the fragments of viral RNA and then the complete genome (Taubenberger, Reid, Krafft et al. 1997; Taubenberger, Reid, Laurens et al. 2005). It characterized the pathogen's extraordinary virulence (Tumpey et al. 2005). Statistical analyses of the best available data confirm a peculiar mortality pattern: in contrast to annual epidemics characterized by typical U-shaped mortality patterns, the 1918–1919 pandemic inflicted high mortality on people aged 15–35; years; 99% of all deaths were in people younger than 65 years (WHO 2005). Many of these deaths were due to viral pneumonia, which caused extensive hemorrhaging of the lungs, with death taking place within 48 hours.

There is little certainty regarding the total global death toll of the 1918–1919 influenza pandemic. Perhaps the most commonly cited worldwide aggregate 20–40 million, but a key World Health Organization document refers to "upwards of 40 million people" (WHO 2005), and the best updated account puts the total at 50 million (Johnson and Mueller 2002). Even the lowest estimate is higher than all military and civilian casualties of World War I (~15 million). The total of 50 million

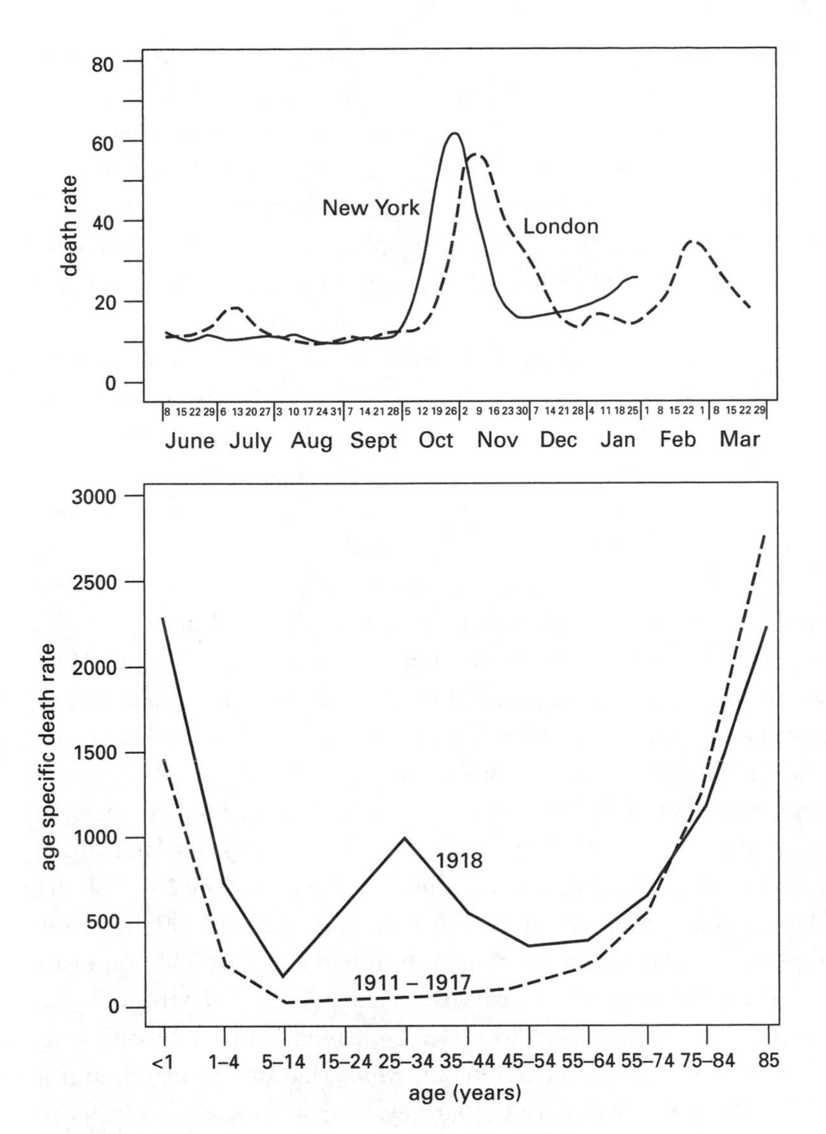

Fig. 2.17

Upper, total mortality in New York and London, 1918–1919. Based on image 3143, National Museum of Health and Medicine, Washington, D. C. *Lower*, influenza and pneumonia mortality in the United States, 1911–1917 and 1918. Plotted from data in Linder and Grove (1943).

deaths from the influenza pandemic would be much higher than the global deaths from the great 1347–1351 plague and almost equal the uncertain grand total of deaths among the populations of the world's two largest Communist regimes of the twentieth century, the Stalinist USSR and Maoist China (White 2003). By any standard, the 1918–1919 influenza pandemic was the deadliest in history. The fairly reliably documented U.S. death toll of 675,000 people (Crosby 1989) was higher than all the deaths sustained by the country's servicemen in all twentieth-century wars.

During the late 1990s, two decades after the last and relatively mild pandemic, new concerns arose because of the emergence of new avian influenza viruses that besides infecting birds and pigs could be transmitted to people. In December 1995 a meeting in Bethesda, Maryland, on pandemic influenza heard from one of the world's leading experts that "at this time, there is no evidence for or against the direct spread of avian influenza viruses to humans" (Webster 1997, S18). By the time this presentation was published, the subtype H5N1 mutated in Hong Kong's poultry markets to a highly pathogenic form, first identified in April 1997, that could kill virtually all affected chickens within two days, and in May 1997 came the first human death, of a three-year-old boy (fig. 2.18) (Sims et al. 2002).

The virus was eventually transferred to at least 18 people, causing six deaths and bringing about the slaughter of 1.6 million birds (Snacken et al. 1999). This episode showed for the first time that avian influenza viruses could infect humans directly, without passing through pigs or other intermediate hosts. Two years later Hong Kong had two poultry-to-people transfers of subtype H9N2, and in 2003 H5N1 strains were isolated from two of the city's SARS patients. Late in 2003 a highly pathogenic subtype H5N1 began to appear again in poultry in East and Southeast Asian countries. During the next three years the virus was found repeatedly in domestic poultry and wild birds in Japan, South Korea, China, Taiwan, Hong Kong, Vietnam, Laos, Cambodia, Philippines, Indonesia, Malaysia, and Thailand, and it spread westward to Mongolia, Kazakhstan, Turkey, and a number of European countries.

Between December 2003 and the end October 2007 there were 329 laboratory-confirmed cases of human H5N1 infection and 201 deaths, indicating a highly virulent pathogen with mortality of 61% (WHO 2007). The highest number of infections and deaths were in Indonesia (107, 86), Vietnam (100, 46), and Thailand (25, 17). Fortunately, the H5N1 virus circulating between 2003 and 2006 was not easily transmissible to humans, and the main impact of its spread was economic,

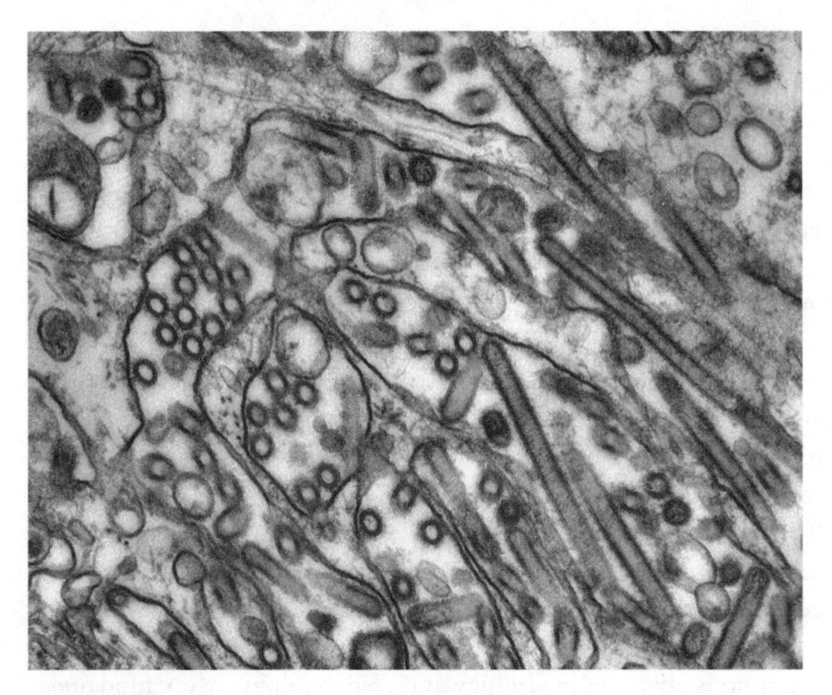

Fig. 2.18
Transmission electron micrograph of avian influenza H5N1 virus. From Centers for Disease Control and Prevention, Atlanta, Ga.

necessitating mass slaughter of infected poultry and the stockpiling of vaccines and antiviral medicines. The Thai outbreak in 2004 was particularly widespread, requiring a slaughter of 40 million chickens in 41 provinces (Chotpitayasunondh et al. 2005).

There is no way to eliminate the natural reservoirs of this virus. South China's high densities and ubiquitous proximities of people, poultry, and pigs make the region a perennial source of new viruses, and studies show that domestic ducks in China's southern provinces are the key reservoir of H5N1 (H. Chen et al. 2004). The serotype is now also widely present in wild migratory fowl; ducks, geese, and swans were credited with spreading it across much of Eurasia. However, that assumption may not be correct because there was no such transmission during the eight years after the virus was identified in Hong Kong in 1997, with billions of birds using the same flyways. Poultry shipment and the movement of contaminated materials and wastes may be the primary routes (Fergus et al. 2006).

Because the H5N1 serotype is highly pathogenic and has become ineradicable throughout large parts of Asia, it clearly has a pandemic potential (Li et al. 2004). Heightened awareness of the risks posed by H5N1 led epidemiologists to predict a high probability of pandemic influenza in the not-too-distant future. The following realities indicate the imminence of the risk. The typical frequency of influenza pandemics was once every 50–60 years between 1700 and 1889 (the longest known gap was 52 years, between the pandemics of 1729–1733 and 1781–1782) and only once every 10–40 years since 1889. The recurrence interval, calculated simply as the mean time elapsed between the last six known pandemics, is about 28 years, with the extremes of 6 and 53 years. Adding the mean and the highest interval to 1968 gives a span between 1996 and 2021. We are, probabilistically speaking, very much inside a high-risk zone.

Consequently, the likelihood of another influenza pandemic during the next 50 years is virtually 100%, but quantifying probabilities of mild, moderate, or severe events remains largely a matter of speculation because we simply do not know how pathogenic a new virus will be and what age categories it will preferentially attack. Assessing the most likely extent of morbidity and mortality is even more challenging. Despite the enormous advances in virology and epidemiology, many fundamental scientific questions concerning the origins, virulence, and diffusion of influenza remain unanswered (Taubenberger and Morens 2006). The origin of the 1918–1919 pandemic remains unidentified, and that virulent strain itself was genetically unlike any other known virus examined since that time. Consequently, the current concern about avian H5N1 may be entirely misplaced, whereas a new strain may turn out to be pandemic. We also do not know how influenza A viruses switch to new avian hosts and, most important, what induces them to change so that they can propagate among humans.

The isolated cases of known human-to-human transmissions of H5N1 do not answer that key question. Certainly the most encouraging fact is that the widespread human exposure to infected poultry has produced only a few hundred human infections. This means that the species barrier to H5N1 diffusion is considerable (Hou and Shu 2006). Moreover, innate immune response (including elevated levels of inflammatory mediators among patients who died) may have been responsible for some of the human disease clusters.

Optimistic assessments of the next influenza pandemic put the infection rate at 20% of the world's population, with 1 in every 100 ill people requiring hospitalization (providing the beds will be available) and 7 million deaths in a few months

(Stöhr and Esveld 2004). But the morbidity rate may actually be 25%–30%, and the World Health Organization believes that a new pandemic may affect 20%–50% of the world's population. Its toll, however, cannot be definitely predicted because we have no way of knowing the eventual virulence of new infectious strains. What is certain is that the appearance of subtype H5N1 has brought us closer to the next pandemic and that, whatever its actual magnitude, we are not adequately prepared for it and for its consequences (WHO 2005).

If the eventual death toll were to resemble those of the last two pandemics, with just a few million excess deaths, there would be no global consequences. If it were merely a repeat of the 1918–1919 event, with mortality of no less than 20–25 million and no more than 50 million people, the relative global toll would be obviously much smaller (only about one-fourth as large) than it was four generations ago. But the overall mortality could also be a proportionally potentiated replica of 1918–1919. In the early 2000s we had a 3.4 times larger global population, and an at least eight times larger (nearly 20 billion compared to less than 3 billion) worldwide inventory of poultry (the main reservoir of lethal viruses), with a large share of these birds in large feeding facilities housing tens of thousands of birds.

There are other obvious reasons that could make the next pandemic more costly even if the virulence of the pathogen and the relative mortality rate were no greater than in the 1918–1919 episode. By 2007 the world's cities, the environment that affords much faster spread of infection, housed 50% of all people (76% in affluent countries), compared to only about 30% in 1918 (Brockerhoff 2000). Moreover, the global economy is incomparably more integrated, and modern travel and traffic in goods (including live animals and other agricultural products) are several orders of magnitude faster and more voluminous than nine decades ago, a near-ideal condition to spread infections around the world.

In 1918 it took six days to cross the Atlantic on a liner; now it takes six hours on a jetliner, and there is no doubt that air travel plays an important role in the diffusion of annual epidemics (Grais et al. 2004). The spread of SARS from Hong Kong to Toronto illustrated how unpredictably and rapidly such diffusions can take place (Abraham 2005). The extent and intensity of global links also make it impractical to adopt rapid and effective quarantine measures. Additional human factors facilitating the spread of infectious diseases include the rising demand for wild meat (common in both Africa and parts of Asia), drug addiction (including intravenous injections), and mass urban prostitution.

Even assuming "only" 25 million deaths in 1918–1919, we could see a proportionally increased global mortality surpassing 80–100 million people; with 50 million deaths in 1918–1919, the proportional total would rise to 150–200 million. With a slightly more than 5% mortality rate (that was the well-documented U.S. mean in 1918–1919) there could be at least 1.5–2 billion people contracting the infection, and it would be clearly beyond the capacity of health services to cope effectively with such a sudden burden of mass sickness. On the other hand, there are positive factors of generally better nutrition, much better hospital care, and incomparably greater virological understanding. Even so, the overall enormity of ubiquitous morbidity and greatly multiplied mortality would pose challenges unseen in most countries for generations.

In addition, many specific impacts would complicate our ability to deal with immediate challenges and would have long-term consequences. A smaller herald wave of infections several months ahead of the main event (as happened in 1918 in the United States) might not be helpful: rather than giving us more time to prepare, it might actually cause more helplessness and fear because any development of a new vaccine, which would begin only once the pandemic virus started its diffusion, would not be completed before the virus covered the world (Stöhr and Esveld 2004). But if the pandemic resembled the 1918–1919 episode, then the event might not be over in six months, substantial mortality could continue during the second season, and many cities and countries might find it particularly difficult to cope with the second wave.

That phenomenon was illustrated in the case of Toronto's second wave of SARS in May 2003, minuscule in terms of total numbers but extremely taxing due to mental burdens and logistical problems cause by quarantined hospitals where none but emergency operations were done and no visits, not even to terminally ill patients, were allowed. The mortality burden might shift, as it did so stunningly during the 1918–1919 pandemic, to younger people (see fig. 2.17). Repetition of this pattern could strain the availability and effectiveness of caregivers (health professionals ranging from physicians to staff at retirement homes) and substantially worsen dependency ratios, particularly in Europe's aging populations, where they have already risen to unprecedented levels.

The massive mortality of people in their prime would also strain the life insurance industry and depress real estate values. And what would the 24-hour news media, so adept at flogging a few accidental deaths in all-day marathons of despair, do with so many deaths that would just keep coming, day after day, week after week?

How would the financial markets react to this massive and indiscriminate dying? More important, what would be the long-term economic cost in fear and depression on top of the immediate social and economic insults to the previously insulated Western way of life?

To what extent would Europe's ravaged countries become open to virtually unchecked Muslim immigration, speeding up the conversion of the continent into Eurabia? What would the collapse of global trade do to the lives of hundreds of millions of factory workers in Asia whose wages depend on exports to affluent economies? How would it affect inequality among the already highly economically polarized populations of Africa and Latin America? Until a new pandemic unfolds in its unique way, we will have more uncertainty than assurance about our chances of coping with what might be a truly unprecedented public health and socioeconomic challenge.

Violent Conflicts

While trying to assess the probabilities of recurrent natural catastrophes and catastrophic illnesses, we must remember that the historical record is unequivocal: these events, even when combined, did not claim as many lives and have not changed the course of world history as much as the deliberate fatal discontinuities that Rhodes (1988) calls man-made death, the single largest cause of non-natural mortality in the twentieth century. Violent collective death has been such an omnipresent part of the human condition that its recurrence in various forms conflicts lasting days to decades, homicides to democides, is guaranteed. Long lists of the past violent events can be inspected in print (Richardson 1960; Singer and Small 1972; Wilkinson 1980) or in electronic databases (White 2003; IISS 2003; PRIO 2004).

Even a cursory examination of this record shows yet another tragic aspect of that terrible toll: so many violent deaths had no or only a marginal effect on the course of world history. Others, however, contributed to outcomes that truly changed the world. Large death tolls of the twentieth century that fit the first category include the Belgian genocide in the Congo (began before 1900), Turkish massacres of Armenians (mainly in 1915), Hutu killings of Tutsis (1994), wars involving Ethiopia (Ogaden, Eritrea, 1962–1992), Nigeria and Biafra (1967–1970), India and Pakistan (1971), and civil wars and genocides in Angola (1974–2002), Congo (since 1998), Mozambique (1975–1993), Sudan (since 1956 and ongoing), and Cambodia (1975–1978). Even in our greatly interconnected world, such conflicts can cause more than 1 million deaths (as did all of the just listed events) and go on for decades without having any noticeable effect on the cares and concerns of the remaining 98%–99.9% of humanity.

By contrast, the modern era has seen two world wars and interstate conflicts that resulted in long-lasting redistribution of power on global scales, and intrastate (civil) wars that led to the collapse or emergence of powerful states. I call these conflicts transformational wars and focus on them next. I then examine the most threatening category of asymmetrical conflicts, terrorist attacks used by small groups or loosely connected networks to challenge even the most powerful nation-states, and in so doing, changing the course of the world history. Determined, protracted terrorist activities on the local or national level are not new, but after 9/11 there can be no doubt about their impact on global history.

Transformational Wars

There is no canonical list of transformational wars of the nineteenth and twentieth centuries. Historians agree on the major conflicts that belong in this category but differ as to others. My own list is fairly restrictive; a more liberal definition of worldwide impacts could extend the list. A long-lasting transformational effect on the course of world history is a key criterion. And most of the conflicts I have called transformational share another characteristic: they are mega-wars, claiming the lives of more than 1 million combatants and civilians. By Richardson's (1960) definition, based on the decadic logarithm of total fatalities, most would be magnitude 6 or 7 wars (fig. 2.19). Their enumeration starts with the Napoleonic wars, which began in 1796 with the conquest of Italy and ended in 1815 in a refashioned, and for the next 100 years also remarkably stable, Europe. This stability was not basically altered, either by brief conflicts between Prussia and Austria (1866) and Prussia and France (1870–1871) or by repeated acts of terror that killed some of the continent's leading public figures while others, including Kaiser Wilhelm I and Chancellor Bismarck, escaped assassination attempts.

The next entry on my list of transformational wars is the protracted Taiping war (1851–1864), a massive millennial uprising led by Hong Xiuquan (Spence 1996). This may seem like a puzzling addition to readers not familiar with China's modern history, but the Taiping uprising, aimed at achieving an egalitarian, reformist kingdom of heaven on earth, exemplifies a grand transformational conflict because it fatally undermined the ruling Qing dynasty, enmeshed foreign actors in China's politics for the next 100 years, and brought in less than two generations the end of

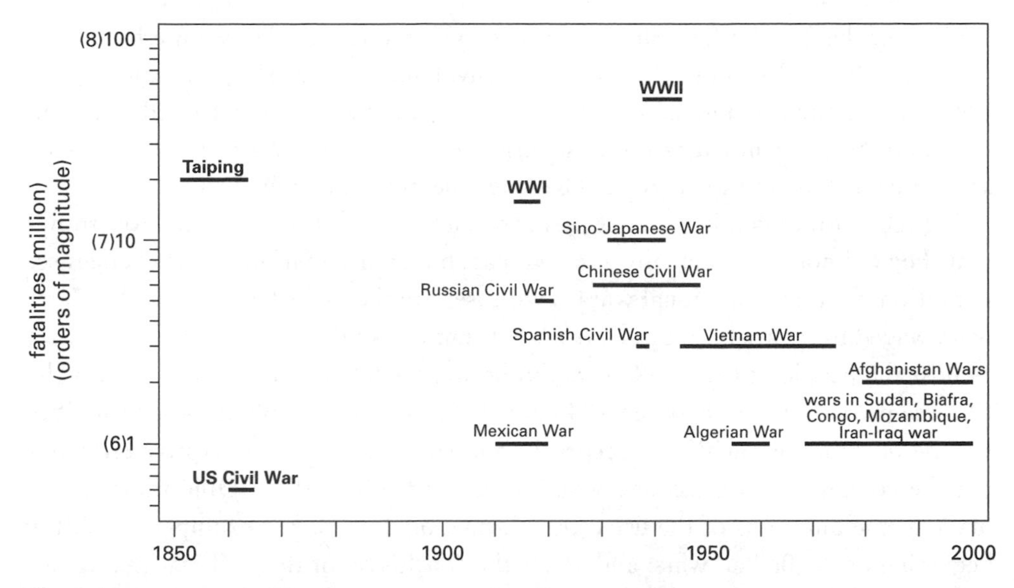

Fig. 2.19
Wars of magnitude 6 or 7, 1850–2000. Boldface font indicates wars that the author considers transformational. Plotted fatalities are minimal to average (heavily rounded) estimates from sources cited in the text.

the old imperial order. With about 20 million fatalities, its human costs were higher than the aggregate losses of combatants and civilians in World War I.

The American Civil War (1861–1865) should be included because it opened the way toward the country's rapid ascent to global economic primacy (H.M. Jones 1971). U.S. GDP surpassed that of Great Britain by 1870; by the 1880s the United States became the technical leader and the world's most innovative economy, firmly set on its rise toward superpower status.

World War I (1914–1918) traumatized all European powers, utterly destroyed the post-Napoleonic pattern, ushered in Communism in Russia, and brought the United States into global politics for the first time. And—a fact often forgotten—it also began the destabilization of the Middle East by dismembering the Ottoman Empire and creating the British and French Mandates whose dissolution led eventually to the formation of the states of Jordan (1923), Saudi Arabia and Iraq (1932), Lebanon (1941), Syria (1946), and Israel (1948) (Fromkin 2001).

World War II (1939–1945) is, of course, the quintessential transformational war, not only because of the sweeping changes it brought to the global order but also because of the decades-long shadows it cast over the rest of the twentieth century. Virtually all the key post-1945 conflicts involving that war's protagonists—the USSR, the United States, and China in Korea; France and the United States in Vietnam; the USSR in Afghanistan; superpower proxy wars in Africa—can be seen as actions designed to maintain or challenge the outcome of WW II.

Arguably, other conflicts might seem to qualify, but closer examination shows that they did not fundamentally alter the past but rather reinforced the changes set in motion by transformational wars. Two cases are the undeclared but no less fatal wars waged by a variety of means from outright mass killings to deliberate famines against the people of the USSR by Stalin between 1929 and 1953, and against the Chinese people by Mao between 1949 and 1976. The actual toll of these brutalities will be never known with any accuracy, but even the most conservative estimates put the combined death toll at above 70 million (White 2003). Objections can be made to the durations of the listed transformational wars. For example, 1912, the beginning of the Balkan wars, and 1921, the conclusion of the civil war that established the Soviet Union, may be more appropriate dating of World War I. And one could say that World War II started with Japan's invasion of Manchuria in 1933 and ended only with the Communist victory in China in 1949.

Even a rather restrictively defined list of transformational wars adds up to 42 years of conflicts in two centuries, with conservatively estimated total casualties (combatant and civilian) of about 95 million (averaging 17 million deaths per conflict). The mean recurrence rate is about 35 years, and an implied probability of a new conflict of that category stands at roughly 20% during the next 50 years. All these numbers could be reduced by including the wars of the eighteenth century, a time of remarkably lower intensity of all violent conflicts than the two preceding and two subsequent centuries (Brecke 1999). On the other hand, most of that century belonged distinctly to the preindustrial era, and most of the then major powers (e.g., Qing China, enfeebled Mughal India, and weakening Spain) were nearing the end of their influence. Thus the exclusion of eighteenth-century wars makes sense.

Three important conclusions emerge from the examination of all armed conflicts of the past two centuries. First, until the 1980s there was an upward trend in the total number of conflicts starting in each decade; second, there was an increasing share of wars of short duration (less than 1 year) (Kaye, Grant, and Emond 1985). Implications of these findings for future transformational conflicts are unclear. So is the fact that between 1992 and 2003 the worldwide number of armed conflicts

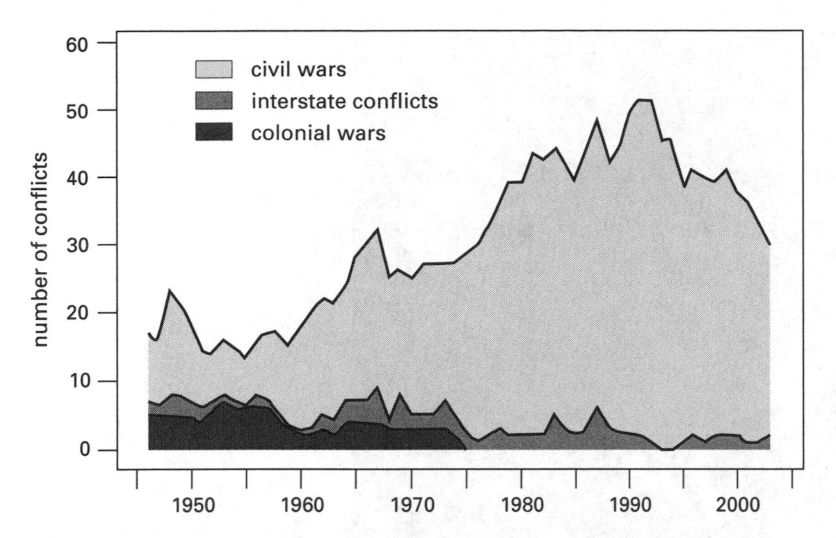

Fig. 2.20
Worldwide conflicts, 1946–2003. From Human Security Centre (2006).

declined by 40%, and the number of wars with 1,000 or more battle deaths dropped by 80% (fig. 2.20) (Human Security Centre 2006). These trends were clearly tied to declining arms trade and military spending during the post–Cold War era; it is thus unclear if the decade was a welcome singularity of reduced violence or a brief aberration.

The most important finding regarding the future likelihood of violent conflicts comes from Richardson's (1960) search for causative factors of war and his conclusion that wars are largely random catastrophes whose specific time and location we cannot predict but whose recurrence we must expect. That would mean that wars are like earthquakes or hurricanes, leading Hayes (2002,15) to speak of warring nations that "bang against one another with no more plan or principle than molecules in an overheated gas." At the beginning of the twenty-first century it could be argued that new realities have greatly diminished the recurrence of many possible conflicts, thus, to continue the metaphor, greatly reducing the density and the pressure of the gas.

The European Union is widely seen as a near-absolute barrier to armed conflicts involving its members. America and Russia may not be strategic partners, but they surely do not take the same adversarial positions that they held for two generations preceding the fall of the Berlin Wall in 1989. The Soviet Union and China came

Fig. 2.21
Napoleon Bonaparte, First Consul of France, 1799–1802. Painting (ca. 1802) by Antoine-Jean Gros.

very close to a massive conflict in 1969 (a close call that prompted Mao's rapprochement with the United States), but today China buys the top Russian weapons and would gladly buy all the oil and gas that Siberia could offer. And Japan's very constitution prevents it from attacking any country. This reasoning would negate, or at least severely undercut, Richardson's (1960) argument, but it would be a mistake to use it when thinking about long spans of history. Neither short-term complacency nor understandable reluctance to imagine the locale or cause of the next transformational was a good argument against its rather high probability.

In 1790 no Prussian high officer or Czarist general could suspect that Napoleon Bonaparte (fig. 2.21), a diminutive Corsican from Ajaccio, who became known to his troops as *le petit caporal*, would set out to redraw the map of Germany before embarking on a mad foray into the heart of Muscovy (Zamoyski 2004). In 1840 the Emperor Daoguang could not have dreamt that the dynastic rule that had lasted for millennia would come close to its end because of Hong Xiuquan, a failed candidate of the state Confucian examination who came to think of himself as a new Christ and who led the protracted Taiping rebellion (Spence 1996). And in 1918 the victorious powers, dictating a new European peace at Versailles, would not have credited that Adolf Hitler, a destitute, neurotic would-be artist and gassed veteran of the trench warfare, would within two decades undo their new order and plunge the world into its greatest war.

New realities may have lowered the overall probability of globally transformational conflicts, but they have not surely eliminated their recurrence. Causes of new conflicts could be found in old disputes or in surprising new developments. During 2005–2007 the probabilities of several new conflicts rose from vanishingly low to decidedly non-negligible as the North Korean threat led Japan to raise the possibility of an attack across the Sea of Japan; as the chances of a U.S.-Iran war (nonexistent during the Pahlavi dynasty, very low even after the Revolutionary Guards took the U.S. embassy hostages) were widely discussed in public; and as China and Taiwan continued their high-risk posturing regarding the fate of the island.

Richardson's (1960) reasoning and the record of the past two centuries imply that during the next 50 years the likelihood of another armed conflict with potential to change world history is no less than about 15% and most likely around 20%. As in all cases of such probabilistic assessments, the focus is not on a particular figure but rather on the proper order of magnitude. No matter whether the probability of a new transformational war is 10% or 40%, it is 1–2 OM higher than that of the globally destructive natural catastrophes that were discussed earlier in this chapter.

Before leaving this topic I must note the risks of an accidentally started, transformational mega-war. As noted, we have lived with this frightening risk since the early 1950s and during the height of the Cold War. Casualties of an all-out thermonuclear exchange between the two superpowers (including its lengthy aftermath) were estimated to reach hundreds of millions (Coale 1985). Even a single isolated miscalculation could have been deadly. Forrow et al. (1998) wrote that an intermediate-sized launch of warheads from a single Russian submarine would have killed nearly instantly about 6.8 million people in eight U.S. cities and exposed millions more to potentially lethal radiation.

On several occasions we came perilously close to such a fatal error, perhaps even to a civilization-terminating event. Nearly four decades of the superpower nuclear standoff were punctuated by a significant number of accidents involving nuclear submarines and long-range bombers carrying nuclear weapons, and by hundreds of false alarms caused by malfunctions of communication links, errors of computerized control systems, and misinterpretations of remotely sensed evidence. Many of these incidents were detailed in the West after a lapse of time (Sagan 1993; Britten 1983; Calder 1980), and there is no doubt that the Soviets could have reported a similar (most likely, larger) number.

The probabilities of such mishaps escalating out of control rose considerably during periods of heightened crisis, when a false alarm was much more likely to be interpreted as the beginning of a thermonuclear attack. A series of such incidents took place during the most dangerous moment of the entire Cold War, the October 1962 Cuban Missile Crisis (Blight and Welch 1989; Allison and Zelikow 1999). Fortunately, there was never any accidental launch, either attributable to hardware failure (crashing nuclear bomber, grounded nuclear submarine, temporary loss of communication) or to misinterpreted evidence. One of the architects of the Cold War regime in the United States concluded that the risk was small because of the prudence and unchallenged control of the leaders of the two countries (Bundy 1988).

The size of the risk depends entirely on the assumptions made in order to calculate cumulative probabilities of avoiding a series of catastrophic mishaps. Even if the probability of an accidental launch were just 1% in each of some 20 known U.S. incidents (the chance of avoiding a catastrophe being 99%), the cumulative likelihood of avoiding an accidental nuclear war would be about 82%, or, as Phillips (1998,8) rightly concluded, "about the same as the chance of surviving a single pull of the trigger at Russian roulette played with a six-shooter." This is at once correct reasoning and a meaningless calculation. As long as the time available to verify the real nature of an incident is shorter than the minimum time needed for a retaliatory strike, the latter course can be avoided and the incident cannot be assigned any definite avoidance probability. If the evidence is initially interpreted as an attack under way, but a few minutes later this is entirely discounted, then in the minds of decision makers the probability of avoiding a thermonuclear war goes from 0% to 100% within a brief span of time. Such situations are akin to fatal car crashes avoided when a few centimeters of clearance between the vehicles makes the difference between death and survival. Such events happen worldwide thousands

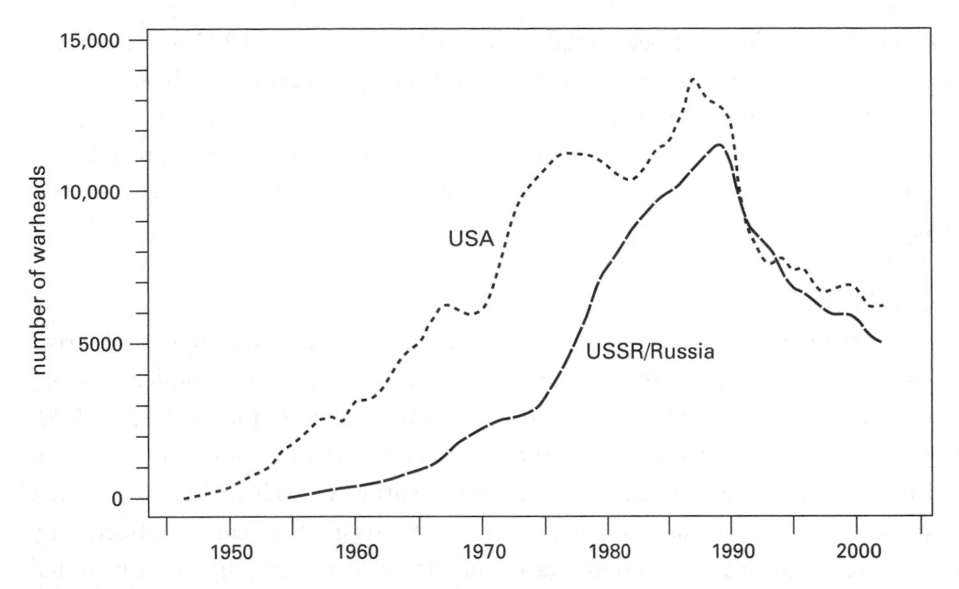

Fig. 2.22
Number of U.S. and Soviet/Russian strategic warheads, 1945–2005. Plotted from data in NRDC (2006).

of times every hour, but an individual has only one or two such experiences in a lifetime, so it is impossible to calculate the probabilities of any future clean escape.

The demise of the USSR had an equivocal effect. On the one hand it undoubtedly diminished the chances of accidental nuclear war thanks to a drastic reduction in the number of warheads deployed by Russia and the United States. In January 2006, Russia had approximately 16,000 warheads compared to the peak USSR total of nearly 45,000 in 1986, and the United States had just over 10,000 warheads compared to its peak of 32,000 in 1966 (Norris and Kristensen 2006). Totals of strategic offensive warheads fell rapidly after 1990 to less than half of their peak counts (fig. 2.22), and the Strategic Offensive Reductions Treaty, signed in May 2002, envisaged further substantial cuts. On the other hand, it is easy to argue that because of the aging of Russian weapons systems, a decline in funding, a weakening of the command structure, and the poor combat readiness of the Russian forces, the risk of an accidental nuclear attack has actually increased (Forrow et al. 1998).

Moreover, with more countries possessing nuclear weapons, it is reasonable to argue that chances of accidental launching and near-certain retaliation have been increasing steadily since the beginning of the nuclear era. Since 1945 an additional nation has acquired nuclear weapons roughly every five years; North Korea and Iran have been the latest candidates. Even as the concerns about nuclear proliferation have been rising, an old kind of violence has assumed new and unprecedented prominence: terrorist attacks are rising to a level of globally transformative events.

Terrorist Attacks

I should start by emphasizing that this discussion is not about state-operated terror aimed at a ruling regime's domestic enemies, a bloody ingredient of modern history that was elevated to governmental policy in revolutionary France (Wright 1990) and perfected during the twentieth century in many countries on four continents, but surely to the greatest extent and with the most ruthless reach in Stalinist Russia (Conquest 1990). My focus is on global transformations that can be effected by relatively small groups of internationally operating terrorists, that is, on global impacts arising from asymmetrical violence whereby a few can inflict serious damage on the many. Although possessing enormous arsenals of armaments and astonishing technical prowess, states find it very difficult to eliminate such threats or even to keep them within acceptable bounds.

There has been no shortage of terrorist actions ("politics by other means") in modern history (Lacquer 2001; Carr 2002; Maxwell 2003; Sinclair 2003; Parry 2006), but as with classical armed conflicts, most of them have not risen to the level of globally transforming events. Following Rapoport's (2001) division of the history of terrorism into four waves, it is clear that neither the first wave, begun in 1879 and dominated by nearly four decades of Russia's *narodnaya volya* (the people's will) assassinations (Geifman 1993; Hardy 1987), nor the second wave, extending from the 1920s to the 1960s and characterized primarily by terror in the service of national self-determination, had globally transformative effects.

More important, in the long run those terrorist actions did not make a great deal of difference even to the collective fortunes of the afflicted societies. In Russia, the *narodnik* killings were less important in the demise of the Czarist Empire than tardiness of internal reforms, imperial overreach leading to conquest of the Muslim populations in Central Asia and in the Caucasus region between 1817–1864 (Yemelianova 2002), and the socioeconomic impacs of World War I.

As for the process of decolonization, it was driven primarily by the colonizers themselves, not by terrorist groups. And in Israel it was the *Haganah*, the predecessor of the Israel Defense Forces, rather than terrorist groups (Stern Gang, Irgun) that created modern Israel (Allon 1970).

Rapoport's (2001) third terrorist wave, in the 1960s and 1970s, was much more far-reaching. It included the PLO (Palestine Liberation Organization), PFLP (Popular Front for the Liberation of Palestine), IRA (Irish Republican Army), Basque ETA, Italian *Brigatte Rosse*, French *Action Directe*, German *Rotte Armee*, and the U.S. Weather Underground (Alexander and Myers 1982; Lacquer 2001; Parry 2006). These terrorists shared the rhetoric of struggle and violence and a predilection for airline hijacking and for obtaining weapons and explosives from the Soviet Union and its allies. Their exploits included such well-publicized actions of international terrorism as the PFLP's hijackings of planes to Amman in September 1970, the Munich Olympics massacre of Israeli athletes in 1972, and the kidnapping of OPEC ministers in Vienna in 1975.

Yet, these terrorist actions did not have any collective global impact. They did not inflict irreparable economic damage on Ireland, the UK, Germany, Spain, or Italy (where the country's Prime Minister, Aldo Moro, was kidnapped and murdered in 1979); did not bring about drastic social or political shifts within these societies; and did not prevent the countries' successful integration into the European Union. The PLO was actually weakened by its radical stance; what looked like the pinnacle of its influence—hijacked airplanes, Arafat's UN speech with holstered gun—was actually the beginning of its road to negotiations in Oslo and to the Madrid meeting of 1991.

Rapoport (2001) sees 1979 as the beginning of a fourth, still unfolding, wave of modern terrorism. That year saw the downfall of the Pahlavi dynasty and the rise of Ayatollah Khomeini's theocracy in Iran as well as, symbolically, the beginning of a new century in the Islamic calendar—year 1400 since the *hijra* began at sundown of November 19, 1979, fifteen days after the Khomeini-inspired mob took over the U.S. embassy in Teheran. These events radicalized other *shī'a* societies and led directly to the establishment of Hizbullah in 1982, to its mass murder of U.S. marines in Beirut in 1983, and to its campaign of assassinations, kidnappings, and prolonged hostage takings.

And there was no lack of religiously motivated *sunnī* terrorism, beginning with a violent temporary takeover of the Grand Mosque of Mecca in 1979 and extending to bombings in North Africa, the Middle East (above all, in Egypt), the Philippines, and Indonesia. This wave was greatly reinforced during the decade of the struggle to remove the Soviet forces from Afghanistan. During the 1990s religiously inspired terror was behind the most frequent kind of indiscriminate attacks, suicide bombings in Israel (perpetrated usually by young men whose designation *shahīd* conveys both "witness" and "religious martyr"). Al Qaeda's first suicide attacks, on U.S. embassies in Kenya and Tanzania, killed 301 people and injured more then 5,000 on August 7, 1998.

But, once again, by the year 2000 it would have been difficult to argue that these terrorist attacks had changed (or had a major role in changing) the fundamental ways of the Western world. I have always found it extraordinarily remarkable that Hizbullah's suicidal attack on the U.S. marines was met with no retaliation whatsoever, had no political domestic consequences, and led to no court-martials; the U.S. response was only a rapid retreat. (In retrospect, if that event had been seen as the opening blast of modern anti-Western Islamic violence, history might have unfolded differently). Similarly, the reaction to the 1993 World Trade Center bombing was near-instant forgetting (despite the discovery of manuals detailing future attacks). And, remarkably, Israeli society has shown an enormous resilience through the years of random suicide bombings; if the U.S. population were attacked with such frequency (proportionately) more than 70,000 people would have died between 1968 and 2004 (fig. 2.23).

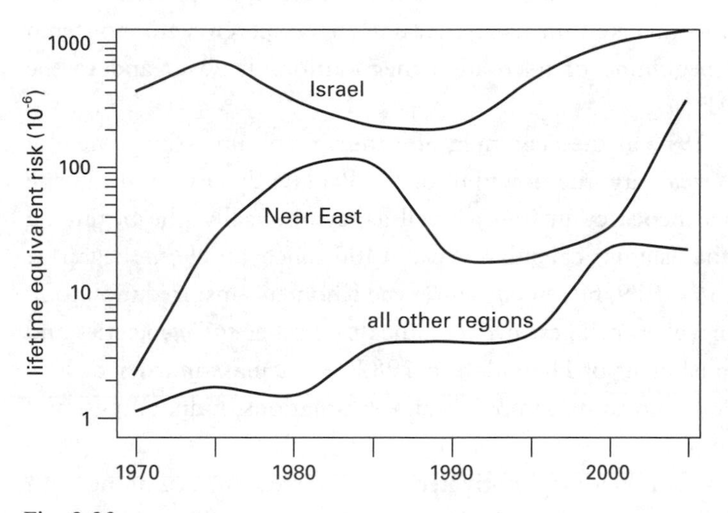

Fig. 2.23
Equivalent lifetime (70-year) risk of casualty or death from terrorist attacks in Israel, the Near East (excl. Israel and Palestine), and the rest of the world, 1968–2004. Based on Bogen and Jones (2006).

The attacks of September 11, 2001, when 19 Islamic terrorists hijacked four Boeing 767s and steered two of them into the Twin Towers of the World Trade Center and one into the Pentagon (the fourth crashed because of passenger resistance), changed everything. They elevated terrorism to the class of global catastrophic events. Through the world's omnipresent visual media, they produced a horrific (and endlessly replayed) spectacle whose impact resides primarily in unforgettable impressions created by the attack's execution rather than in total number of deaths or lasting economic impact. (I address comparative fatalities and economic consequences in some detail in chapter 5).

Similarly, the main reason for the post-9/11 response was not the ensuing economic damage, not even the tragic human toll, but the shock of the experience (the first attack on the U.S. mainland since the British raids during the War of 1812); the unmistakable symbolism of the attacks (striking at the seat of the global economy as well at the power centers of the superpower); and the attendant fear of possible repeats and even more deadly strikes. The consequences of 9/11 can be understood only when looking far beyond the immediate casualties and near-term economic losses, and the same kind of broader perspectives are needed to assess future threats. Unfortunately, only a single conclusion can be reached with certainty, namely, that the oft-repeated post-2001 aspiration to eliminate terrorism ("winning the war on terror") is unachievable. As Rapoport (2001,424) put it, "Terrorism is deeply rooted in modern culture. Even if the fourth wave soon follows the path of its three predecessors, another inspiring cause is likely to merge unexpectedly, as it has too often in the past."

The real questions thus concern the likely extent, frequency, and impact of continuing terrorist attacks. Put differently, what is the likelihood that terror will become a recurrent factor (albeit at irregular and relatively lengthy intervals) in shaping global affairs or perhaps even a dominant preoccupation of the modern world? In the wake of the 9/11 attack, the modalities and consequences of future terrorist actions that would fall into the same class have been explored—from different perspectives, at once excessively yet insufficiently, with exaggeration as well as with inadequate appreciation of possibilities (Carr 2002; Silvers and Epstein 2002; Calhoun, Price and Timmer 2002; Bennett 2003; Flynn 2004; Fagin 2006)—but their probability remains beyond useful quantification.

While nobody can assign meaningful numerical probabilities regarding the extent, frequency, and impact of future attacks, there is a great deal of historical and statistical evidence that offers some comforting and some disturbing conclusions. At the outset of such an assessment is the troubling array of choices available to determined terrorists: cyberattacks on modern electronic infrastructures, poisoning of urban water or food supplies, decapitation of national leaders, dirty bombs, and release of old or new pathogens. Public policy and precaution dictate that none of these incidents, no matter how low their probabilities might seem, can simply be dismissed as unlikely. This very reality makes it very difficult to assess the relative likelihood of these different modes of attack.

But a skeptical appraisal must point out at least two important facts. Many of these attacks are not as easy to launch as media coverage would lead us to believe, and many of them, even if successfully executed, would have relatively limited impacts and would not rise to the level of global transformational events. A notable example is the use of nerve gas. As the fanatics of Japan's Aum Shinrikyō discovered, it is not easy to disperse a nerve gas (sarin in their case) and kill a large number of people even in such a densely populated setting as Tokyo's subway system, even if the people selected for dispersing the gas were well trained. A total of 12 people died as the result of the March 20, 1995, attack (Murakami 2001).

And the U.S. anthrax scare of October and November 2001 was primarily a matter of irresponsibly exaggerated fears. Indeed, many experts argue, the entire threat of bioterrorism has been vastly overblown (Enserink and Kaiser 2005). Given the large number of pathogens that could be used by terrorists (smallpox, plague, anthrax, botulism, tularemia, Venezuelan equine encephalitis, transmissible by mosquitoes from horses to humans) and the large number of ways in which the diseases can be spread (spraying viruses in shopping malls, using crop duster planes over cities) means that a continuous effort to reduce such risks to a negligible level is beyond our health and security resources.

And given the enormous amount of money that is now pouring into U.S. research on anthrax and smallpox (more than a half billion dollars in FY 2004 and FY 2005), the work on common pathogens that annually claim millions of lives worldwide is already being shortchanged. There is a much greater danger (counterintuitive but credible) that a bioterror agent could be released by a disgruntled employee of one of the 14 new research superlabs built to handle the most dangerous pathogens than that it would be carried to the United States in a suitcase by a fundamentalist zealot from the caves of Waziristan.

Worries about pathogens deployed by terrorists could be multiplied by considering the attacks with plant and animal diseases that offer perfect opportunities for low-tech, high-impact (in terms of economic cost) terrorism (Wheelis, Casagrande, and Madden 2002; Gewin 2003). Again, there is no shortage of possible pathogens, from *Phytophthora* fungus to *Brucella suis* to foot-and-mouth disease, but it might be more rewarding to ensure first that our food supply is free of unnecessary, preventable risks caused by dubious commercial choices (e.g., turning herbivorous cattle into cannibalistic carnivores, an insanity that brought us mad cow disease).

In general, the arguments about the high risks of terrorist actions, which are supposedly much easier to launch than spectacular mass murders but which might ultimately prove costlier and deadlier—attacks that Homer-Dixon (2002) defined as ingredients of complex terrorism—should be seen with a great deal of skepticism. Rather than killing people, terrorists would exploit the growing complexities of modern societies and deploy their "weapons of mass disruption" in attacking non-redundant nodes of critical infrastructures like electricity networks, other energy supply systems, chemical factories, or communication links. A skeptical riposte to this scenario is, given the fact that these attacks are so easy to launch, why are we not seeing scores of them every month?

After all, it is impossible to safeguard against explosives every one of hundreds of thousands of steel towers that carry a large nation's high-voltage lines; or to protect round-the-clock every one of tens of thousands of transforming substations; or to detect every poisoned kilogram among millions of tonnes of harvested crops; or to secure thousands of reservoirs and rivers that furnish drinking water.

A relatively rich experience with accidental large-scale electricity supply outages (caused by weather, human error, or technical problems) demonstrates that similar or even larger failures would not rise to the level of historic milestones. For example, who now remembers the great U.S. blackouts of 1965 and 1977? And the great U.S.-Canadian blackout of August 2003 provided a remarkable illustration of technical and social resilience. Caused by a series of preventable technical failures, it affected some 50 million people in the northeastern United States and Ontario, and it extinguished all traffic lights and stopped all subway trains in New York (Huber and Mills 2004). But the trading on Wall Street continued, the blackout did not lead to any catastrophic disruption of the region's business or economic growth, and hospitals were able to maintain adequate care (Huber and Mills 2004). And despite hot weather, the crime rate had actually dropped. In general, people made the best of their often taxing experiences, and if they looked upward, they were rewarded by a rare sight, dark Manhattan under a starry sky.

Use of the weapons of mass disruption would most likely amount to nothing but a spatially and temporarily limited (though fairly expensive) nuisance, which would remain far below the threshold of events capable of modifying the course of global history. This may be the most important reason why terrorists have little interest in such attacks. The globally reverberating shock they seek can hardly be achieved by a temporary disruption of a city's electricity supply or by killing a few thousand pigs. Launching new spectacular mega-attacks is clearly not easy; a detailed summary of worldwide terrorist attacks during the fourth year after 9/11 reinforces the conclusion that the global impact of terrorism in a "normal" year is at best marginal.

According to the National Counterterrorism Center (NCC 2006), the year saw 11,111 terrorist attacks, but in 5,980 of them (54%) there were no fatalities, 2,884 had a single fatality, and 1,617 had two to four fatalities. This means, fortunately, that in nearly 95% of all terrorist events the human impact was akin to the level of moderate to serious car accidents. The country with the second highest number of fatalities (after Iraq) was India, with 1,357 deaths, but this toll had no effect on the country's overall economic performance and led to no discernible changes in its policies. Western Europe's most deadly attack targeted three subway trains and a bus in London on July 7, 2005, killing 52 people, but except for some new policing and counterterrorism measures, it had a surprisingly small effect on the country's affairs. Finally, given the prominence of the United States and its citizens as the targets of Islamic terror, it is notable that in 2005 only 0.4% (56 people) of all deaths due to terror attacks were U.S. citizens.

An extended perspective also does not support an image of terrorism as an extraordinarily risky, world-changing phenomenon. Analysis of 40 years of about 25,000 worldwide terrorist events shows about 34,000 deaths and 82,000 nonfatal injuries (Bogen and Jones 2006), a small fraction of the nearly 20 million fatalities caused by traffic accidents, and an annual average close to deaths from volcanic eruptions or airline accidents (fig. 2.24). But this comparison could change with a single new terrorist attack using a nuclear weapon or other effective means of mass destruction. Here we must confront several uncomfortable realities. The first is that even the most assiduous deployment of the best available preventive measures (smart policing, clever informants, globe-spanning electronic intelligence, willingness to undertake necessary military action) will not be able to thwart all planned attacks.

The second reality is that the most dangerous form of terrorist attacks cannot be deterred because the political and ideological motivations for terrorist attacks that characterized Rapoport's (2001) first three waves of terror have blended with reli-

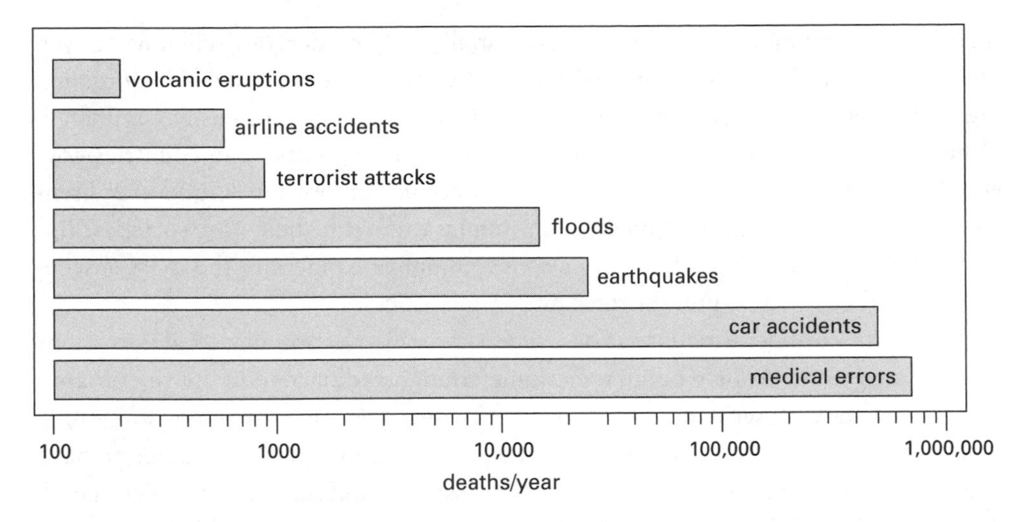

Fig. 2.24
Global fatalities due to terrorist attacks compared to mortality from traffic and airline accidents, major natural disasters, and errors during hospitalization. Annual averages for 1970–2005 calculated from data in Bogen and Jones (2006), WHO (2004a), Boeing (2006), Swiss Re (2006b), and Kohn, Corrigan, and Donaldson (2000).

gious zealotry and become one with the Muslim concept of martyrdom, providing the perpetrators with an irresistible reward: instant access to paradise. Murder by suicide has deep roots in Muslim history, going back to the *shī'a* Nizari state of the eleventh through thirteenth centuries, which perfected the practice of skilled sacramental suicidal assassins (Lewis 1968; Andriolo 2002). Modern revival of the practice first sent thousands of Iranian boys and young men to their deaths during the war with Iraq in the early 1980s (Taheri 1986); at the same time it was adopted by the Lebanese *shī'ī* Hizbullah as the principal tool of its terrorist attacks; and soon it was copied by the *sunnī* Hamas and Fatah in their fight with Israel; and by al-Qaeda in its quest for a new caliphate. Fighting this kind of commitment to terror is exceedingly difficult: "The suicide bomber's thumb pressing the detonator simultaneously clocks him into paradise" (Andriolo 2002, 741).

The third sobering consideration is that neither personal instability nor an individual's hopelessness or overt personal defects, factors that come immediately to mind as the most likely drivers, are dependable predictors of candidates for suicidal martyrdom (Atran 2003; 2004). Nor are such indicators as poverty, level of education (a typical *shahīd* is not an illiterate simpleton), or religious devotion (before their indoctrination, many youngsters are initially only moderately religious or even secular-minded). Institutional manipulation of emotional commitment (by organizing fictive kin groupings) seems to be a key factor, and one not easily eliminated. Other obvious contributions are rapidly rising youth populations in countries governed by dictatorial regimes with limited economic opportunities, and the disenchantment of second-generation Muslim immigrants with their host societies. But none of these factors can offer any selective guidance to identify the most susceptible individuals and to prevent their murderous suicides.

The fourth consideration is that our understandable fixation on suicidal missions may be misplaced. A dirty bomb containing enough radioactive waste to contaminate several downtown blocks of a major city and cause mass panic (as anything invisible *and* nuclear is bound to do) can be positioned in a place calculated to have a maximum impact (a building roof, a busy crossroad) and then remotely detonated. And while Hizbullah's more than 30 days of rocket attacks on Israel in the summer of 2006 were not particularly deadly, they paralyzed a large part of the country and demonstrated how more conventional weapons could be used in the service of terrorism.

Imaginable Surprises

This is a nebulous category of events whose number is limited only by one's inventiveness—and by a mesh of the sieve used to separate plausible events from imaginary constructs. Still, this category should not be dismissed as unhelpfully speculative but probed, within reason, in order to go beyond the appraisal of well-appreciated risks. Rather than following Joy (2000) or Rees (2003), with his concerns about the annihilation of the Earth through particle physics experiments, I engage in a much more plausible stretching of past events and do so, as with the just completed appraisals of probable catastrophic discontinuities, within two distinct categories of natural and human-made events.

I believe that our understanding of natural catastrophes leaves little room for events that were not already covered earlier in this chapter. Perhaps the only important exception is unusually rapid climate change—*rapid* in climatic terms, lasting years rather than centuries or millennia, and *unusually* because major uncertainties exist in the prevailing conclusions regarding past abrupt climate changes, and today's biospheric conditions are not conducive to similarly abrupt shifts.

When the $\delta^{18}O$ (normalized ratios of two oxygen isotopes, ^{18}O and ^{16}O) derived from Greenland ice cores are taken as proxies for local temperature change (they are so only within a factor of ~2), there is an unmistakable pattern of large, positive spikes correlated with episodes of rapid warming during the last glacial period (Stuiver and Grootes 2000). A widely held assumption is that these Dansgaard-Oeschger (D-O) events were hemispheric or global in extent and that they originated because of shifts in North Atlantic ocean circulation (Broecker 1997). But this genesis is dubious, and there is little solid evidence that they were anything more than local phenomena restricted to Greenland (most likely due to the interaction of the wind field with continental ice sheets) that ceased abruptly about 10,000 years ago with the progressing deglaciation of the Northern Hemisphere (Wunsch 2006). Ever since that time, $\delta^{18}O$ variations have been remarkably stable, and there appears to be virtually no possibility that another D-O warming could take place during the next 50 years.

But is a reverse shift, rapid global cooling, any more plausible? The last rapid temporary cooling of the Younger Dryas period took place 12,700–11,500 years ago at the end of the last glacial cycle as the Earth was warming (Dansgaard, White, and Johnsen 1989). Because we are not sure how this cooling was triggered we should include its possibility among imaginable natural discontinuities that could change the course of global history. For decades consensus posited a massive flood of fresh water (~9500 km^3) pouring from the glacial Lake Agassiz via the Great Lakes to the North Atlantic and disrupting the normal thermohaline circulation (see fig. 5.1) (Johnson and McClure 1976; Teller and Clayton 1983).

This cool period, with temperatures ~15°C lower than today's, began and ended abruptly, with warming of ~5°C–10°C spread over ~40 years, most of it over 5 years (Alley 2000). There is today no similarly massive body of fresh water that could suddenly spill into the Atlantic, but there is also no clear proof of that past massive eastward outflow from Lake Agassiz and no clear conclusion about the cooling's trigger (Broecker 2003; 2006). There is no obvious geomorphic evidence for the postulated outlet toward the Great Lakes; a northern route (via Athabasca) appears very unlikely, as does a massive escape beneath the ice. Another explanation finds the source of fresh water in a precipitous melting of massed icebergs, and yet another rejects that trigger entirely and attributes the sudden cooling to a shift of wind patterns caused by a tropical temperature anomaly.

In the absence of any reliable explanation, it is easy to imagine that an as-yet-unidentified trigger might act once more to cause cooling of 5°C–10°C over several decades. This change would pose many challenges for the functioning of a modern civilization whose centers of economic activity and large shares of affluent populations are between 35°N and 55°N. As with the D-O-type warming episodes, such cooling must be seen as highly unlikely during the next 50 years. Our major preoccupation with climate will remain focused on a much slower (though rapid in geological terms) process of global warming.

Of imaginable catastrophic human-made surprises, none is as worrisome as the accidental or deliberate use of nuclear weapons. Simple logic dictates that the probabilities of nuclear accidents should rise as the number of nuclear nations increases and some of them fail to enforce strict precautions to avoid such mishaps. Although the probabilities of such accidents remain uncertain, they may surpass by several orders of magnitude the likelihood of any known global natural catastrophes. It is the same with the probabilities of deliberate nuclear war among minor (or future) nuclear states. Inevitably, India and Pakistan come to mind.

Additional frightening nuclear scenarios can be imagined: Pakistani bombs fall into the hands of extremists bent on realizing Usama bin-Lādin's favorite project of a Muslim caliphate extending from Spain to Indonesia; or the Russian nuclear arsenal (perhaps after a collapse of global energy prices, prolonged worldwide recession, and a drastic pauperization of Russia that would be simultaneously terrorized to an unprecedented extent by attacks from its Muslim fringe) is controlled by a reckless, nationalist, anti-Western regime (remember Vladimir Zhirinovsky?). But the most worrisome among the imaginable futures of deliberate nuclear use is the Iranian determination, voiced in no uncertain terms by Ayatollah Khomeini (cited in Lewis 2006):

> I am decisively announcing to the whole world that if the world-devourers wish to stand against our religion, we will stand against the whole world and will not cease until the annihilation of all of them. Either we all become free, or we will go to the greater freedom which is martyrdom. Either we shake one another's hands in joy at the victory of Islam in the world, or all of us will turn to eternal life and martyrdom. In both cases, victory and success are ours.

This could be dismissed as just another example of hyperbolic, apocalyptic preaching, but it may be prudent to take it seriously. After all, Hitler's *Mein Kampf* (1924) turned out to be a programmatic statement. Khomeini's version of freedom means turning the whole world into a medieval theocracy, and his definition of

Islamic victory leaves no space for compromise arrangements or the fear of mutually assured destruction that has restrained the two superpowers. Even Stalin did not court death by U.S. bombs. But to Khomeini and the president of Iran, Mahmoud Ahmedinejad, assured death in a retaliatory nuclear strike appears to be a shortcut to martyrdom.

A new threat may come from accidental escapes or unintended mutation of bacteria or yeast that will be engineered from life's fundamental genetic components in order to create artificial species with superior abilities for energy generation, enzymatic processing, or food and drug production. As an editorial in *Nature* (Futures of artificial life 2004) put it, "This is no longer a matter just of moving genes around. This is shaping life like clay," and such manipulations must evoke unease and concern, Craig Venter's boisterous pronouncements (about writing and not just reading the code of life) notwithstanding.

Finally, musings on imaginable catastrophes should not fail to note that given the multitude of viruses and their inherent mutability, there is always the possibility of a new pathogen as potent as HIV (or even more virulent) and as easily transmissible as influenza. And given the history of bovine somatotrophic encephalopathy, or mad cow disease, it is also possible to posit new prions (proteinaceous infectious particles) that might spread a new version of transmissible spongiform encephalopathy. The consequences of such eventualities are truly frightening to contemplate.

3

Unfolding Trends

Augescunt aliae gentes, aliae minuuntur, inque brevi spatio mutantur saecla animantum et quasi cursores vitai lampada tradunt.
(Some nations rise, others diminish, the generations of living creatures are changed in short time, and like runners carry on the torch of life.)
Titus Lucretius Carus, *De Rerum Natura*, II, 7S

Fundamental changes in human affairs come both as unpredictable discontinuities and as gradually unfolding trends. Discontinuities are more common than is generally realized, and chapter 2 addressed those natural and anthropogenic catastrophes that have the greatest potential to affect the course of global civilization. But another category of discontinuities deserves at least a brief acknowledgment, that of epoch-making technical developments. Incremental engineering progress (improvements in efficiency and reliability, reduction of unit costs) and gradual diffusion of new techniques (usually following fairly predictable logistic curves) are very much in evidence, but they are punctuated by surprising, sometimes stunning, discontinuities.

By far the most important concatenation of these fundamental advances took place between 1867 and 1914, when electricity generation, steam and water turbines, internal combustion engines, inexpensive steel, aluminum, explosives, synthetic fertilizers, and electronic components created the technical foundations of the twentieth century (Smil 2005a). A second remarkable saltation took place during the 1930s and 1940s with the introduction of gas turbines, nuclear fission, electronic computing, semiconductors, key plastics, insecticides, and herbicides (Smil 2006). The history of jet flight is a perfect illustration of the inherently unpredictable nature of these rapid technical shifts. In 1955 it did not require extraordinary imagination to see that intercontinental jet travel would become a large-scale enterprise, but no one could have predicted (a year after the third fatal crash of the pioneering British Comet) that by 1970 there would be an intercontinental plane capable of carrying more than 400 people. Introduction of the Boeing 747 was an unpredictable step, a result of Juan Trippe's (PanAm's chairman) vision and William Allen's (Boeing's CEO) daring, not an inevitable outcome of a technical trend (Kuter 1973).

Some political discontinuities of the second half of the twentieth century were equally stunning. The year 1955 was just six years after the Communist victory in China's protracted civil war, two years after China's troops made it impossible for the West to win the Korean War and forced a standoff along the 38th parallel, and three years before the beginning of the worst (Mao-made) famine in history (Smil 1999). At that time China, the legitimacy of its regime unrecognized by the United States, was an impoverished, subsistence agrarian economy, glad to receive a few crumbs of Stalinist industrial plant, and its annual per capita GDP was less than 4% of the U.S. average.

Fifty years later China, still very much controlled by the Communist party, had become a workshop for the world, an indispensable supplier of goods ranging from pliers to cell phones. In 2005 its per capita GDP (expressed in PPP values) was at about $4,100 (comparable to that of Syria or Namibia), and the country was providing key support for the United States' excessive spending through its purchases of U.S. Treasury bills. How could anyone have anticipated all these developments in 1955, or for that matter in September 1976, right after Mao's death, or in 1989, after the Tian'anmen killings?

Demographic discontinuities offer perhaps the best illustration of a continuum between abrupt changes and gradually unfolding trends. Clear directional changes (rather than mere fluctuations) in fertility, mortality, or marriage rates become obvious only after one or two decades of reliable records. A major change accomplished in a decade is an abrupt shift on the demographic time scale (where one generation, 15–25 years, is the basic yardstick) but a rapidly unfolding trend by any other temporal metric. Declines in fertility (leaving aside China's one-child policy) have best exemplified these relatively rapid shifts/unfolding trends.

Two generations ago Europe's Roman Catholic South had total fertility rates close to or above 3.0, but by the year 2000 Spain and Italy shared the continent's record lows with Czechs, Hungarians, and Bulgarians (Billari and Kohler 2002; Rydell 2003). In Spain's case, most of this decline took place in just a single decade, between the late 1970s and the late 1980s (fig. 3.1). In Canada a similar phenom-

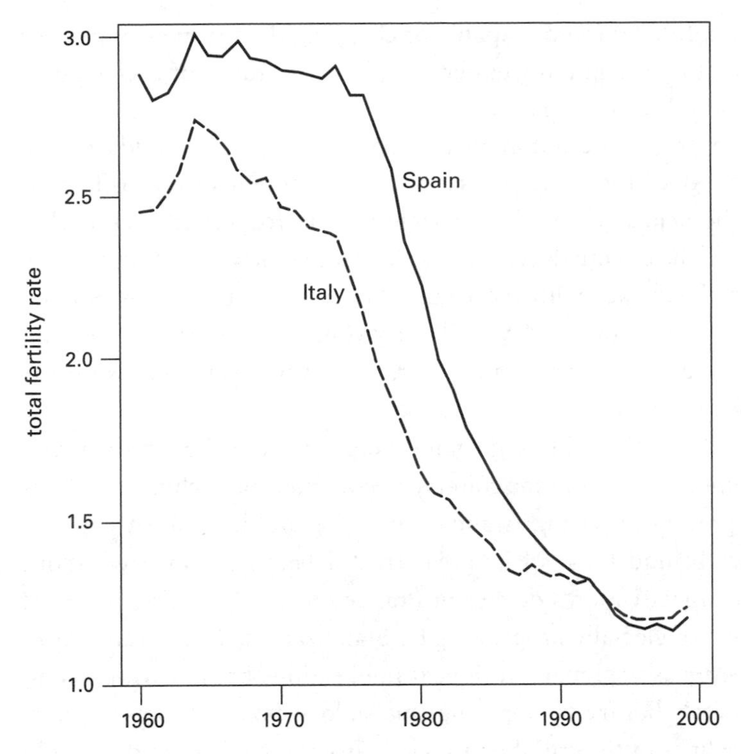

Fig. 3.1
Steep decline in total fertility rate in Italy and Spain, 1960–2000. Based on Billari and Kohler (2002).

enon was seen not only in Roman Catholic Quebec, whose fertility during the mid-1920s was nearly 60% higher than Ontario's, but also among the francophones outside the province, whose fertility fell from nearly 5.0 during the late 1950s to below 1.6 by the mid-1990s (Rao 1974; O'Keefe 2001).

Energy transitions provide another illustration of a continuum between abrupt changes and gradually unfolding trends. Substitution of a dominant source of primary energy by another fuel or by primary electricity (hydro and nuclear) has been slow. It usually takes decades before a new source captures the largest share of the total supply. But in some countries these primary energy transitions have been relatively fast. An outstanding example is France's successful strategy to produce most of its electricity from nuclear fission; the country quintupled its nuclear generation during a single decade, the 1980s (Dorget 1984).

The probabilities of unfolding trends capable of changing the fortunes of nations and reshaping world history are not any easier to assess than those of catastrophic discontinuities. There are three reasons for this.

First, many long-term trends are not explicitly identified as they unfold and are recognized only ex post once they end in discontinuities. Recent history offers no better example than the demise of the Soviet Union. In retrospect it is clear that during the second half of the twentieth century the Soviet regime was falling steadily behind in its self-proclaimed race with the United States. Yet the West—spooked by Nikita Khrushchev's famous threat, "We will bury you," awed by a tiny Sputnik, and willing to believe in a huge missile gap—operated for two generations on the premise of growing Soviet might.

In reality, any Soviet domestic and foreign gains were outweighed by internal and external failures, and improvements in the country's economic and technical abilities were not sufficient to prevent its average standard of living and its military capability from falling further behind those of its great rival. I became convinced about the inevitability of the Soviet Union's demise in Prague during the spring of 1963, when as yet another sign of glacially progressing de-Stalinization, *Time*, *Newsweek*, and *The Atlantic Monthly* as well many technical publications became available in the Carolinum University's library. Comparing this sudden flood of facts, figures, and images with the realities of the crumbling society around me, I could draw only one conclusion: contrary to Khrushchev's boasts, the gap between the East and the West was widening and the Communist regimes would never catch up. But I, like others, was not bold enough to imagine the collapse coming within a single generation, and thought that the Soviet Union would endure into the next century.

Second, we cannot foresee which trends will become so embedded as to seem immune to any external forces and which ones will suddenly veer away from a predictable course or never come to pass. The history of nuclear fusion is an example of an unrealized trend. Ever since the late 1940s nuclear physicists have been anticipating that we need 30–40 years to make the first breakthroughs in controlled fusion and to follow them with relatively rapid commercialization of fusion-powered electricity generation. Skeptics should be forgiven for concluding that this ever-receding horizon will not be reached soon, perhaps not even by 2100 (Parkins 2006).

Third, what follows afterwards is often equally unpredictable: a new long-lasting trend or a prolonged oscillation, a further intensification or an irreversible weakening? This challenge can be illustrated by concerns about the rapidly approaching end of the oil era. Are we facing only more unpredictable price fluctuations that will eventually be moderated by transition to natural gas and renewable energies, or is this the real beginning of the end of cheap oil as the most important energy source of modern civilization? In turn, the latter could be seen as fairly catastrophic (we do not now have an equally flexible and affordable alternative) or as a tremendous opportunity for technical innovations and social adjustments that would actually improve the world's economic fortunes and environmental quality (Smil 2003; 2006).

My assessment of globally important demographic, economic, political, and strategic trends is done in three ways. First, I address the most fundamental future shift in the global economy. It is not, as one might think, further globalization but rather the coming epochal energy transition. Second, I look at long-term trends as they affect the major protagonists on the world scene, the United States, the European Union, China, Japan, Russia, and the militant political Islam. Given the number and complexity of these factors, my approach is to address only those key trends that will most likely shape the fortunes of the world's leading economies. Third, I close the chapter with some musings about who is on top. That section also includes some conclusions about the equivocal effects of continuing globalization of the world's economy, above all, rising inequality.

Energy Transitions

Energy flows and conversions sustain and delimit the lives of all organisms, and hence also of such superorganisms as societies and civilizations. No human action can take place without harnessing and transforming energies. From a fundamental biophysical (thermodynamic) perspective, the fortunes of nations are not determined primarily by strategic designs or economic performance but by the magnitude and efficiency of their energy conversions (Smil 2008). Global consumption of coal (supplemented by small amounts of crude oil and natural gas) surpassed that of phytomass (wood, charcoal, crop residues) during the 1890s (Smil 1994). Coal's share in the world's total primary energy supply (TPES), excluding the phytomass, stood at 95% in 1900; it slid below 50% during the early 1960s, but in 2005 it was still at about 28%. Crude oil accounted for 4% of global TPES in 1900, 27% in 1950, and about 46% in 1975. By 2005 it was about 36%, while the importance of natural gas kept rising; natural gas supplied nearly 24% of global TPES in 2005 (fig. 3.2).

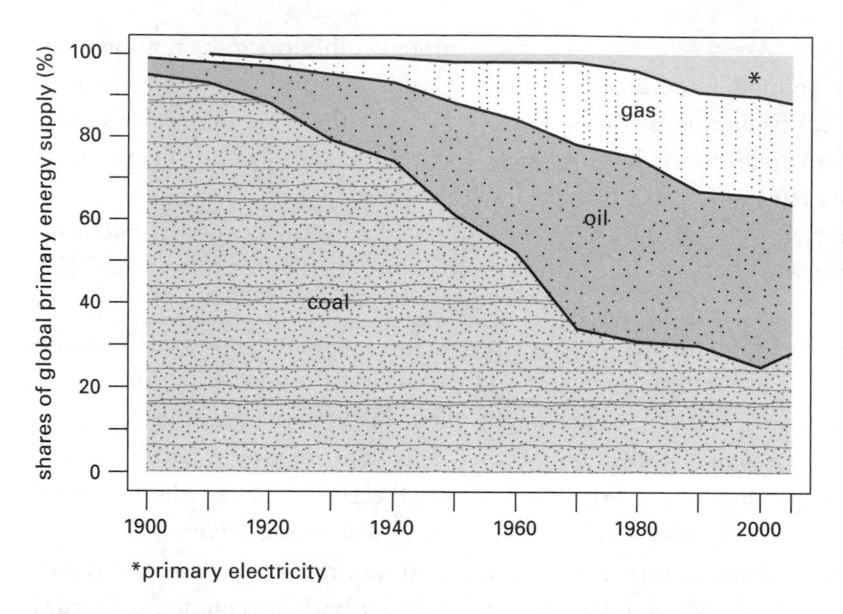

Fig. 3.2
Shares of fossil fuels and primary electricity in global energy consumption, 1900–2000. Based on Smil (2006) and BP (2006).

Given the need to put in place new infrastructures and the difficulty of rapidly discarding existing capital investments, the substitution process is slow, with a new energy source taking about a century after its initial introduction to capture half of the market share. Until the early 1970s the process was highly regular, leading Marchetti and Nakićenović (1979) to conclude that the entire energy system operates as if driven by a hidden clock. In reality, the pace and timing of the transition is far less predictable. OPEC's oil price hikes (1973–1974 and 1979–1981) clearly disrupted the previously orderly energy substitutions. Coal has remained important (holding the same share in 2005 as it did in 1975, 28%); oil's share fell from 46% to 36%, and natural gas gained only half the expected share (25% by 2005). Altogether, fossil fuels supplied about 88% of global TPES in 2005, down just 5% compared to 1975, whereas new nonfossil alternatives (other than well-established hydro and nuclear electricity generation) are still only marginal contributors.

Although coal has lost its traditionally large transportation, household, and industrial markets, it generates about 40% of the world's electricity (WCI 2006). It is the principal fuel for the rapidly rising production of cement and coke, and it remains the key energizer and reducing agent of iron smelting. The energy density of natural gas is typically around 34 MJ/m^3 compared to 34 GJ/m^3 for oil, or a 1,000-fold difference. This limits the use of gas as transportation fuel; gas is also more costly to transport and store. Consequently, modern economies will do their utmost to stay with the more convenient liquid fuels, and contrary to alarmist claims about the end of oil production, the transition to renewable energies will be a protracted affair.

Dominant Fuels, Enduring Prime Movers

Many comments on energy futures (ranging from catastrophic predictions of an imminent oil drought to unrealistic forecasts of future biomass energy uses) betray the fundamental lack of understanding of the nature and dynamics of the global energy system. The three key facts for this understanding are these: We are an overwhelmingly fossil-fueled civilization; given the slow pace of major resource substitutions, there are no practical ways to change this reality for decades to come; high prices have concentrated worldwide attention on the availability and security of the oil supply, but coal and natural gas combined provide more energy than do liquid fuels.

As coal's relative importance declined, its absolute production grew roughly sevenfold between 1900 and 2005, to more than 4 billion t of bituminous coal and nearly 900 million t of lignites. Yet this prodigious production was only about 0.6% of that year's fuel reserves, the share of mineral resources in the Earth's crust that can be recovered profitably with existing extraction techniques. This means that coal's global reserves/production (R/P) ratio is nearly 160 years, and new extraction techniques could expand the reserves. Coal production is thus constrained not by resources but by demand, and by old (particulate matter and sulfur and nitrogen oxides) and new (emissions of CO_2, the leading anthropogenic greenhouse gas) environmental considerations.

Natural gas produces the least amount of CO_2 per unit of energy (less than 14 kg C/GJ, compared to about 25 kg C/GJ for bituminous coal and 19 kg C/GJ of refined fuels); hence it is the most desirable fossil fuel in a world concerned about global warming. But until recently long-distance pipeline exports were limited to North America; Russian supplies of Siberian gas to Europe; and pipelines from Algeria and Libya to Spain and Italy. Rising prices have led to new plans for long-distance pipelines and to renewed interest in liquefied natural gas shipments. As with coal, the fuel's resources are abundant, and in 2005 the reserves (at 180 Tm^3) were twice as large as in 1985.

A small army of experts has disseminated an alarmist notion of imminent global oil exhaustion followed by economic implosion, massive unemployment, breadlines, homelessness, and the catastrophic end of industrial civilization (Ivanhoe 1995; Campbell 1997; Laherrère 1997; Deffeyes 2001). Their alarmist arguments mix incontestable facts with caricatures of complex realities, and they exclude anything that does not fit preconceived conclusions in order to issue obituaries of modern civilization.

Their conclusions are based on a lack of nuanced understanding of the human quest for energy. They disregard the role of prices, historical perspectives, and human inventiveness and adaptability. Their interpretations are anathema to any critical, balanced scientific evaluation, but, precisely for that reason, they attract mass media attention. These predictions are just the latest installments in a long history of failed forecasts but their advocates argue that this time the circumstances are really different and the forecasts will not fail. In order to believe that, one has to ignore a multitude of facts and possibilities that readily counteract their claims. And, most important, there is no reason that even an early peak to global oil production should trigger any catastrophic events.

The modern tradition of concerns about an impending decline in resource extraction began in 1865 with William Stanley Jevons, a leading economist of the Victorian era, who concluded that falling coal output must spell the end of Britain's national greatness because it is "of course . . . useless to think of substituting any other kind of fuel for coal" (Jevons 1865, 183). Substitute *oil* for *coal* in that sentence, and you have the erroneous foundations of the present doomsday sentiments about oil. There is no need to elaborate on how wrong Jevons was. The Jevonsian view was reintroduced by Hubbert (1969) with his "correct timing" of U.S. oil production, leading those who foresaw an early end to oil reserves to consider Hubbert's Gaussian exhaustion curve with the reverence reserved by Biblical fundamentalists for Genesis.

In reality, the Hubbert model is simplistic, based on rigidly predetermined reserves, and ignoring any innovative advances or price shifts. Not surprisingly, it has repeatedly failed (fig. 3.3). Hubbert himself put the peak of global oil extraction between 1993 and 2000. The Workshop on Alternative Energy Strategies (WAES 1977) forecast the peak as early as 1990 and most likely between 1994 and 1997; the CIA (1979) believed that global output must fall within a decade; BP (1979) predicted world production would peak in 1985 and total output in the year 2000 would be

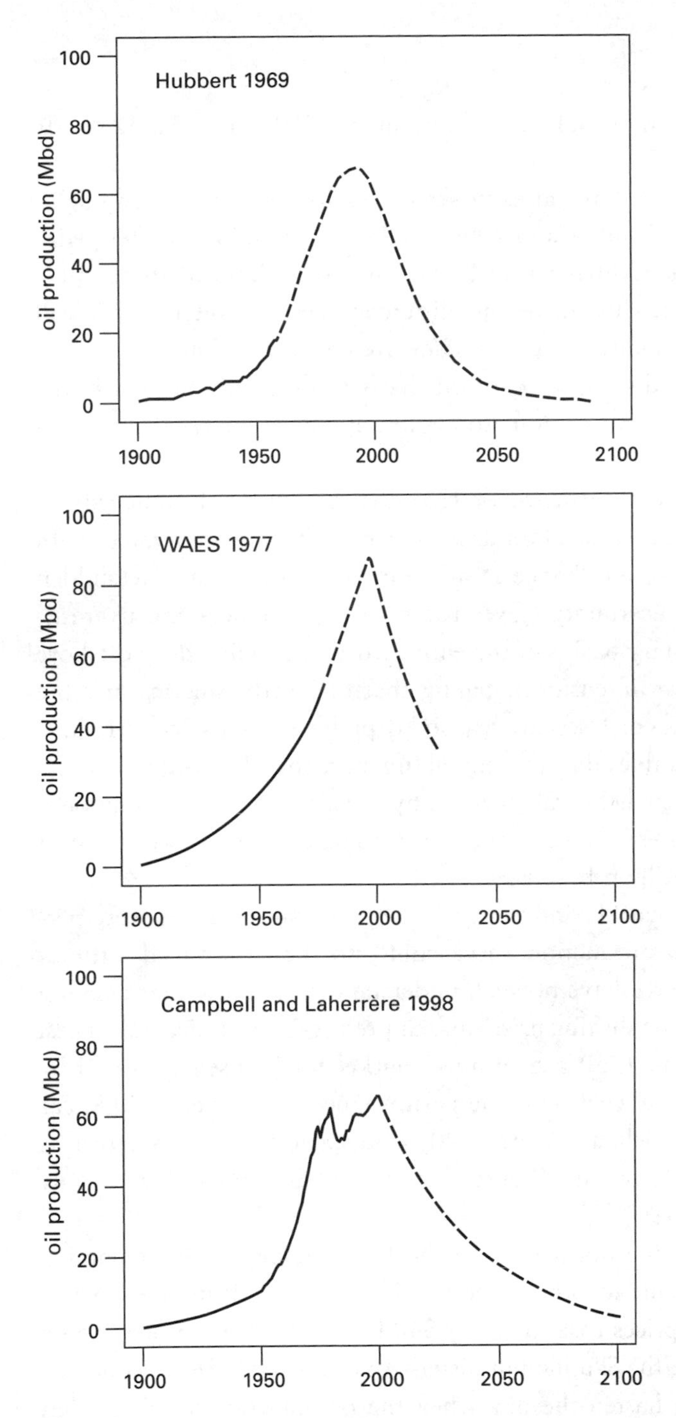

Fig. 3.3
Failed forecasts of global peak oil production: Hubbert (1969), WAES (1977), and Campbell and Laherrère (1998). From Smil (2003).

nearly 25% below that maximum (global oil output in 2000 was actually nearly 25% above the 1985 level).

Some peak-of-oil proponents have already seen their forecasts fail. Campbell's first peak was to be in 1989, Ivanhoe's in 2000, Deffeyes's in 2003 and then, with ridiculous specificity, on Thanskgiving Day 2005. But the authors of these failed predictions would argue that this makes no difference because oil reserves will inevitably be exhausted in a matter of years. They are convinced that exploratory drilling has already discovered some 95% of oil that was originally present in the Earth's crust and that nothing can be done to avoid a bidding war for the remaining oil.

True, there is an unfortunate absence of rigorous international standards in reporting oil reserves, and many official totals have been politically motivated, with national figures that either do not change at all from year to year or take sudden suspicious jumps. But this uncertainty leaves room for both under- and overestimates, and until the sedimentary basins of the entire world (including deep offshore regions) are explored with an intensity matching that of North America and the U.S. sector of the Gulf Mexico, I see no reason to prefer the most conservative estimate of ultimately recoverable conventional oil (no more than 1.8 trillion barrels) rather than the substantially higher totals favored by other geologists, although not necessarily the highest values estimated by the U.S. Geological Survey, whose latest maximum is in excess of 4 trillion barrels.

Even if the amount of the world's ultimately recoverable oil resources were perfectly known, the global oil production curve could not be determined without knowing future oil demand. We have no such understanding because that demand will be shaped, as in the past, by shifting prices and unpredictable technical advances. Who would have predicted in 1930 a new huge market for kerosene, created by commercial jets by 1960, or in 1970 that the performance of an average U.S. car would double by 1985? As Adelman (1992, 7–8), who spent most of his career as a mineral economist at MIT, put it, "Finite resources is an empty slogan; only marginal cost matters."

Steeply rising oil prices would not lead to unchecked bidding for the remaining oil but would accelerate a shift to other energy sources. This lesson was learned painfully by OPEC after oil prices rose to nearly $40/bbl in 1981, and it led Sheikh Ahmed Zaki Yamani (2000), the Saudi oil minister from 1962 to 1986, to conclude that high prices would only hasten the day when the organization would be left with untouched fuel reserves because new efficient techniques would reduce the demand for transport fuels and leave much of the Middle East's oil in the ground forever.

And yet, as noted, price feedbacks are inexplicably missing from all accounts of coming oil depletion and its supposedly catastrophic consequences. Instead, there is an assumption of demand immune to any external factors. In reality, rising prices do trigger powerful adjustments. Between 1973 and 1985 the U.S. CAFE (corporate automobile fuel efficiency) was doubled to 27.5 mpg, but further improvements were not pursued largely because of falling oil prices. A mere resumption of that rate of improvement (technically easy to do) would have automobiles averaging 40 mpg by 2015, and a more aggressive adoption of hybrids could bring the rate to 50 mpg, more than halving the current U.S. need for automotive fuel and sending oil prices into a tailspin.

And although oil prices are still relatively low (adjusted for inflation and lower oil intensity of modern economies, even $100/bbl is at least 25% below the 1981 peak), they have already reinvigorated the quest for tapping massive deposits of nonconventional oil as well as the development of new gas fields aimed at converting the previously "stranded" reserves into a massively traded global commodity (liquefied natural gas). Technical advances will also make possible the conversion of that gas (and coal) into liquids, and increasing recoveries of coalbed methane and extraction of methane from hydrates will supply more hydrocarbons. But even if the global extraction of conventional crude oil were to peak within the next two decades, this would not mean any inevitable peak of overall global oil production, and even less so the end of the oil era, because very large volumes of the fuel from traditional and nonconventional sources would remain on the world market during the first half of the twenty-first century.

As oil becomes dearer, we will use it more selectively and efficiently and intensify the shift from oil to natural gas and to renewable and nuclear alternatives. Finally, it must be stressed that fossil fuels will retreat only slowly because the dominant energy converters depend on their supply. The evolution of modern energy systems has shown a great deal of inertia following the epochal commercial introduction of new prime movers. All those overenthusiastic, uncritical promoters of new energy techniques would do well to consider five fundamental realities.

First, the steam turbine, the most important continuously working high-load prime mover of the modern world, was invented by Charles Parsons 120 years ago, and it remains fundamentally unchanged; gradual advances in metallurgy simply made it larger and more efficient. These large (up to 1.5 GW) machines now generate more than 70% of our electricity in fossil fueled and nuclear stations; the rest comes from gas and water turbines and from diesels.

Second, the gasoline-fueled internal combustion engine, the most important transportation prime mover of the modern world, was first deployed (based on older stationary models) during the same decade as the Parsons machine, and it reached a remarkable maturity in a single generation after its introduction.

Third, Diesel's inherently more efficient machine followed shortly after the Benz-Daimler-Maybach design, and it matured almost as rapidly. As I explain later, it is entirely unrealistic to expect that we could substitute most of the gasoline or diesel fuel by fuels derived from biomass within a few decades.

Fourth, the gas turbine, the most important prime mover of modern flight, is now entering the fourth generation of service after a remarkably fast progression from Frank Whittle's and Pabst von Ohain's conceptual designs to high-bypass turbofans (Smil 2006). Again, conversion of biomass could not supply an alternative aircraft fuel at the requisite scale for decades (even if that conversion were profitable).

Fifth, Nikola Tesla's induction electric motor, commercialized during the late 1880s, diffused rapidly to become the dominant prime mover of industrial production as well as of domestic comfort and entertainment. Renewable conversions should eventually be capable of supplying the needed electricity for these motors by distributed generation.

Solar (Nuclear?) Civilization

There are five major reasons that the transition from fossil to nonfossil supply will be much more difficult than is commonly realized: scale of the shift; lower energy density of replacement fuels; substantially lower power density of renewable energy extraction; intermittence of renewable flows; and uneven distribution of renewable energy resources.

The scale of the transition is perhaps best illustrated by comparing it to the epochal shift from biomass to fossil fuels. By the late 1890s, when phytomass slipped below 50% of the world's total primary energy supply (TPES), coal (and a small amount of oil) was consumed at the rate of 600 GW, whereas in 2005 the world used fossil energies at a rate of 12 TW, a 20-fold difference. Of course, phytomass was never totally displaced. During the twentieth century its use (now mainly in poor countries) roughly doubled, but it now provides only about 10% of global TPES.

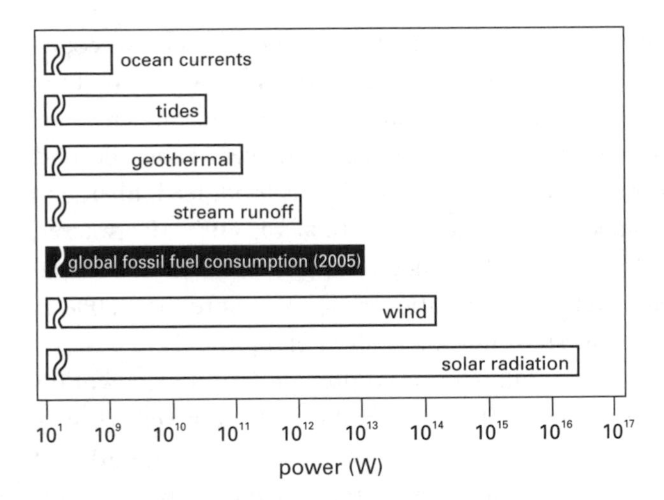

Fig. 3.4
Annual global flows of renewable energies compared to the world's total commercial energy consumption in 2005. From Smil (2008). Technically feasible conversions are much lower.

The magnitude of the needed substitution also runs into some important resource restrictions. At 122 PW the enormous flux of solar radiation reaching the Earth's ground is nearly 4 OM greater than the world's TPES of nearly 13 TW in 2005 (fig. 3.4). But this is the only renewable flux convertible to electricity that is considerably larger than the current TPES; no other renewable energy resource can provide more than 10 TW. Generous estimates of technically feasible maxima are less than 10 TW for wind, less than 5 TW for ocean waves, less than 2 TW for hydroelectricity, and less than 1 TW for geothermal and tidal energy and for ocean currents. Moreover, the actual economically and environmentally acceptable rates may be only small fractions of these technically feasible totals.

The same disparity applies to the production of phytomass that yields solid fuels directly and that can be converted to liquids and gases. The Earth's net primary productivity (NPP) of 55–60 TW is more than four times as large as was global TPES in 2005. Its mass is dominated by the production of woody tissues (boles, branches, bark, roots) in tropical and temperate forests, and recent proposals of massive biomass energy schemes are among the most regrettable examples of wishful thinking and ignorance of ecosystemic realities and necessities. Their proponents are either unaware of (or deliberately ignore) some fundamental findings of modern biospheric studies.

As the Millennium Ecosystem Assessment (2005) demonstrated, essential ecosystemic services (without which there can be no viable economies) have already been modified, reduced, and compromised to a worrisome degree, and any massive, intensive monocultural plantings of energy crops could only accelerate their decline. Second, humans already appropriate 30%–40% of all NPP as food, feed, fiber, and fuel, with wood and crop residues supplying about 10% of TPES (Rojstaczer, Sterling, and Moore 2001). Moreover, highly unequal distribution in the human use of NPP means that the phytomass appropriation ratios are more than 60% in East Asia and more than 70% in Western Europe (Imhoff et al. 2004).

Claims that simple and cost-effective biomass could provide 50% of the world's TPES by 2050 or that 1–2 Gt of crop residues can be burned every year (Breeze 2004) would put the human appropriation of phytomass close to or above 50% of terrestrial photosynthesis. This would further reduce the phytomass available for microbes and wild heterotrophs, eliminate or irreparably weaken many ecosystemic services, and reduce the recycling of organic matter in agriculture. Finally, nitrogen is almost always the critical growth-limiting macronutrient in intensively cultivated agroecosystems and in silviculture. Mass production of phytomass for conversion to liquid fuels, gases, or electricity would necessitate a substantial increase in continuous applications of this element (Smil 2001). Proponents of massive bioenergy schemes appear to be unaware of the fact that human interference in the global nitrogen cycle has already vastly surpassed anthropogenic changes in the carbon cycle (see chapter 4).

The transition to fossil fuels introduced fuels with superior energy densities, but the coming shift will move us in an opposite, less desirable direction. Ordinary bituminous coal (20–23 GJ/t) contains 30%–50% more energy than air-dried wood (15–17 GJ/t); the best hard coals (29–30 GJ/t) are nearly twice as energy-dense as wood; and liquid fuels refined from crude oil (42–44 GJ/t) have nearly three times higher energy density. With this transition we are facing the reverse challenge: replacing crude oil-derived fuels with less energy dense biofuels. Moreover, this transition would also require 1,000-fold and often 10,000-fold larger areas under crops than the land claimed by oil field infrastructures, and shifting from coal-fired to wind-generated electricity would require at best 10 times and often 100 times more space (fig. 3.5) (Smil 2008). In order to energize the existing residential, industrial, and transportation infrastructures inherited from the fossil-fueled era, a solar-based society would have to concentrate diffuse flows to bridge power density gaps of 2–3 OM (Smil 2003).

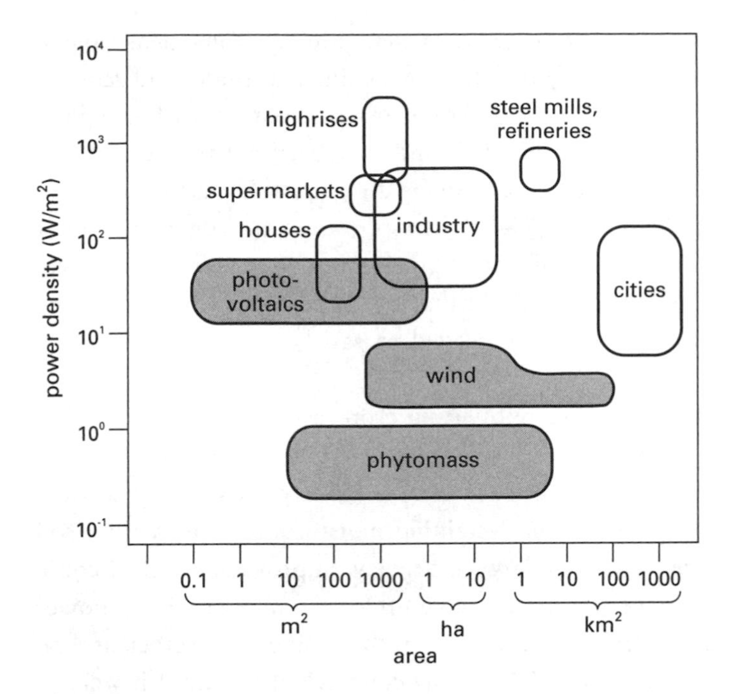

Fig. 3.5
Mismatch between power densities of renewable energy conversions and common energy uses. From Smil (2008).

The mismatch between the inherently low power densities of renewable energy flows and the relatively high power densities of modern final energy uses means that a solar-based system would require profound spatial restructuring, with major environmental and socioeconomic consequences. Direct solar radiation is the only renewable, energy flux available, with power densities of 10^2 W/m^2 (global mean of ~170 W/m^2), which means that increasing efficiencies of its conversion (above all, better photovoltaics) could harness it with effective densities of several 10^1 W/m^2 (the best all-day rates in 2005 were on the order of 30 W/m^2). All other renewables have low (<10 W/m^2), or very low (<1 W/m^2) production power densities. Low extraction power densities would be the greatest challenge in producing liquid fuels from phytomass. Even the most productive fuel crops or tree plantations have gross yields of less than 1 W/m^2, and subsequent conversions to electricity and liquid fuels prorate to less than 0.5 W/m^2.

During the first years of the twenty-first century, global consumption of gasoline and diesel fuel in land and marine transport and kerosene in flying was about 75 EJ.

Even if the most productive solar alternative (Brazilian ethanol from sugar cane at 0.45 W/m^2) could be replicated throughout the tropics, the aggregate land requirements for producing transportation ethanol would reach about 550 Mha, slightly more than one-third of the world's cultivated land, or nearly all the agricultural land in the tropics. There is no need to comment on what this would mean for global food production. Consequently, global transportation fuel demand cannot be filled by even the most productive alcohol fermentation. Corn ethanol's power density of 0.22 W/m^2 means that about 390 Mha, or slightly more than twice the country's entire cultivated area, would be needed to satisfy the U.S. demand for liquid transportation fuel.

The prospect does not change radically by using crop residues to produce cellulosic ethanol (from cereal straw, crop stalks, and leaves). Only a part of these residues could be removed from fields so as to maintain the key ecosystemic services of recycling organic matter and nitrogen, retaining moisture, and preventing soil erosion (Smil 1999). Moreover, even large efficiency improvements in alcohol fermentation or car performance would not make up for the inherently low power densities of cropping. The U.S. transportation sector, three times more efficient than it was in 2000, would still claim some 75% of the country's farmland if it were to run solely on ethanol produced at rates prevailing in 2005.

The intermittence of renewable energy flows poses a particularly great challenge to any conversion system aimed at a steady, reliable supply of energy required by modern industrial, commercial, and residential infrastructures. Solar radiation, wind, waves, and plant harvests fluctuate daily or seasonally, but the base load's needed share of the maximum power load has been increasing in all affluent societies. Easily storable high energy density fossil fuels and thermal electricity generating stations capable of operating with high load factors (>75% for coal-fired stations, >90% for nuclear plants) meet this need. By contrast, the two most important renewable flows seen as the key future generators of electricity, wind and direct solar radiation, are intermittent and far from perfectly predictable. Photovoltaic generation is still so negligible that it is impossible to offer any meaningful averages, but annual load factors of wind generation in countries with relatively large capacities (Denmark, Germany, Spain) are just 20%–25%.

Much is made of the uneven distribution of fossil fuels, particularly of the hydrocarbon anomaly of the Persian Gulf Zagros Basin, which contains nearly two-thirds of the world's known oil reserves. But renewable flows are also highly unevenly distributed. The equatorial zone has much reduced direct solar radiation: the peak midday flux in Jakarta (6°S) is no different from the summer fluxes in Canada's sub-Arctic Edmonton (nearly 55°N). Large areas of all continents are not sufficiently windy or are seasonally too windy: think of large turbines on top of tall towers in any areas with strong cyclonic flows, or what another Katrina hurricane would do to the Gulf of Mexico wind farms. Sites with the best potential for geothermal, tidal, or ocean energy conversions are even less common. Some densely populated regions have low potential for both solar and wind conversions, for instance, Sichuan, China's most populous province. Also, many windy or sunny sites are far from major load centers, and their exploitation will require entirely new extra-high-voltage lines (from southern Algeria to Europe or from North Dakota to New York).

Another myth concerning renewable conversions is the expectation of a near-automatic decline in their cost with increasing volume. This trend is common for most techniques undergoing commercialization, but not inevitable; if it were, we would already have had inexpensive electric cars for decades. The costs of many renewable techniques have actually been increasing. Photovoltaic silicon prices have more than doubled, and the cost of structural steel, aluminum, and plastics for wind turbines and of ethanol fermentation from corn have both risen because all these techniques depend on large inputs of more costly fossil energies.

For all these reasons (and because of other nontrivial technical problems) the global transition to a nonfossil fuel world will be gradual and uneven, driven as much by external political and strategic considerations as by autonomous technical advances. The entire process could be speeded up by aggressively pushing nuclear electricity generation. A general agreement among energy experts is that nuclear electricity must have a significant place in a nonfossil fuel world. This conclusion is based on the high capacities of nuclear power plants and on the high load factor of fission-powered generation.

Nuclear plants have been among the largest electricity generation facilities for some three decades. The largest stations surpass 5 GW, the largest turbogenerators are about 1.5 GW, yet new reactor designs would make units as small as a few hundred megawatts commercially profitable. By contrast, typical large wind turbines rate below 5 MW, and by 2005 the largest PV assemblies were on the order of 4–6 MW of peak power. Well-run nuclear power plants can operate 95% of the time, compared to typical rates of 65%–75% for coal-fired stations, 40%–60% for hydrogeneration, and 25% for wind turbines. A reliable, predictable, high-capacity, high-load mode of electricity generation would be a perfect complement to various renewable conversions that share the attributes of low-capacity, moderate-load, and often unpredictable intermittent operation.

Concerns about global warming have made the pronuclear sentiment much more widespread than it was a generation ago. Some new advocates now include prominent former adversaries, including Patrick Moore, the co-founder of Greenpeace, and James Lovelock, the originator of Gaia theory (Moore 2006; Lovelock 2006). Both now believe that nuclear generation may be the best choice to save the Earth from a catastrophic climate change. Nevertheless, the realities of large-scale nuclear expansion are otherwise. The industry still suffers because of its rushed post-1945 development, and public acceptance of nuclear generation and the final disposal of radioactive wastes remain the key obstacles to massive expansion. No other mode of primary electricity production was commercialized as rapidly as the first generation of water reactors (most of them operating with a pressurized water loop, hence pressurized water reactors, PWRs). Only some 25 years elapsed between the first sustained chain reaction, which took place at the University of Chicago on December 2, 1942, and the exponential rise in orders for new nuclear power plants after 1965.

Consequently, it was widely expected that by the year 2000 worldwide electricity generation would be dominated by inexpensive fission. Instead, the industry experienced stagnation and retreat. In retrospect, it is clear that the industry was far too rushed and that too little weight was given to public acceptance of fission (Cowan 1990). The economics of fission generation has been always arguable because the accounts have excluded both the enormous government subsidies for nuclear R&D and the costs of decommissioning the plants and safely storing radioactive wastes permanently. Accidental core meltdown and the release of radioactivity during the Chernobyl disaster in Ukraine in May 1986 made matters even worse (Hawkes 1987). Although the Western PWRs with their containment vessels and tighter operating procedures could not experience such a massive release of radiation, the Ukrainian accident only reinforced the common public perception that all nuclear power plants are inherently unsafe.

Another problem to overcome is the final disposal of a small volume of highly radioactive wastes that must be sequestered for thousands of years. Ancient stable rock formations provide such repositories, but public distrust of these plans, objections to chosen sites, and bureaucratic procrastination have prevented the activation of any of these sites. To this must be added the serial failure of fast breeder reactors that were designed to use limited supplies of fissionable fuel more efficiently. During the 1970s it was widely predicted that by the year 2000 they would dominate global electricity generation. Instead, the three major national programs were soon abruptly shut down, first in the United States (1983), then in France (1990), and finally in Japan (1995).

If there is widespread expert agreement regarding the desirability of a major nuclear contribution to the energy picture, there is also clear consensus that any new major wave of reactor building must be based on new designs. There is no shortage of these new, more efficient, more reliable and safe designs, including reactors that could be entirely buried underground and that would not have to be refueled during the entire life of their operation. Edward Teller, one of the great pioneers of the nuclear era, detailed the technical parameters of this ingenious solution (Teller et al. 1996). But the likelihood of their early large-scale commercialization is very low, and longer-term prospects remain highly uncertain.

And it is also extremely unlikely that nuclear fusion can be a part of an early (before 2050) or indeed any solution. The engineering challenges of a viable plant design (heat removal, size and radiation damage to the containment vessel, maintenance of vacuum integrity) mean that the technique has virtually no chance to make any substantial contribution to the global primary energy supply of the next 50 years (Parkins 2006). Yet this *fata morgana* of energy techniques keeps receiving enormous amount of taxpayer monies; U.S. spending on fusion has averaged about a quarter billion dollars per year for the past 50 years.

A miracle of a new generation of inexpensive, safe, and reliable fission reactors (or significant breakthroughs in efficiency and cost of photovoltaics) would provide an essential foundation for a transition to a hydrogen-based energy system, but even then its realization would be a protracted affair. Undeniably, energy transitions have been steadily decarbonizing the global supply as average atomic H/C ratios rose from 0.1 for wood, to 1 for coal, 2 for crude oil, and 4 for methane. As a result, the logistic growth process points to a methane-dominated global economy after 2030, but a hydrogen-dominated economy, requiring production of large volumes of the gas without fossil energy, could take shape only during the closing decades of the twenty-first century (fig. 3.6) (Ausubel 1996).

I agree with those who say that hopes for an early reliance on hydrogen are just hopes (Mazza and Hammerschlag 2004). There is no inexpensive way to produce this high energy density carrier and no realistic prospect for the hydrogen economy to materialize for decades (Service 2004). In any case, a methanol economy may be a better, although also very uncertain, alternative (Olah et al. 2006). And there will be no rapid massive adoption of fuel cell vehicles because they do not offer a major

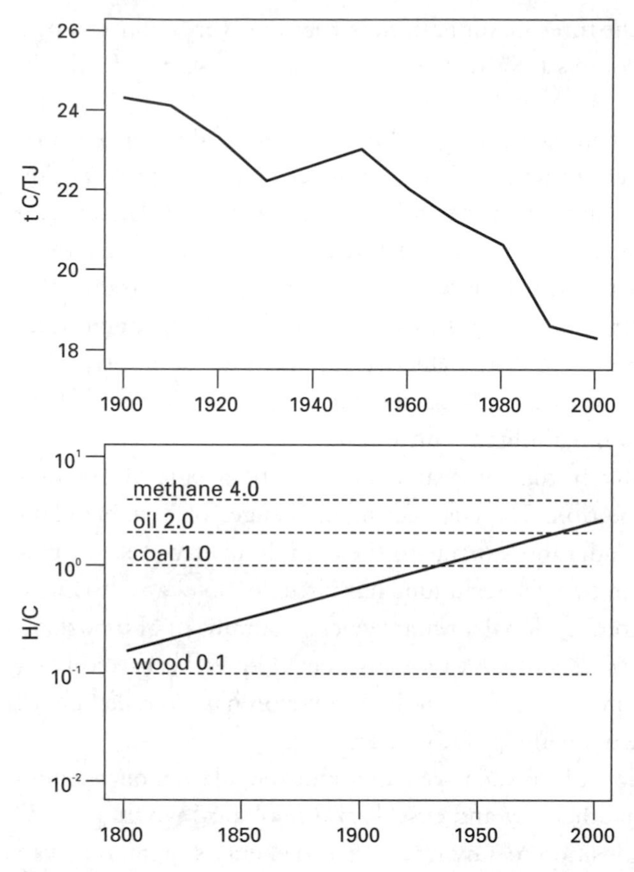

Fig. 3.6
Decarbonization of global primary energy supply, 1900–2000: *upper*, declining carbon content of fossil fuels; *lower*, rising H/C ratio of fuels. From Smil (2003) and Ausubel (1996).

efficiency advantage over hybrid cars in city driving (Demirdöven and Deutsch 2004).

Three key factors drove the transition to fossil fuels: declining resource availability (deforestation), the higher quality of fossil fuels (higher energy density, easier storage, greater flexibility), and the lower cost of coals and hydrocarbons. The coming transition will be entirely different. There is no urgency for an accelerated shift to a nonfossil fuel world: the supply of fossil fuels is adequate for generations to come; new energies are not qualitatively superior; and their production will not be substantially cheaper. The plea for an accelerated transition to nonfossil fuels results almost entirely from concerns about global climate change, but we still cannot quantify its magnitude and impact with high confidence.

Given the complications outlined in this saction, the coming energy transition will be even more challenging than were past shifts. Wishful thinking is no substitute for recognizing the extraordinary difficulty of the task. A nonfossil fuel world may be highly desirable, and determination, commitment, and persistence could accelerate its arrival, but the transition would be difficult and prolonged even if it were not complicated by specific national conditions and trends creating a new constellation of world power.

New World Order

Using the term *new world order*, judging if rise or retreat best describes a nation's (or a continent's or a religion's) recent past and imminent future, and assessing a global power ranking to conclude who is "on top" (as I do here)—such analysis is both useful and problematic. Useful because it offers effective shortcuts for appraising the dynamics of the global power system. Unfilled niches are as rare in ecosystems as they are in the now undoubtedly global quest for influence, affluence, and power. Problematic in that any such ranking must be suspect because all the "data" have complex, multiattribute dimensions and eventual outcomes do not always conform to the expectations of a zero-sum game. Rising global interdependence on resources (mineral or intellectual) and a shared dependence on the biosphere's habitability preclude that.

Consequently, the following appraisals respect the multifaceted nature of rising or falling national fortunes, do not attempt quantitative international comparisons, do not aim at definite rankings, and do not suggest time frames. Multifactorial reviews of complex realities may inform us clearly enough about a nation's relative trajectory, imminent prospects, or comparative position vis-à-vis its principal competitors, but they do not suffice to predict its overall standing after decades of dynamic global contest. As for relative trajectories, how can there be any doubt about China's post-1980 rise?

The post–WW II history of the world's four largest economies provides many illustrations of these complex and elusive realities and profoundly changing trends. I have already noted the curiously taken-for-granted demise of the Soviet Union. As for China, less than four years after Mao Zedong's death in September 1976, Deng

Xiaoping, his old revisionist comrade, launched the modern world's most far-reaching national reversal. He has transmuted the country, stranded for two generations in the role of an autarkic Stalinist underperformer capable of providing only basic subsistence to its people, into a global manufacturing superpower closely integrated into the international economy.

At the same time, Japan, the world's most dynamic large economy of the 1960s, 1970s, and 1980s, widely predicted to become the world's leading economy by the beginning of the twenty-first century, has suddenly lost its momentum and has spent nearly a generation in retreat and stagnation.

U.S. economic and strategic fortunes seemed rather bleak during most of the 1980s, but during the 1990s the country recovered in a number of remarkable ways. This rebound was short-lived, however, and was followed by worrisome fiscal and structural reversals accompanied by unprecedented strategic, military, and political challenges.

I argue that two key trends, China's rise and U.S. retreat, will continue during the coming generation, but I also caution against overestimating the eventual magnitude of the Chinese ascent or the speed of the U.S. decline. At the same time, I do not see how the united Europe or a reenergized Japan can regain their former dynamism. Russia, despite its many problems, is reemerging as a more influential power than it appeared to be after the unraveling of the USSR. Certainly the most difficult challenge is to appraise the likely influence of Islam on world history in the next 50 years: feelings of despair, fear of violence, and expectations of continued failure abound, but a number of factors suggest possibilities of better trends.

I am sure I will be criticized for not devoting a separate account to India in this brief examination of the "new world order," but I defend this decision and comment on the country's key strengths and weaknesses in the closing section of this chapter, where I survey the modern history of global leadership. Who is on top matters in many tangible and intangible ways, and the next two generations will almost surely see an epochal shift, the end of the pattern that has dominated world history since WW II. India will definitely be a part of this rearrangement. I also comment on increasing intra- and international inequality, which I see as the most important long-term destabilizing consequence of globalization.

Europe's Place

Several recent publications have been quite euphoric about Europe's prospects, leaving little room for doubts about the continent's future trajectory. The director of foreign policy at the Centre for European Reform predicts, astonishingly, that Europe will economically dominate the twenty-first century (Leonard 2004). The former London bureau chief of the *Washington Post* maintains that the rise of the United States of Europe will end U.S. supremacy (Reid 2004). And Rifkin (2004) is impressed by the continent's high economic productivity, the grand visions of its leaders, their risk-sensitive policies and reassuring secularism, and the ample leisure and high quality of life provided by caring social democracies.

Such writings make me wonder whether the authors ever perused the continent's statistical yearbooks, read the letters to editors in more than one language, checked public opinion polls, walked through the postindustrial wastelands and ghettos of Birmingham, Rotterdam, or Milan, or simply tried to live as ordinary Europeans do. A perspective offered by the author, a skeptical European who understands the continent's major languages, who has lived and earned money on other continents, and who has studied other societies should provide a more realistic appraisal. Of course, this does not give me any automatic advantage in appraising Europe's place, but it makes me less susceptible to Euro-hubris and gives me the necessary *Abstand* to offer more realistic judgments.

Russia, too, is part of my Europe. Arguments about Russia's place in (or outside of) Europe have been going on for centuries (Whittaker 2003; McCaffray and Melancon 2005); I have never understood the Western reluctance or the Russian hesitancy to place the country unequivocally in Europe. Of course, Russia has an unmistakable Asian overlay—there must be a transition zone in such a large land mass, and centuries of occupation by and dealing with the expansive eastern nomads had to leave their mark—but its history, music, literature, engineering, and science make it quintessentially European. On the other hand, its size, resources, and past strategic posture make it a unique national entity, and there is a very low probability that Russia will be integrated into Europe's still expanding union during the coming generation. For these reasons, I deal with Russia's prospects in a separate section.

The argument about Europe being the leading economy of the twenty-first century is inexplicably far off the mark. The reality, illustrated by Maddison's (2001) millennial reconstruction of Western Europe's GDP and population shares, shows an unmistakable post-1500 ascent that culminates during the nineteenth century and is followed by a gradual descent that is likely to accelerate during the coming decades (fig. 3.7). In 1900, Europe (excluding Russia) accounted for roughly 40% of global economic product; 100 years later it produced less than 25% of global

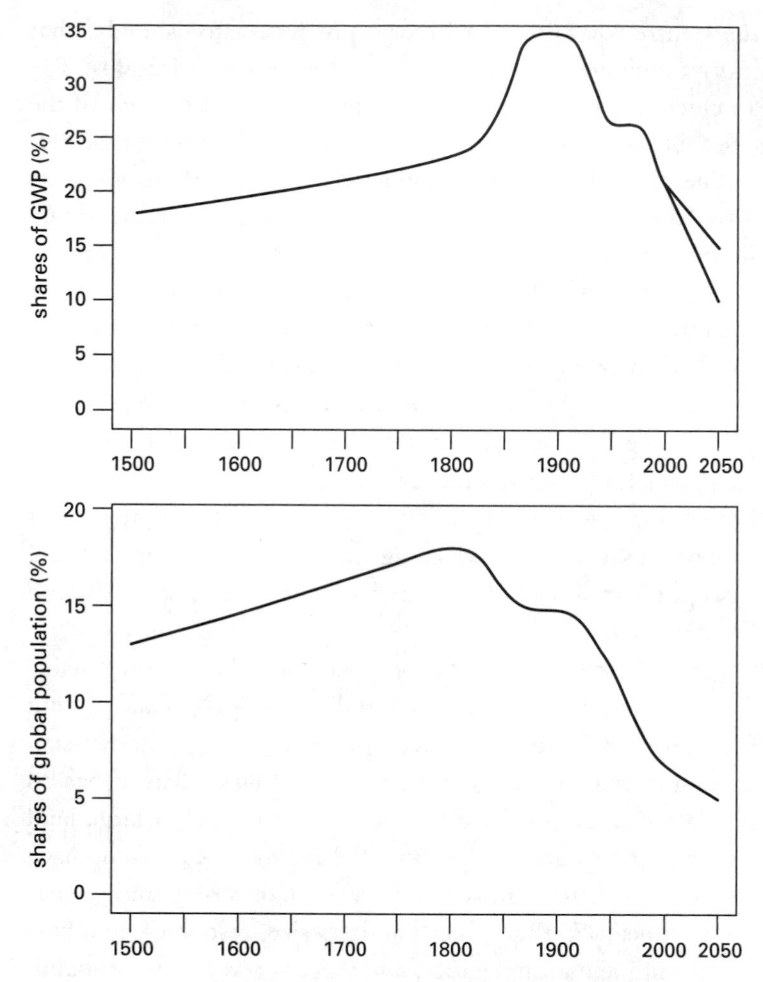

Fig. 3.7
Western Europe's shares of (*upper*) economic product and (*lower*) global population, 1800–2050. Plotted from data in Maddison (2001) and United Nations (2005).

output, and by 2050, depending above all on growth in the GDPs of China and India, its share of global economic product may be as low as 10%. By 2050, Europe's share of global economic product may be lower than it was before the onset of industrialization, hardly a trend leading toward global economic dominance. In addition, the continent has no coherent foreign policy or effective military capability. As Zielonka (2006) argues, the current European Union is simply too large and unwieldy to ever act like state; rather than a coherent actor on an international scene, he sees a "maze Europe."

The European Union's member states do not see eye to eye even on a major issue whose excesses and burdens simply cannot continue: the bloated, trade-distorting agricultural subsidies that have been swallowing about 40% of the EU's annual budget. Europeans, so eager to sermonize about their superior economic and social policies and higher moral standards, and so ready to voice anti-Americanism, could do with some introspection in all of these respects. Europe's labor productivity and ample leisure time have been bought by mass (and in some countries, persistent) unemployment, roughly twice the U.S. rate for the entire work force (~10% vs. ~5%). In some parts of the continent more than 25% of people younger than 24 years are jobless, and in 2005 the peaks were above 50% in three regions in Italy, two in France, and Poland (Eurostat 2005b).

Just two months after the self-congratulatory comments on the United States' ineptitude in dealing with the hurricane Katrina ("something like that could never happen here"), *banlieus* were burning all across France, the flames of thousands of torched cars and properties illuminating segregation, deprivation, and neglect no less deplorable than the reality of New Orleans' underclass exposed by Katrina's breached levees. (The torchings of about 100 cars a day have continued ever since.) As for the EU's morally superior risk-sensitive policies (as opposed to what Europeans see as unsophisticated, brutal, and blundering U.S. ways), they allowed EU member states to watch the slaughter of tens of thousands of people and the displacement of millions of refugees in the Balkan wars of the 1990s. Only the U.S. interventions, on behalf of Muslim Bosniaks and Kosovars, prevented more deaths (Cushman and Meštrovic 1996; Wayne 1997).

As for the grand visions of professional politicians, they have been rejected in a referendum even by France, the European Union's pivotal founding nation. Many managers as well as ordinary citizens would characterize the EU's modus operandi as bureaucratic paralysis instead of caring social democracy. Additionally, and sadly, the continent's ever-present anti-Semitism is undeniably resurgent (requiring the repeated assurances of political leaders that it is not); it surfaced in some stunningly direct comments during the Israeli-Hizbullah war of August 2006. Finally, a multitude of national problems with European integration will not go away.

The presence of a supranational entity like the EU has the effect of weakening ancient national entities. Spain has its Euskadi (Basque) and Catalunyan (and Galician) aspirations. Combined challenges from devolution (Scotland, Wales, Northern Ireland), European integration, multiculturalism in general, and large Muslim populations add up to a trend that may see the end of Britain (Kim 2005).

Germany, France, and Italy, the continent's three largest nations whose unity is not threatened by serious separatist movements (episodic Corsican violence or Lega Nord are not going to dismember France or Italy), have their own deep-seated challenges. The German and French economies underperformed during the last two decades of the twentieth century. In the German case there was the partial excuse (or terrible miscalculation) of the economic cost of absorbing the former East Germany (DDR) (Bentele and Rosner 1997; Zimmer 1997; Larres 2001; Bollmann 2002). The French situation is a textbook illustration of failures arising from an overreaching dirigisme, economic planning and control by the state. Neither country offers a model for a dynamic reinvention of Europe during the coming two generations.

As for Italy, the EU's third-largest economy, I can do no better than to quote an acute observer of his *patria* (Severgnini 2005, 77–78): "Life in Italy is so pleasant it becomes narcotic. . . . Italy, it would seem, suffers from a 'squirrel syndrome': everybody finds a comfortable hole, and hunkers down. The problem is that there are so many holes in the national tree that it may topple unless something is done soon."

But this comfort cannot last. Italy's economy suffers due to the combined impact of rapid aging (rivaled only by Japan), precipitous destruction of traditional small- and mid-scale artisanal manufacturing by Chinese imports, and growing numbers of immigrants. The deep economic and cultural breach between the *Nord* and *Centro* on one hand, and *Mezzogiorno* on the other, has not diminished. Decades of massive (and largely wasteful) investment have not lowered chronic unemployment. And the Mafioso culture of violence and corruption still operates (Gambetta 1992; Cottino 1998 and 1999; P. Schneider and J. Schneider 2003).

Even if one were inclined to see the post–WW II path of Western European as an undisputed success story and assumed that recent economic problems could be addressed fairly rapidly by suitable reforms, there is one inescapable factor that will determine the future of Europe's most affluent western and central parts and poorer eastern regions: a shrinking and aging population. After many generations of very slow demographic transition (Gillis, Tilly, and Levine 1992), Western Europe's total fertility rates slid below replacement level (2.1 children per mother) by the mid-1970s. A generation later, by the mid-1990s, the total fertility level of EU-12 was below 1.5; the new members did nothing to lift it: by 2005 the average fertility of EU-25 was 1.5 (Eurostat 2005b). Europe's population implosion (Douglass 2005) now appears unstoppable.

Naturally, the reliability of long-term population projections declines as the projection's final date advances, and a new trend of increased fertility cannot be categorically excluded. However, it is unlikely that it would last very long. The last notable regional rebound lifted the Nordic countries from an average of about 1.7 in 1985 to almost 2.0 by 1990, but by the end of the decade fertility was back to the mid-1980s level. More important, it is unlikely that a meaningful rebound can even begin once the rate had slipped below 1.5. That is the main argument advanced by Lutz, Skirbekk, and Testa (2005).

Once the total fertility rate reaches very low levels, three self-reinforcing mechanisms can take over and result in a downward spiral of future births that may be impossible to reverse. First, delayed childbirths and decades of low fertility shrink the base of the population pyramid and produce sequentially fewer and fewer children. Second, as fertility plummets, its norms will be even lower during subsequent generations, and even the perception of ideal family size (the number of children wished for under ideal conditions) shifts below the replacement level (already the case in Germany and Austria). Third, low fertilities, aging population, and shrinking labor force lead to cuts in social benefits, higher taxes, and lower expected income, working against higher fertility.

The economic consequences of population aging have been examined in considerable detail, and none of them can be contemplated with equanimity (Bosworth and Burtless 1998; England 2002; McMorrow 2004). Many of them are self-evident. Older populations reduce the tax base, and hence they lower average per capita state revenues and increase the average tax burden. Falling numbers of employed people push up the average dependence (pensioners/workers). Europe's already high pensioner/worker ratios (Britain being the only exception among the continent's largest economies) mean that old-age dependence ratios will typically double by 2050 (Bongaarts 2004). And in some countries most of this rise will happen during the next generation; for example, in 2001, Germany had 44 retirees (60+ years old) per 100 persons of working age, but the Federal Statistical Office (2003) predicts that the rate will jump to 71 by 2030.

As most countries finance the current retirement costs of their workers by current contributions from the existing labor force (pay-as-you-go arrangement), increasing retiree/worker ratios will bankrupt the entire system unless current contributions are sharply raised, pensions substantially cut, or both. Older workers may be more knowledgeable, but they still tend to be less productive because of physical or mental restrictions, higher disease morbidity, and a greater tendency toward workplace injuries and hence more frequent absenteeism. Higher dependence ratios and a higher share of very old people (>80 years old)—for example, every eighth person in Germany will be that old by 2050—will put unprecedented stresses on the cost and delivery of health care.

But as health care and pension expenditures rise, the average savings rates of the aging population will fall. This will affect capital formation, change the nature of the real estate market, and shift retail preferences for commodities ranging from food to cars. Despite the tightening labor market, many younger people may find their choice of jobs limited as some companies prefer to relocate their principal operations to areas with plentiful and cheap labor. Most new companies are started by individuals 25–44 years of age, and the shrinking share of this cohort will also mean less entrepreneurship and reduced innovation.

In addition to these consequences, which aging Europe will share with low-fertility societies in other parts of the world, the continent faces a specific problem whose resolution may crucially determine its economic and political future. As Demeny (2003) has noted, the process of moving toward a smaller and older population could be contemplated with equanimity only if Europe were an island, but instead "it has neighbors that follow their own peculiar demographic logic" (4). This neighborhood—Demeny calls it the European Union's southern hinterland—includes 29 states (counting Palestine and Western Sahara as separate entities) between India's western border and the Atlantic Ocean, all exclusively or predominantly Muslim (fig. 3.8).

By 2050, EU-25 is projected to have 449 million people (after losing some 10 million from the present level and an assumed net immigration of more than 35 million, 2005–2050), half of them older than 50 years. The population of its southern hinterland is projected to reach about 1.25 billion by 2050. Immigration to the continent from this hinterland is already the greatest in more than 1,000 years.

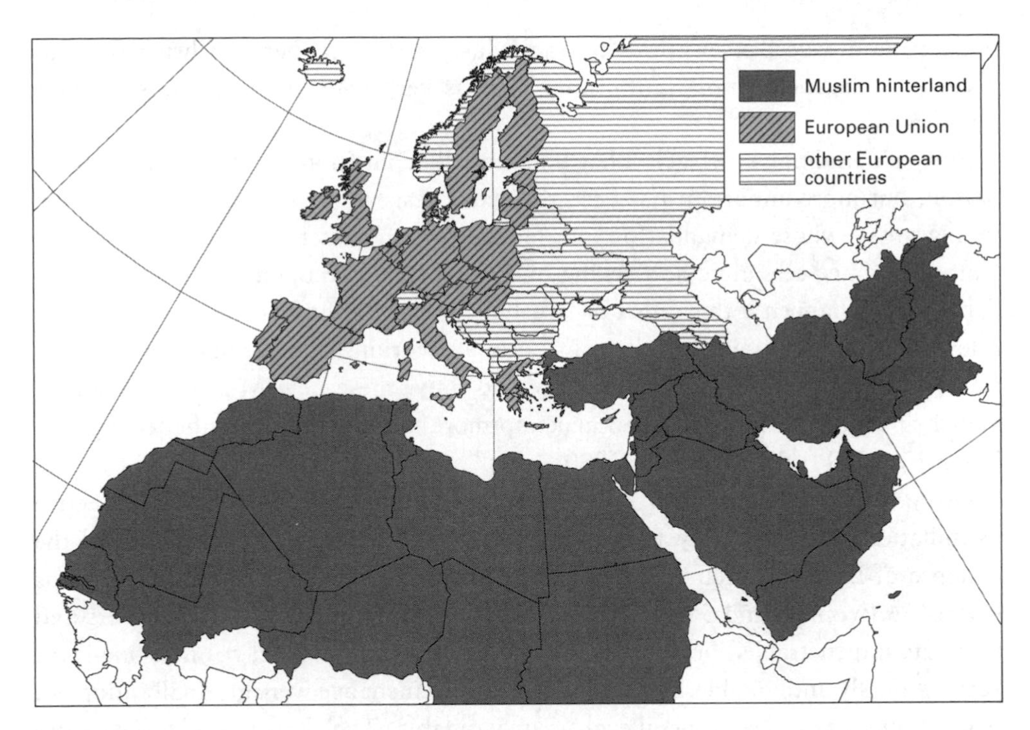

Fig. 3.8
Europe's Muslim hinterland. The population ratio of Muslim countries to EU-25 will rise from 1.4 in 2005 to 2.75 by 2050.

During the previous period of mass incursions, intruders such as Goths, Huns, Vikings, Bulgars, and Magyars destroyed the antique order and reshaped Europe's population. So far, the modern migration has been notable not for its absolute magnitude but for five special characteristics.

First, as is true for immigrants in general, the Muslim migrants are much younger than the recipient populations. Second, the migrants' birth rate is appreciably (approximately three times) higher than the continent's mean. Third, the immigrants are disproportionately concentrated in segregated neighborhoods in large cities: Rotterdam is nearly 50% Muslim; London's Muslim population has surpassed 1 million, and Berlin has nearly 250,000 Muslims. Fourth, significant shares of these immigrants show little or no sign of second-generation assimilation into their host societies. A tragically emblematic illustration of this reality is that three of the four suicide bombers responsible for the July 7, 2005, attacks in London's underground were British-born Pakistani Muslims. Fifth, whereas Christianity has become irrelevant to most Europeans, Islam is very relevant to millions of these immigrants.

Europe's traditional ostracism has undoubtedly contributed to the lack of assimilation, but more important has been the active resistance by many of the Muslim immigrants—whose demands for transferring their norms to host countries range from segregated schooling and veiling of women to the recognition of *sharī'a* law. What would happen if this influx of largely Muslim immigrants were to increase to a level that would prevent declines in Europe's working-age population? In many European countries, including Germany and Italy, these new Muslim immigrants and their descendants would then make up more than one-third of the total population by 2050 (United Nations 2000).

Given the continent's record, such an influx would doom any chances for effective assimilation. The only way to avoid both massive Muslim immigration and the collapse of European welfare states would be to raise the retirement age—now as low as 56 (women) and 58 years in Italy and 60 years in France—to 75 years and to create impenetrable borders. The second action is impossible; the first one is (as yet) politically unthinkable. But even if a later retirement age were gradually adopted, mass immigration, legal and illegal, is unavoidable. The demographic push from the southern hinterland and the European Union's economic pull produce an irresistible force.

Two dominant scenarios implied by this reality are mutually exclusive: either full integration of Muslim immigrants into European societies, or a continuing incompatibility of the two traditions that through demographic imperatives will lead to an eventual triumph of the Muslim one, if not continentwide, then at least in Spain, Italy, and France. I do not think that the possibility of a great hybridization, akin to the Islamo-Christian syncretism that prevailed during the earliest period of the Ottoman state (Lowry 2003), is at all likely. The continent's Christians are now overwhelmingly too secular-minded to be partners in creating such a spiritual blend, and for too many Muslims, any dialogue with "nonbelievers" is heretical.

Other fundamental problems will prevent Europe from continuing to act as a global leader. Europe cannot act as a cohesive force as long as its internal divisions and disagreements remain as acute as they have been for the past three decades despite the continent's advances toward economic and political unification. Yet the ruinous agricultural subsidies, national electorates alienated from remote bureaucracies, Brussels's rule by directive, and inability to formulate common foreign policy and military strategy are, in the long run, secondary matters compared with the eventual course of the EU's enlargement. Even an arbitrarily permanent exclusion of Russia from the EU leaves the challenge of dealing with the Balkans, Ukraine, and Turkey. The EU's conflicting attitudes toward Turkey—among some leaders an eager or welcoming, economics-based embrace, among others a fearful, largely culture-based, rejection—capture the complexity of the challenge.

Turkey's exclusion would signal an unwillingness to come to terms with the realities of the southern hinterland. And, as the Turkish Prime Minister said, Turkey's achieving membership in the EU "will demonstrate to the world at large that a civilizational fault-line exists not among religions or cultures but between democracy, modernity, and reformism on the one side and totalitarianism, radicalism, and lethargy on the other" (Erdoğan 2005, 83). Admirable sentiments, but only if one forgets a number of realities. The wearing of *hijāb* has become a common act in Turkey, overtly demonstrating the rejection of Turkey's European destiny (even Erdoğan's wife, Emine, would not appear in public without it and hence cannot, thanks to Atatürk's separation of Islam from the state power, take part in official functions in Ankara or Istanbul). The Turkish police and courts habitually persecute writers and intellectuals who raise the taboo topic of Armenian genocide and question the unassailability of "Turkishness." The Kurds, some 15% of Turkey's population, are still second-class citizens. So much for "democracy, modernity, and reformism."

And how could one posit a rapid cultural harmonization (*integration* would be the wrong word here) of what would be EU's largest nation with the rest of the Union when Turkish immigrants have remained segregated within Islamic islands in all of Europe's major cities? Perhaps the only quality that might endear the Turkish public to Europeans is the fact that the former's share of very or somewhat favorable opinion of the United States is even lower (<20%) than in Pakistan (Pew Research Center 2006). Europe's anti-U.S. elites would thus have a multitude of new allies. But if the EU admits Turkey, why not then the neighboring ancient Christian kingdoms of Georgia and Armenia?

And if the EU, as Erdoğan says, is not a Christian club, why not admit Iraq, one of the three largest successor states of the Ottoman Empire, ancient Mesopotamia, a province of the *Imperium Romanum*? And, to codify the inevitable, why not make the EU's southern borders coincidental with those of the Roman Empire? Why not embrace all the countries of the Arab *maghrib* and *mashriq*, that is, North Africa from the Atlantic Morocco (Roman *Mauretania Tingitana*) to the easternmost Libya

(*Cyrenaica*), and the Middle East from Egypt (*Aegyptus*) to Iraq? Their populations will be providing tens of millions of new immigrants in any case.

No matter how far the EU expands, what lies ahead is highly uncertain except for one obvious conclusion. An entity so preoccupied with its own makeup, so unclear about its eventual mission, and so imperiled in terms of its population foundations cannot be a candidate for global leadership. But it already is the planet's foremost destination of tens of millions of tourists. And many more are poised to come. When one sees the endless procession of travelers in today's Rome, Prague, Paris, or Madrid, one can imagine a not-too-distant future (2020?) when the intensity of Chinese Europe-bound travel will surpass today's U.S. rate (12.5 million visitors in 2005), and Europe will see every year more than 20 million Chinese tourists. This is perhaps the likeliest prospect: Europe as the museum of the world.

Japan's Decline

Japan's rise, more phenomenal than Europe's recovery after WW II, lasted less than two generations, between 1955, when the country finally surpassed its prewar GDP, and the late 1980s, when it was widely seen as an unbeatable economic Titan. This surge was all the more remarkable considering its near-total dependence on imported energy and the OPEC-driven oil price shocks of 1973–1974 and 1979–1980. At that time its admirable dynamism and enviable economic performance earned it widespread admiration and generated apprehension and outright fear regarding its future reach. Its rise was not derailed even by the Plaza Accord of September 22, 1985, by the G-5, the group of five nations with leading economies, which eventually led to near halving of the yen/dollar exchange rate (from 254 by the end of 1984 to 134 by the end of 1986) and to a spree of foreign acquisitions by Japanese companies and record purchases by the country's art collectors (Funabashi 1988).

As Japan's high-quality exports kept rising, Ezra Vogel, Harvard University's leading expert on Japan, published *Japan as Number One* (1979). Japan's expansive trend defied the revaluation of the yen in 1985 and accelerated during the next four years; the Nikkei index was at just over 13,000 by the end of 1985 and peaked at nearly 39,000 in December 1989 (fig. 3.9). But right after that, Japan's bubble economy burst in spectacular fashion (Wood 1992; Baumgartner 1995). History has no other example of a country whose standing switched so rapidly from that of a globally admired technical and manufacturing superpower to that of a deeply ailing economy.

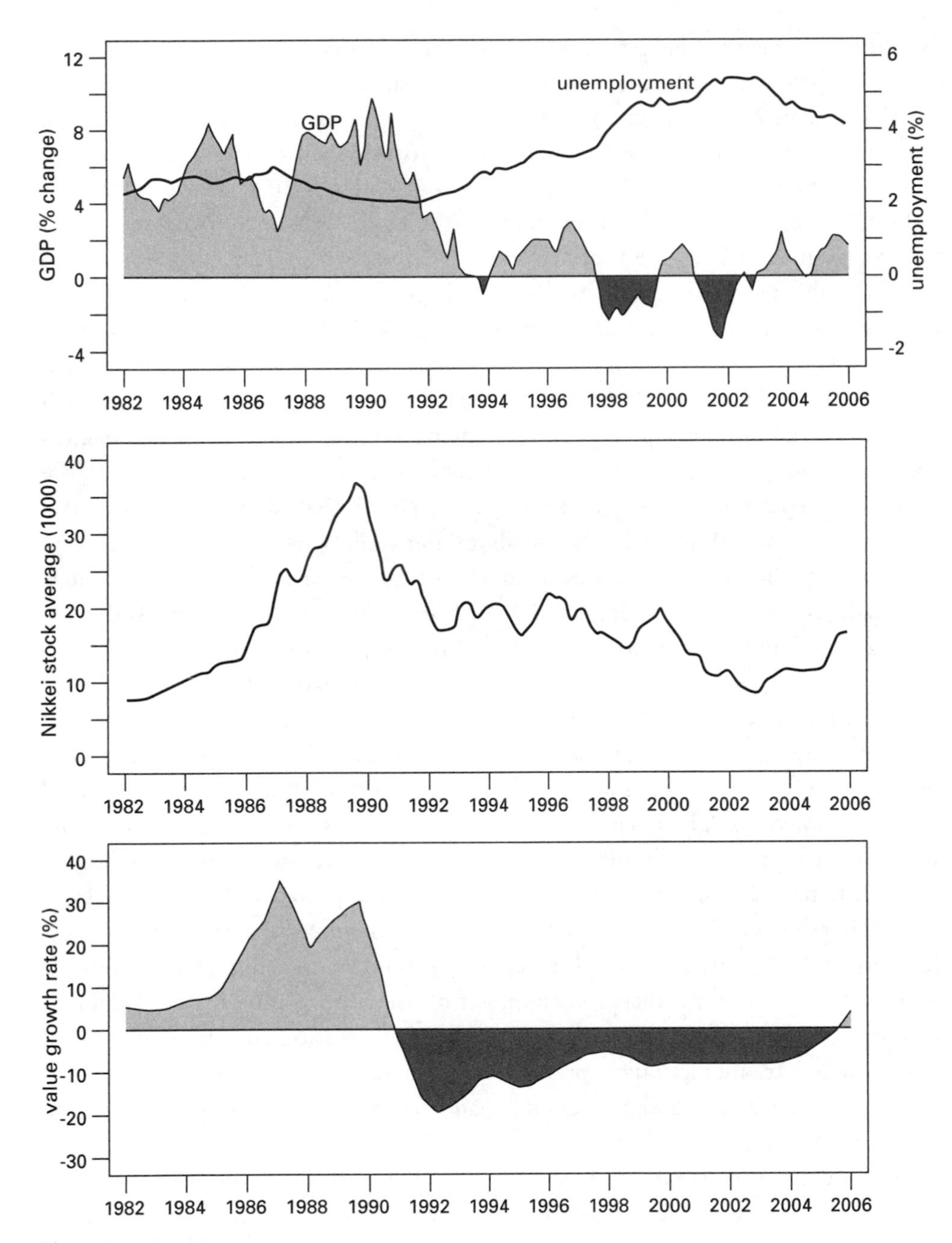

Fig. 3.9
Japan's GDP growth, unemployment, Nikkei stock index, and urban land prices, 1982–2006. Plotted from data in JREI (2006).

As just about everything began to unravel, the critics of Japan's bubble, ridiculed during the 1980s, became the new prophets. By the end of 1990, as Nikkei index fell to less than 24,000, many experts still foresaw an imminent recovery. But the decline continued, and by 2002 the index fell to less than 8,600. Then it rose to about 11,400 by the end of 2004, and by the beginning of 2008, it was at 15,308, still 60% below its record level. In contrast, the Dow Jones index reached 13,624 by the beginning of 2008, 16% above its pre-9/11 peak of 11,722. Because so much of Japan's inflated stock market was propped up by a real estate price bubble, the slipping Nikkei index devastated property values. By 1995 the index of urban land prices in Japan's six largest cities fell to half of its peak 1990 level, and then it continued to decline; by 2005 it was 25% of its top bubble value (see fig. 3.9; JREI 2006). More important, Japan, previously the paragon of value-added manufacturing, began losing jobs to other East Asian countries. In 1989, Japan derived more than 27% of its GDP from manufacturing; by 2005 that share fell below 20% (Statistics Bureau 2006). Complaints about the hollowing-out of the economy, heard strongly in the United States for the first time because of the country's huge trade deficits with Japan during the 1980s, became common in Japan. And every passing year has failed to arrest, much less reverse, Japan's profound and long-lasting retreat from its place as a top economy and a leading technical innovator (Yoda 2001; Callen 2003; Hutchison and Westermann 2006).

Japan's stagnation has given rise to many unaccustomed sights, such as homeless men living in cardboard boxes in railway stations, parks, and side streets, and dismal statistical indicators. The unemployment rate, which was mostly 2%–2.5% during the 1980s, rose to 5.5% by the end of 2001; the suicide rate, traditionally higher in Japan than in Europe or North America, increased from 16.4/100,000 in 1990 to 25.5/100,000 by 2003 (Statistics Bureau 2006). And in 2007 the first large cohort of elderly baby boomers launched the country's mass retirement wave (typically at age 60). At the same time, increasing numbers of young people (more than 1 million) have opted out of the labor market. This NEET generation (not in employment, education, or training), which prefers just hanging out in strange clothes and hairdos, can be seen as a sign both of Japan's national decline and its continuing personal affluence.

But some things have not changed. Japanese women still live longer than women elsewhere (their average life expectancy at birth surpassed 85 years in 2003, compared to 83 in France and 82 in Canada), and the mean per capita GDP is (in terms of PPP) only marginally behind the French or Canadian level. There have even been some welcome gains. After two generations of very high savings, people began spending more freely, on air conditioners, bathrooms, cars, and flights to Thailand or Europe. To be sure, the savings rate plummeted, but more Japanese enjoy life in greater comfort at home and more of them spend their vacations (still short, even when compared to U.S. vacations) abroad. In 2005 more than 17 million Japanese tourists (or nearly every seventh person) left the archipelago.

Japan's prospects are discouraging. Despite the prolonged economic shock, the country still has not made sufficient adjustment to its peculiar banking, management, and decision-making systems, which are generally considered the preconditions of a new beginning (Carlile and Tilton 1998; Lincoln 2001; Grimond 2002; Tandon 2005). Prolonged recovery has become much harder because of a combination of economic and political factors: the relentless rise of China and its confrontational style of foreign policy, increasingly precarious dependence on the grossly overextended United States, and the perennial danger of irrational North Korea. By 2005 there were many signs of a real turnaround, and a key question seemed to be: If Japan's rise during the 1980s was uncritically hyped by most of the country's admirers, are not the country's detractors now repeating the same mistake in reverse by degrading Japan to a category of lasting underperformer?

The editor of *The Economist* argued that this is precisely the case and concluded that the country "is at last ready to surprise the world how well it does, not how badly" (Emmott 2005, 3). There has been no shortage of statistics to buttress an optimistic case. By 2003 Japan's annual GDP growth rose again to more than 2%, and many large companies became profitable again (some because of their links with manufacturing in, China, others thanks to growing worldwide demand for Japan's well-known brands of manufactures). By 2005 newly available jobs nearly matched the number of applicants (the ratio was below 0.5 in 1998), and in July 2006 seven years of deflation (as high as –0.9% of the consumer price index in 2002) came to an end as the Bank of Japan raised its interest rate from 0 to 0.25% (the rate was 6% in 1990).

There are at least three major reasons that I do not foresee Japan's regaining a status comparable to its position during the 1980s. The first reason is perfectly captured in Donald Richie's perceptive Japan diaries, in his entry for February 12, 1999. When Karel van Wolferen, who authored a book on the enigma of Japanese power (Wolferen 1990), remarked that the only way out of Japan's dilemma was some kind of revolt that he could not imagine, Richie (2004, 429) told him "that Nagisa Oshima had said that this occurred only three times in Japan's history; the

Tempo Reforms, the beginning of Meiji, and in 1945. And each time the structure recrystallized, and petrified." This may be dismissed as too deterministic, but no diligent student of history can deny the existence of national predilections.

The second reason is that the signs of Japan's domestic renaissance have been accompanied by unsettled relations with its three western neighbors, by their deepening distrust and dislike of Japan, whose manifestations include mass demonstrations in China's cities, South Korea's frequent diplomatic protests, and the undisguised hostility of North Korea that provoked the government to contemplate a future possibility of a preventive defensive strike. These seemingly intractable external factors are the main reason that even if a widely discussed change in its constitution removed the restrictions on Japan's military actions (Nippon Keidanren 2005), the country would remain no less dependent on its strategic ties with the United States.

The third reason is Japan's declining population. By far the most fundamental obstacle to Japan's reincarnation as a great power of the twenty-first century is the fact that the country's partial economic recovery came so late that it merged with the onset of Japan's depopulation and with a globally unprecedented aging of its people. Two generations of decreasing total fertility rate—from a post–WW II peak of 2.75 in the early 1950s to about 1.3 (well below the replacement level of 2.1) by the early 2000s—have made it inevitable that Japan's total population will decline. Only a massive, Canadian- or Australian-style immigration that would admit at least half a million people per year could reverse this, but such a policy change is most unlikely. Consequently, the only uncertainty concerning Japan's aging is the pace. Its many socioeconomic consequences should be similar to those that will be affecting other countries (England 2002; McMorrow 2004; MacKellar et al. 2004).

The medium variant of the best Japanese projections of the early 2000s expected peak population in 2006, at 127.74 million (NIPSSR 2002), but the preliminary counts of the 2005 census (held on October 1) showed that the total population of 127.76 million was about 19,000 below the estimate for October 2004. Apparently, Japan has already entered a long period of depopulation. If there are no dramatic changes in Japan's fertility rate, the country's population would decline slowly at first, to about 121 million by 2025, then to only about 100 million people by the year 2050 (NIPSSR 2002). For comparison, the latest United Nations (2005) forecast sees only a marginal decline by 2025 (nearly 125 million) and the total of about 112 million by 2050. But these differences matter less than what the absolutes hide:

it is virtually certain that by the middle of this century Japan will become the most aged of all aging high-income societies.

According to the medium variant of the latest Japanese projections (NIPSSR 2002), the country's median age will reach 50 years by 2025. Whereas in 2005 one out of five people was 65 years or older (the highest share worldwide), by 2025 the share will be nearly 30%, and it will reach 35% by 2050. Japan's age-gender population structure will assume a cudgel-like profile, in contrast to today's barrel-shape and the classical pyramid of the early 1950s (fig. 3.10). The share of adults of economically active age will drop from 66% in 2005 to 53% by 2050, when astonishingly, about one out of seven people will be 80 years or older. This would mean that there would be more very old people (80+ years) than children (0–14 years; their 2050 share is projected to be less than 11%), a situation that would create the world's first truly geriatric society (United Nations 2005).

Japan's depopulation is already starkly evident in many of the country's rural areas, where entire villages are abandoned and shrinking municipalities are merging to boost their tax base. This trend is evident not only in remote areas. A few years ago I was offered (if I'd move in) a large, well-built house overlooking a lovely valley in a small village only a little more than one hour north of Kyōto that had already lost its post office, its school, and all but a few of its inhabitants. The implications of this continuing depopulation and the aging trend would be far-reaching, and some are difficult to imagine. There has been never a society, much less a major nation, where octogenarians outnumbered children.

The absolute drop would push Japan from being the world's tenth most populous nation in 2005 (after Nigeria and ahead of Mexico) to thirteenth place by 2025 and to seventeenth place by 2050, behind Vietnam and ahead of Turkey (United Nations 2005). And because most Japanese companies still have a minimum mandatory retirement age at 60, there will be a wave of retirees between 2010 and 2020 as the high-fertility cohort of the 1950s quits working. A new law passed in 2004 raised the minimum mandatory retirement age to 65 by 2013. While many people will want to work past 60, none of these adjustments will provide more workers for occupations that require more demanding physical exertions, which will be in greater demand as both Japan's infrastructure and its population age rapidly.

Reconstruction of crumbling highways (air hammers, concrete pouring, laying down reinforcing steel bars), repairs of buildings damaged by earthquakes, or the care of bedridden patients cannot be done by octogenarians. Japan's much admired robotization has often been offered as a partial solution of the aging challenge:

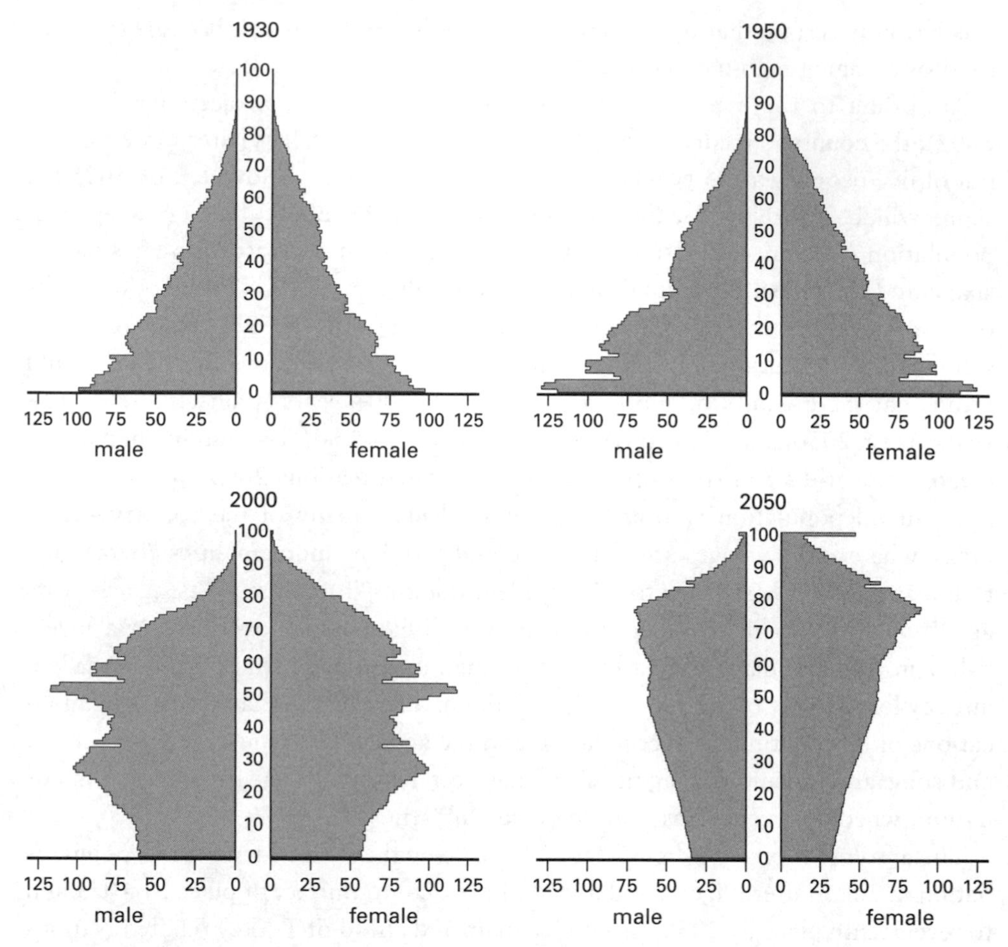

Fig. 3.10
From a pyramid to a cudgel: Japan's age and gender population structure in 1930, 1950, 2000, and 2050. From NIPSSR (2002).

instead of importing foreign labor Japan leads the world in using industrial robots. By 2005 the country had about 356,000 robots, more than 40% of the worldwide total and nearly 90% of the combined stock of these machines installed in Europe and North America (IFR 2005). The country's many makers of robots include such leading producers as FANUC, Fujitsu, Kawasaki, Mitsubishi, Muratec, Panasonic, and Yaskawa. But the actual gap between Japan and the rest of industrialized world is not that large because Japanese statistics also include data on simple manipulators that are controlled by mechanical stops, and these machines would not pass a stricter definition of industrial robots used in the United States and the EU.

Moreover, Joseph F. Engelberger (who in 1956, with George Devol, established the world's first robot company, Unimation) has been very critical of the direction taken by Japan's leading robot researchers: "Nothing serious. Just stunts. There are dogs, dolls, faces that contort . . ." (cited in Cameron 2005). Instead, he advocates, as he did for the first time in the late 1980s (Engelberger 1989), an intensive development of household service robots to help the elderly and infirm, an advance that would particularly benefit the world's soon-to-be most geriatric nation.

I do not see robots saving Japan, and I do not see, as Pyle (2007) clearly does, the 15 years since the end of the Cold War as a time of transition, with the country now standing on the verge of new triumphs as it reasserts not only its economic power but also its military capacity. Of course, Japan will not—as so dramatically portrayed in a recent sci-fi bestseller (Komatsu and Tani 2006)—become a victim of mighty geotectonic forces and sink into the ocean, and the Japanese will not become refugees in Papua New Guinea, the Amazon, and Kazakhstan, but even if there were a newly assertive Japan, its immediate impact and its long-term global influence will not be anything like its achievements up to the 1980s.

If the odds do not favor either Europe or Japan filling the roles of strong, dynamic, nimbly adaptive great powers of the first half of the twenty-first century, we are left with four choices. Can resurgent Islam realize an oft-stated goal of some of its militant leaders and establish a caliphate even more extensive than that of the Umayyad dynasty (661–750), whose power reached from the Iberian Peninsula to Afghanistan and from Armenia to Yemen? Can Russia, strategically downsized by the collapse of the USSR, regain its superpower status? Can China translate its economic power into a wider strategic dominance? And is it possible that the United States can successfully cope with many signs of advanced decay and retain its place on top for at least another two generations?

Islam's Choice

The question regarding the establishment of a new centrally governed Muslim caliphate extending from Spain to Pakistan that I posed in the preceding section is easy to answer. Because the Muslim world is too heterogeneous (in sectarian, economic, cultural, and political ways), the chances of seeing such an extensive, coherent, and globally powerful political and economic entity before 2050 are vanishingly small. That leaves us with questions that are much harder to answer. Why is the current Muslim world (with a few exceptions)—taken as the entire population of believers, *al-'umma al-islāmīya* or *al-dar al-islām* (fig. 3.11)—so poorly governed and undemocratic? Why does it lack an acceptable human rights record, why is it so stridently intolerant, why does it relegate women to second-class status? Why is

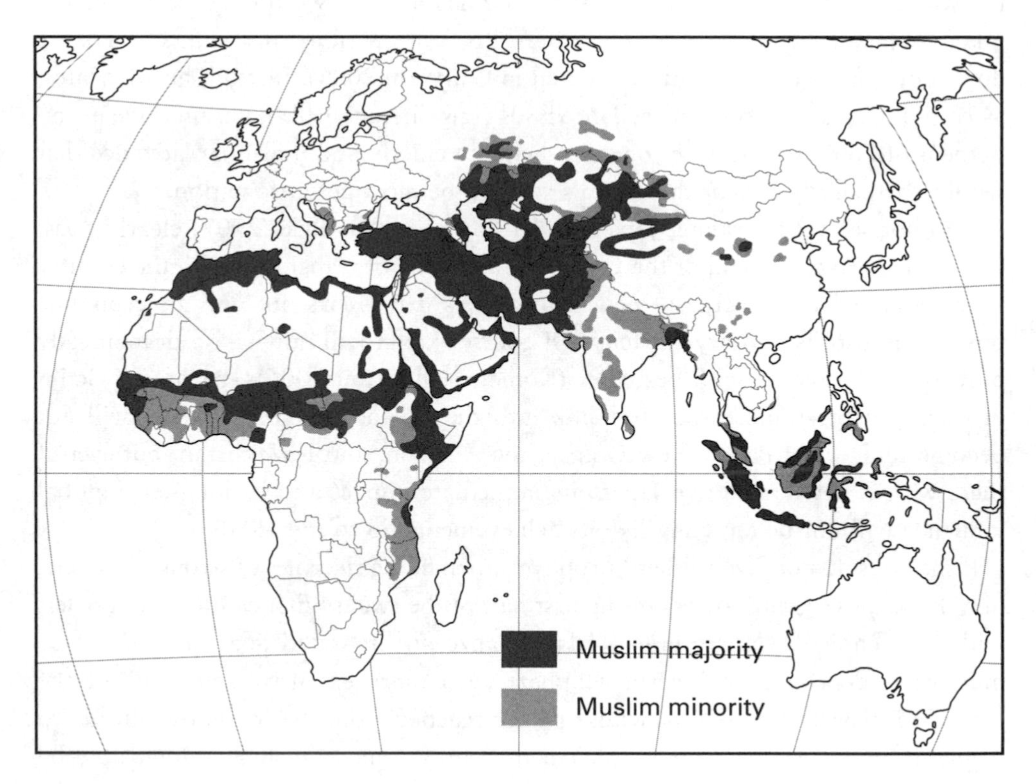

Fig. 3.11
The Muslim world extends from Senegal to the Philippines, and from the Volga region of Russia to coastal East Africa. Simplified from various maps of religious denominations.

it so prone to internal violence, and why is it the leading global perpetrator and exporter of suicide terrorism?

Reasoned answers to these questions should tell us what we should really be worried about in the future, and what is the outlook for defusing and reducing threats. The Muslim faith itself offers at best a limited explanation. The now common Western perception of Islam as a zealous, monolithic faith with a near-hypnotic power over populations imbued with a uniform hatred of the West is a figment of historical ignorance and unhelpful generalization. Islam's comprising so many populations and different traditions makes the faith fairly heterogeneous; its practitioners span a wide continuum of commitments and devotions (B. Lewis 1992; Esposito 1999). More important, a large part of the Muslim world is no different from the Christian world in its adherence to (some or most) major tenets of a traditional religion, observation of principal holidays and ceremonies, participation (by the more devout) in pilgrimages, and pride in the associated literary and artistic heritage.

These Muslims—be they in China, Indonesia, India, Iran, Russia, Bosnia, Egypt, Tunis, Mali, Niger, or Tanzania—are not going to upend the history of the next 50 years any more than will the hundreds of millions of their (lukewarm to earnest) Christian counterparts in Europe, the Americas, and Africa. Nor should the term *Islamic fundamentalism* generate reflexive fears. The term has entirely different connotations when understood as a strictly religious identifier or as a modern call to arms against "infidels" (Sidahmed and Ehteshami 1996; Davidson and Miller 1998; Moaddel and Talattof 2002; Khaled 2005; Margulies 2006). On one hand, it may mean only exemplary piety, but it could also describe a strident insistence that only the Koran and *sunna*, the traditions based on the sayings, actions, and approvals of the Prophet Muhammad incorporated in the books of *hadīth*, are the exclusive guides of the faith. Or it could be a label for the *sunnī* intolerance of their *shī'ī* co-religionists, who do not accept *sunna* as binding.

On the other hand, *Islamic fundamentalism* has come to designate religiously inspired violence perpetrated by Muslims. Most Muslim fundamentalists (*sensu lato*) will never commit any acts of violence, and the actions of Muslim terrorists have been condemned (albeit belatedly) by a *fatwā* authored by the Fiqh Council of North America (2005) as well as by a Saudi *fatwā* (Sandi Committee 2004) issued by the most fundamentalist (*sensu stricto*) body of all, Saudi Arabia's Permanent Committee of Religious Research and Ifta, which is chaired by the kingdom's leading religious authority, Grand Mufti Skaikh Abdulazīz Al-Shaikh. The Saudi *fatwā*, released a day after the Riyādh bombings of April 21, 2004, not only condemned them as "a forbidden and sinful act" but also forbid "to justify the acts of these criminals." Tellingly, no such *fatwā* followed the destruction of the Twin Towers on September 11, 2001.

Islam's guiding texts do not give us, contrary to what is claimed by ill-informed commentators, any unequivocal guidance regarding aggression and compassion. Much like the contradictory instructions and enjoinments of the Torah, and the stark dichotomy between vengeful passages in the Old Testament and the submissive and compassionate preaching of Jesus Christ, the Koran and *hadīth* offer many contradictory messages that a determined exegesis can use to reach opposite conclusions. I juxtapose just two sets of quotes in order to exemplify this contrast between militant and compassionate Islam. The third *sūra* of the Koran (verse 151) puts it unequivocally: "We will put terror into the hearts of the unbelievers. . . . Hell shall be their home." But Abu Daūd (*hadīth* 4923) states with equal clarity: "If you show mercy to those who are on Earth, he who is in the Heaven will show mercy to you."

The fifth *sūra* (verse 32) says, "We laid it down for the Israelites that whoever killed a human being, except as a punishment for murder or other wicked crimes, should be looked upon as though he had killed all mankind; and whoever saved a human life should be regarded as though he had saved all mankind." This eloquent passage was cited in the *fatwā* against terrorism issued in July 2005 by the Fiqh Council of North America, albeit leaving out the preamble "We laid it down for the Israelites." But the very next verse (33) of the same *sūra* carries a very different message, one that could easily be used to justify violence against infidels: "Those that make war against Allah and His apostle and spread disorders in the land shall be put to death or crucified or have their hands and feet cut off on alternate sides, or be banished from the country." Notice that "spreading disorders," a conveniently elastic category, is enough to qualify for such a treatment.

The term *jihād* is now a common synonym for *holy war* (the term for which there is actually no word in classical Arabic), but the noun, derived from the verb *jahāda*, to exert, struggle, or strive, originally signified any sort of exertion to the utmost, striving toward an exalted goal. Muhammad himself called the moral struggle against one's self *al-jihād al-akbar*, the greater *jihād*, in contrast to the lesser *jihād* of the sword. Contrary to the common Western perception that *jihād* is the guiding star of Islam, *hadīth* 48 orders a different sequence:

> I asked Allah's Messenger (may be peace upon him) which deed was the best. He (the Holy Prophet) replied: The Prayer at its appointed hour. I (again) asked: Then what? He (the Holy Prophet) replied: Kindness to the parents. I (again) asked: Then what? He replied: Earnest struggle (jihād) in the cause of Allah.

First piety, then compassion. This sequence could have come straight from the sermons of Jesus.

And there is more. How can any *jihād* be compulsory when the 256th verse of the second *sūra* says quite explicitly that "there shall be no compulsion in religion"? And when in *hadīth* 1009 the Prophet explicitly warns: "O you men, do not wish for an encounter with the enemy. Pray to Allah to grant you security," much as the 190th verse of the second *sūra* admonishes the faithful "to fight for the sake of Allah those that fight against you but do not attack them first. Allah does not love the aggressors." Perhaps I have already belabored the point enough. The problem is not Islam, a religion with tenets as contradictory, as open to diverse interpretation, and as confusing in its totality as are its two great monotheistic inspirations, Judaism and Christianity. The problem is political or politicized Islam, Islam tendentiously interpreted, or not interpreted at all and hence stubbornly anchored in its medieval origins.

But if arguments have no place in dealing with the Koran (because it is God's revealed word), it must either be accepted in its entirety by the faithful or rejected by unbelievers. Doubting parts of it amounts to rejection, as does any adaptation of the text to modern needs. The challenge to literalists is not how to reconcile the revelation with modernity but how to adapt life in the modern society to the revealed teaching. This approach leaves no room for any interfaith dialogue that so many Westerners call for. More important, it leaves little hope for any true modernization of the Muslim world because it prevents any adaptation to challenges ranging from open discussion of key problems facing the Muslim societies to environmental change. It rejects functional fusion of an ancient religious tradition with the needs of the modern world. And it rules out the broad-minded tolerance that is required for international cooperation in a globalized world. In sum, this approach immures the faith.

Not surprisingly, many Muslims feel that their religion has been hijacked by extremists (be they *madrasa*-bound rigid traditionalists or modern descendants of Ibn al-Sabbah's *al-hashashīn* who are bent on murder by suicide) and that it must be reclaimed (Khaled 2005). Perhaps none has outlined this great Muslim dilemma better than Muhammad Shahrūr, by profession a civil engineer, by vocation a passionate advocate of modernized, tolerant, enlightened Islam. In two path-breaking books (Shahrūr 1990; 1994, neither of them in the Library of Congress, a telling commentary on Americans understanding of Islam) and in many presentations and interviews, he argues boldly for a fundamental reinterpretation of the faith: "We cannot go on without radical, religious reform . . . , like the one initiated by Martin Luther. We have reached a dead end; we are stuck in a dark tunnel. . . . If we don't, there is no hope for us, because we will continue living in the past" (Shahrūr 2004).

His most programmatic statement is a proposal for an Islamic covenant based on "interpreting the Book in a contemporary way" that uses the Koran to justify such fundamental tenets of modern, tolerant faith as a full exercise of all personal freedoms ("freedom is the only form in which man's worship of God can be embodied"), no religious coercion ("The fact of having no compulsion in religion is too fundamental to be subject to elections or debate"), no persecution of other religions ("Therefore this distinction does not justify enmity, hatred, and killing"), the need for democracy ("democracy is so far the best relative standard man has achieved"), and rejection of religiously motivated violence (Shahrūr 2000).

Islam thus faces a fundamental, long-lasting, contentious, even outright dangerous internal fight over its modern identity, similar to what Christianity experienced during the centuries between the impassioned preachings of Jan Hus (1369–1415; a reformist Czech priest who preceded Martin Luther (1483–1546) by a century) and the separation of church and state, which in some nations was legally completed only during the twentieth century. I do not claim that what the non-Muslim world does (being considerate, delicately critical, disdainful, indifferent) is irrelevant, but its actions or inactions are not the key determinant of the outcome of Islam's internal conflict.

The contrast between Islam's two choices is as profound as in any previous clash of incompatible ideologies. On the one hand is the extremism of *al-wahhābīya* (a Saudi sect often presented under the label of Salafism, *al-salāfīya al-jihādīya*), which accuses other Muslims of apostasy (*takfīr*) to justify such killings as the *sunnī* massacres of *shīʿī* (Schwartz 2002). Its means of exalting terrorism include widespread distribution of religious cassettes that advocate violent *jihād* by emphasizing sexual rewards for suicide "martyrs" (MEMRI 2005a). On the other side, with Shahrūr, are spokesmen who call for modernization, including Muhammad bin 'Abd Al-Latīf Aal Al-Shaikh, a Saudi writer, who sees the ideology of terror groups as very similar to Nazism (both based on hatred and physical elimination) and asks why, given this similarity, Muslims are not fighting against the foundations of this religiously inspired terror, against "its religious scholars, its theoreticians, and its preachers—just as we deal with criminals, murderers, and robbers?" (MEMRI 2005b).

Of course, not in all countries with Muslim majorities do people feel that this contest is a critical part of their societies or a key determinant of their futures. But certain aspects of this conflict are present in almost all countries, and the outcome, all too uncertain, will determine Islam's global impact during the next 50 years.

Although this internal contest over the meaning and the mission of the faith is important, it is not the only Muslim challenge. There are at least three intertwined problems and trends that are influenced by this contest and whose evolution will, in turn, help to determine its outcome: the lack of secular sources of state legitimacy, Muslim countries' modernization deficit, and a tardy demographic transition.

The lack of secular sources of state legitimacy leads to what Shahrūr (2005, 39) calls "the bizarre combination of ruling regimes" with their default reliance on traditional religious sources and emergence of archaic despotic authorities. Saudi Arabia and Iran (and Afghanistan under the Taliban) are the most extreme examples of this reality. But its persistence is clearly evident even in Turkey eight decades after Kemal Atatürk secularized the country in 1924 (fig. 3.12)—a nation that seeks entry into the EU and but a large part of whose population still imbues a piece of cloth covering a woman's head with transcendent qualities and sees it as a sacred banner of non-negotiable identity. Pamuk's (2004) book dealing with these matters should be required reading.

This default reliance on religious sources of state legitimacy explains the prominence of *sharī'a* (literally, a path, a way), the Islamic code of law and living whose precepts deal with personal affairs, as well as crime, commerce, finance, and education, and thus preempt adoption of laws that form the foundation of modern states. *Sharī'a* reinforces the persistence of an archaic worldview dependent on rigid interpretations of religious texts, and its reach is not weakening under the pressure of modernity. Just the opposite is true, owing to demands to make it a law for Muslim minorities in Western countries (Marshall 2005).

The modernization deficit in the Muslim world is in many ways the result of the situation just outlined. It ranges from repression of individual freedoms to a parlous state of scientific research to a relatively poor quality of life experienced by the majority of Muslim countries. In 2006 only 3 out of 46 countries with Muslim majorities had the highest ranking (1 or 2) in a global comparison of political rights

Fig. 3.12
Mustafa Kemal Atatürk (1881–1938), the creator of a strong secular Turkish state. Photo courtesy of Basin-Yayin ve Enformasyon Genel Müdürlüğü (General Directorate of Press and Information), Ankara.

and civil liberties (Freedom House 2006). The famous Syrian poet Ali Ahmad Sa'īd (living in exile in Paris) noted that in today's Arab world "words are treated as crime" (MEMRI 2006b, 1). This situation is common even in Turkey, where ultranationalist tendencies lead the authorities to repeatedly accuse and prosecute writers who are seen as disrespecting or denigrating "Turkishness."

Nothing spoke as eloquently of the vengeful intolerance of retrograde Islam's usurpation of modern state authority as did the most famous modern *fatwā*, issued on February 14, 1989, by Ayatollah Khomeini. He asked the faithful to kill Salman Rushdie for writing *The Satanic Verses*, a book that neither he nor any but a few score among his tens of millions zealous supporters ever read. (Contrary to widespread reports, this *fatwā* was never formally repealed; indeed it was reaffirmed by Khomeini's successor, Ali Khamenei, in 2005.)

How can any modern state, how can any self-respecting individual engage in a dialogue with those who see death as a fitting verdict for a fictional tale, a beheading for a work of art? Unfortunately, this intolerance is deeper, aimed at all non-Muslims and fully institutionalized. While new mosques (including some large and magnificent buildings constructed with Saudi money) rise in every major city in Europe, North America, and Australia, in July 2000 the Saudi Permanent Council for Scholarly Research and Religious Legal Judgment reaffirmed the ban on construction of churches in any Muslim country because, in its view, all religions other than Islam are heresy and error (MEMRI 2006a).

The modernization deficit is also obvious in the dismal state of learning and research in the Muslim world. Reiterating the glories of medieval Muslim science and engineering (Al-Hassani 2006) will not change the fact that at the beginning of the twenty-first century researchers in Muslim nations (with more than 12% of the world's population) authored only 2% of all scientific publications, and 60% of that minuscule share came from a single country, Turkey (Kagitçibasi 2003; Butler 2006). No Muslim country ranked among the 20 states with the highest overall scientific output (a group that contains such small and resource-poor nations as Belgium, Denmark, and Israel), and the same is true for health-related publications (Paraje, Sadana, and Karam 2005). Countries of North Africa and the Middle East also rank near the bottom of the worldwide record for patents and trademarks.

These realities led Sa'īd (MEMRI 2006b, 2) to compare the achievements of Arabs, resources and great capacities, with what others have done over the past century and to conclude, "I would have to say that we Arabs are in the phase of extinction, in the sense that we have no creative presence in the world."

The modernization deficit is also responsible for a belief in bizarre conspiracy theories that pervades the Muslim world. Even many Western-educated Arab intellectuals believe in the authenticity of *The Protocols of the Elders of Zion*, an anti-Semitic fabrication of the early twentieth century. In 2003, Nigeria's northern Muslim states suspended polio vaccination because of a rumor of deliberate contamination of vaccines with hormones aimed at making young Muslim girls infertile to depopulate the region. And an opinion survey of 13 countries (Pew Research Center 2006) found that large majorities of Muslims do not believe that their coreligionists carried out the 9/11 attacks. In Indonesia, the world's largest Muslim country, as well as in Turkey, only 16% of people thought so, hardly a basis for rational dialogue on responsibility and the need for modernization.

As far as ordinary Muslims are concerned, the modernization gap is most obvious with respect to their quality of life. As a group, Muslim countries, including those fabulously rich, small Persian Gulf states, do not enjoy the high quality of life that could be expected based on their high per capita GDPs. Even among the richest Muslim countries of the Middle East, only the Gulf states have infant mortalities at modern levels (near or below 10/1,000 live births); in the early 2000s the Saudi rate was above 20, and the Iranian and Turkish rates were above 30 (UNDP 2006). And, perhaps most tellingly, in 20 out of 24 Muslim countries of Europe's southern hinterland, the difference between GDP per capita and Human Development Index rank is negative (fig. 3.13).

The third great challenge facing the Muslim world is its tardiness in completing the demographic transition, a reality that becomes particularly important given the just-described lack of secular state authority and adequate scientific advances, two necessary mainsprings of modern economic growth. These deficits will continue to hamper any progress in coping with the aspirations of expanding populations, in particular of assertive young men. At the beginning of the twenty-first century the only countries with Muslim majorities whose total fertility was near replacement level were Iran, Indonesia, and Malaysia. In all populous Muslim countries of North Africa and the Middle East, as well as in Pakistan and Bangladesh, total fertility was 50%–100% above replacement.

Arab countries of the Middle East (from Egypt to Iraq, and from Syria to Yemen, excluding North Africa) had about 190 million people in 2005, but the UN's medium forecast expects the doubling of this population to about 380 million by 2050 (United Nations 2005). Populations of those Muslim societies that have historically been most prone to various forms of fundamentalism (*al-wahhābīya*,

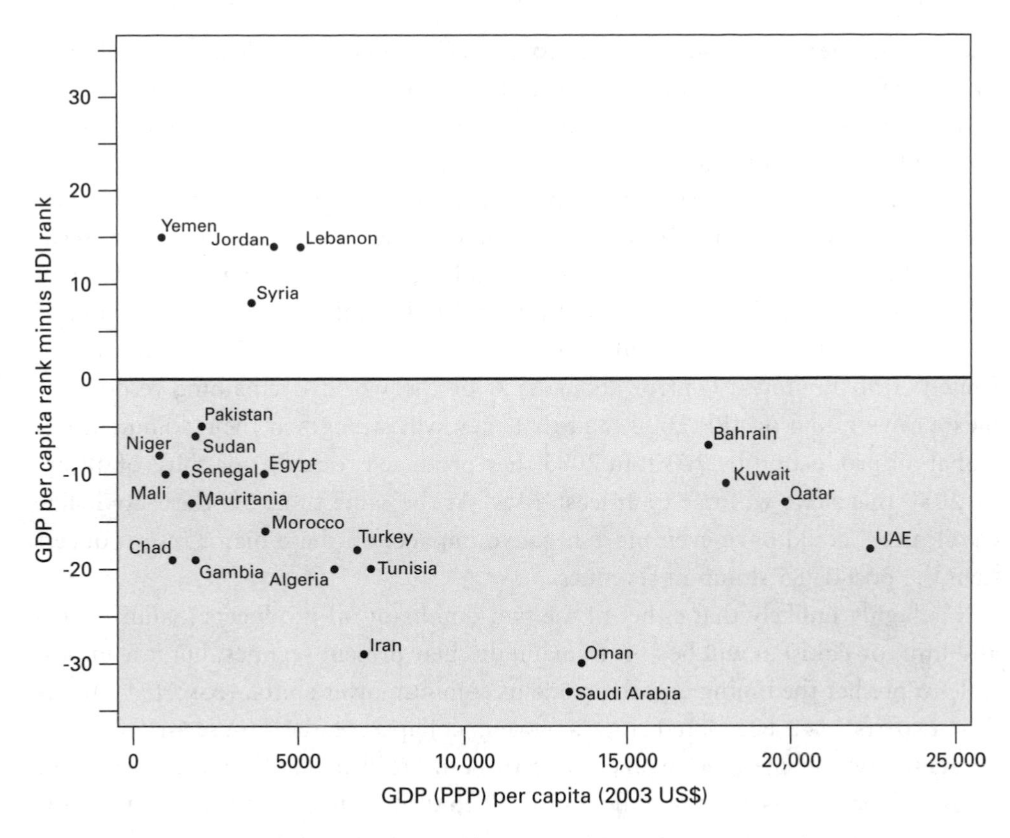

Fig. 3.13
Modernization status of Muslim countries: only 4 out of 24 have positive scores. Plotted as per capita GDP vs. global GDP rank minus HDI rank, from data in UNDP (2006).

militant *shī῾a*) and that harbor deep anti-Western sentiments (Pakistan, Afghanistan, Iran, Saudi Arabia, Yemen) are projected to rise from nearly 300 million in 2005 to about 610 million by 2050, when they could be nearly as large as Europe's declining population (the UN's medium forecast puts it at about 650 million by 2050).

Most important, by 2025 nearly every seventh person in that region will be a man in his late teens or twenties (in Europe it will be every sixteenth person). Young (usually unmarried) men have always been the most common perpetrators of deadly violence. This is not only an impression gathered by watching the images of Somali gangs in Toyota pickups or Taliban attacks in Kandahar. Analyses of demographic and war casualty data show that this variable consistently accounts for more than one-third of the variance in severity of conflicts (Mesquida and Wiener 1996; 1999). Large numbers of young unattached males will thus be a major source of internal tension, instability, and violence in *dar al-islām* that can only be accentuated by politically organized terrorism.

The combination of these trends guarantees decades of worrisome social and political convulsions. The Muslim world would present a considerable global security challenge even if it did not harbor any radical movements engaged in external terrorism. The global repercussions of this internal instability will be compounded by the fact that five Persian Gulf countries (Saudi Arabia, Iran, Iraq, Kuwait, and United Arab Emirates) control about 65% of the world's remaining reserves of inexpensive liquid oil (BP 2006) and that they will strengthen their dominance of global oil production by 2030. In 2005 they produced roughly one-third of all oil; by 2030 this share will rise to at least 40%. At the same time, another rapid slide of oil prices could have even more negative impacts on these major oil producers than the post-1985 slump in revenues.

It is highly unlikely that either of the two dominant oil producers (Saudi Arabia and Iran) or Pakistan will be able to maintain their present regimes, but it is impossible to predict the timing and the mode of coming power shifts. (Assorted Middle East experts have been predicting a violent collapse of the House of Sa'ūd for decades.) Even if the new regimes were to be more democratic and secular, they might not be less assertive and more difficult to deal with. Beyond that, it is unfortunately all too easy to sketch some very scary scenarios, ranging from the emergence of a new militant *shī'a* state in *al-Sharqīya*, the Eastern province of Saudi Arabia (it would control one-quarter of the world's oil reserves) to deliberate nuclear proliferation by Pakistan and (perhaps soon) Iran. For sleepless nights, think of a future nuclear Sudan or Somalia.

Russia's Way

Counting Russia out would be historic amnesia. There is nothing new about Russia's leaving the great power game and then reentering it vigorously decades, even generations, later. The country was out of the contest for great power status in 1805 as Napoleon was installing his relatives as rulers of Europe. But less than a decade later, as Czar Alexander I rode on his light-grey thoroughbred horse through defeated Paris, followed by thousands of his Imperial Guard, Bashkirs, Cossack, and Tartar troops, the country was very much an arbiter of a new Europe (Podmazo 2003). Russia was sidelined again in 1905, convulsed by its first bloody revolution, but 15 years later a victorious revolutionary regime regained control over most of the former Czarist territory and inaugurated seven decades of Communist rule. Then, as the first waves of revolutionary fervor subsided, Russia turned inward, and by 1935, with Stalin plotting murderous purges of his comrades, his army, and millions of innocent peasants, the country's international role seemed once again marginalized.

Fourteen years later Russia was not only a victorious superpower with troops stationed from Korea to Berlin but also the second nuclear power with a successfully tested fission bomb (August 29, 1949). This last episode of Russia's militant role on the global stage lasted nearly 50 years, between 1945 and 1991, cost the West many trillions of dollars in armaments and dubious alliances, prolonged many violent proxy conflicts in far-flung places from Angola to Afghanistan, and brought the world to the brink of thermonuclear war (Freedman 2001). Moreover, it contributed massively to environmental degradation with hundreds of highly contaminated nuclear weapons sites in both the United States and the USSR, and with particularly high levels of radioactive wastes detectable in Russian rivers (including Siberia's main artery, the Yenisei) and in the Arctic Ocean (USDOE 1996).

But any suggestion of Russia's reemergence as a great power seemed far-fetched in the years immediately following the demise of the Soviet Union (officially on December 8, 1991, at a meeting of Russian, Ukrainian, and Belorussian leaders in a hunting lodge at Belovezhskaya Pushcha, the virgin forest reserve near the Polish border). Russia seemed to be assaulted by too many intractable problems. The sudden demise of the command economy that had governed the country since the Stalinist five-year plans of the 1930s and a messy creation of new (often criminalized) business structures were accompanied by years of declining GDP. This shrank by 5% in 1991, 14% in 1992, 9% in 1993, and 13% in 1994; by the time it bottomed out in 1998, its per capita rate (in constant monies) was only 60% of the 1991 level, and real income had fallen to about 55% of the peak Soviet rate.

The country also faced a new, determined challenge from resurgent Islam. The first war in Chechnya (1994–1996) turned the republic's capital into ruins and ended in a fragile truce, and the second war, which began in 1999, continues at a low level, with intermittent large-scale brutalities and terrorist attacks. To this must be added widespread social ills, such as the parlous health of Russian men resulting from widespread alcoholism, smoking, poor diet, low physical fitness, and mental problems (World Bank 2005). From a long-term perspective the most worrisome challenge is a steep decline in Russia's fertility rates and the prospect of considerable depopulation.

But Russia rebounded once again. By 2005 its economic performance had improved to such an extent that it actually became fashionable to speculate about its renewed superpower status. Its economic recovery was driven largely by the exports of mineral riches, in particular oil and gas, and this accelerated with the post-2004 rise of crude oil prices. In addition to ranking among the world's leading producers of such strategically important metals as nickel (used in specialty steels), titanium (for alloys and pigments), and palladium (essential in catalytic converters), Russia is an unmatched treasure house of energy resources (Leijonhielm and Larsson 2004).

The country has the world's second largest coal reserves (after the United States), second-largest potential for hydroelectricity generation (after China), seventh-largest crude oil reserves, and by far the world's largest natural gas reserves (BP 2006; Merenkov 1999). Russia's reserves of natural gas amount to about 27% of the global total, nearly as much as the combined reserves of Iran and Qatar and almost ten times larger than those of the United States.

Russia's oil production peaked at nearly 570 Mt/year in 1987–1988, fell to 303 Mt/year in 1996, and rose to 470 Mt by 2005, accounting for about 12% of the world's total (BP 2006). Concurrently, Russia's crude oil exports rose from about 100 Mt/year to nearly 250 Mt/year, and forecasts envisage sales of more than 350 Mt/year by 2010 (Lee 2004). Russia is Europe's largest supplier of natural gas, delivered through the world's largest and longest pipelines from western Siberia (fig. 3.14). In 2005 its exports of about 151 Gm^3 were largest in the world, with nearly one-quarter destined for Germany and 15% for Italy. Even larger projects are planned to carry Siberian natural gas to Japan and China and to sell liquefied natural gas around the world (Smil 2003; USDOE 2005). The expected boom in natural gas exports is based on the fact that the fuel can substitute for oil in many of today's uses and on its inherent environmental advantage: its combustion generates about 30% less CO_2 per unit of liberated energy than oil.

Russia has another strength in its intellectual capacity. The country has always had many highly creative scientists and engineers, whose fundamental contributions are generally unknown to the Western public. How many people watching a scanner toting up their groceries know that Russian physicists, together with their U.S. colleagues, pioneered masers and lasers. (Nobel prizes in physics were awarded to Nikolai Gennadievich Basov and Aleksandr Mikhailovich Prokhorov in 1964 and

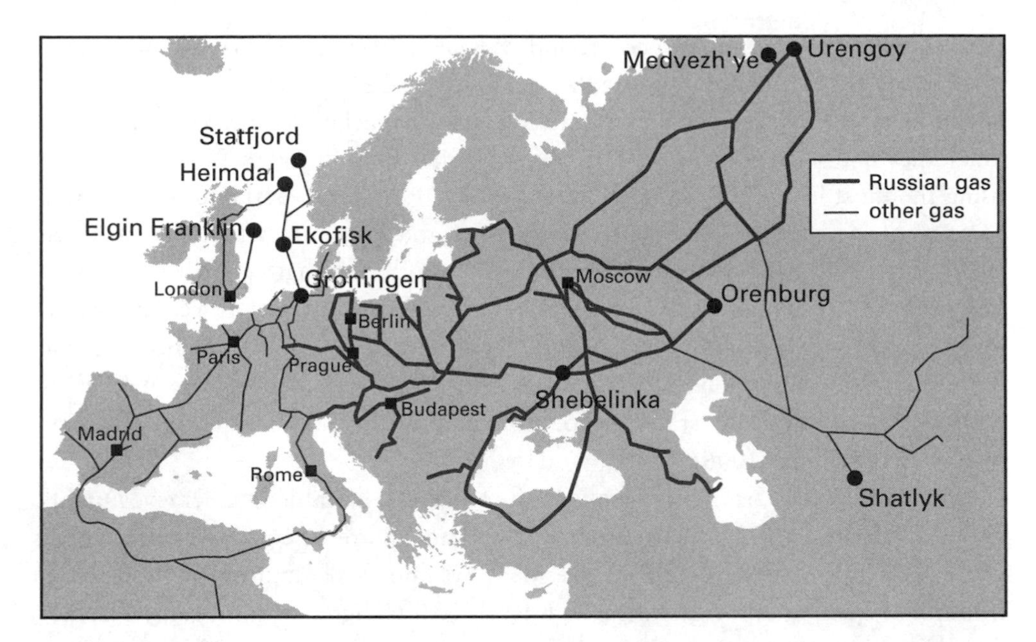

Fig. 3.14
The world's longest and largest-diameter natural gas pipelines transport fuel from Siberia to Western Europe. Simplified networks compiled from maps in various industry publications.

to Zhores Ivanovich Alferov in 2000.) How many people seeing the images of the U.S. Air Force stealth planes know that this class of aircraft began with Piotr Iakovlevich Ufimtsev's (1962) equations for predicting the reflections of electromagnetic waves from surfaces? How many patients appreciate that Russian surgeons pioneered such remarkably unorthodox methods as extending leg bones shortened by bone deficiencies, deformities, and fractures (Gavril Abramovich Ilizarov) and treating nearsightedness by radial keratotomy (Svyatoslav Nikolaevich Fyodorov)?

The Russian economy is reenergized by Siberian gas and oil, profiting from rising exports of hydrocarbons and strategic metals, and drawing on admirable scientific and engineering expertise. The Russian military is being rearmed with a new generation of weapons, and Russian diplomacy is vigorously reengaging in the world's major conflict zones. These factors add up to an undeniable reassertion of the country's global primacy. But are these natural, economic, and intellectual advantages enough to outweigh the country's many problems and elevate it once again to superpower status? A closer look indicates that the prospects for a strong and lasting comeback is not encouraging.

True, the economy has turned around. Between 1998 and 2005, Russia's GDP rose nearly 60% (inflation-adjusted), but it was still some 10% behind the peak Soviet level in 1990. Personal consumption, which reached its low point in 1992, did somewhat better, but in 2005 it was no higher than in 1990. When converted using the most likely PPP, Russia's per capita GDP of about $10,000 in 2005 was comparable to that of Malaysia or Costa Rica and lower than the Spanish level. The country's reliance on its natural resources is both a strength and a great potential weakness. A sudden fall in cyclical commodity prices would deprive the government of nearly two-thirds of its revenue. And there is plenty of evidence that the so-called resource economies, particularly the oil-dependent ones, are neither as efficient nor as politically stable as countries that have to rely on imports of basic resources (Friedman 2006).

Politically, Russia has a long way to go to become a stable democracy. Doubts about its progress and hopes for gradual improvement are reflected in contradictory characterizations of the country's president, Vladimir Vladimirovich Putin, as a traditional Russian autocrat with a high-level KGB pedigree or a *sui generis* democrat (Astrogor 2001; Blanc 2004; Zdorovov 2004; Herspring 2006). But these evaluations might seem quaint in the future if Putin were to be supplanted by a much more inward-looking, xenophobic leader. Strategically, Russian leaders know (despite the reflexive protests) that NATO at its borders is not a mortal threat, but their post-1991 worries about resurgent Islam are likely only to intensify.

The Chechen conflict keeps spilling into neighboring Dagestan, Ingushetia, and Georgia. Oil-rich Azerbaijan and Kazakhstan are Russia's only relatively stable southern neighbors, whereas the former Soviet Central Asian republics with Muslim populations (Uzbekistan, Tajikistan, Kirgizstan) and their neighbor, Afghanistan, have a much more uncertain future. The centuries of migration and the expansion of Czarist Russia brought relatively large resident Muslim nationalities into the Slavic state. They now amount to nearly 15% of its population and are concentrated in such oil-, gas- and mineral-rich regions as Tatarstan, between the Volga and the Kama, and in neighboring Bashkortostan in the southern Urals (Crews 2006; Bukharaev 2000). War in Chechnya, instability in the North Caucasus regions, the post-Soviet Islamic rebirth, and the nationalist aspirations of Muslim minorities worsened the relations between Russians and Muslims both on the state and individual levels, but because Muslims inside Russia are divided along cultural, ethnic, and religious lines, they are not a monolithic power confronting the state (Gorenburg 2006).

Above all, Russia's renewed economic growth and rising resource exports have done little to alleviate the country's social ills and population challenges that became so prominent during the 1990s and that do not appear to have any effective near-term solutions. Russia's decreasing population and poor health statistics have attracted a great deal of attention at home and abroad (Notzon et al. 1998; DaVanzo 2001; Powell 2002; World Bank 2005). The country shares the problem of aging population with the European Union and Japan, but the rapid rate of its population decline and the principal reasons for it have no counterpart elsewhere in industrialized world.

The country has a peculiar aging profile. The decline in its share of children (0–14 years of age) is expected to be similar to that of Japan. By 2025 children will make up less than 16% of the population in Russia and less than 13% in Japan. But in 2025, Russia's share of people 65+ years of age will be much smaller than Japan's: less than 18% compared to about 29%. Unlike in Japan or some European countries, Russia's aging population will not include large numbers of old and very old people because people are simply dying too young. Hence population is shrinking not only because of falling fertility but also because of excessive premature mortality.

Russia's total fertility rate dropped (with a temporary uptick during the 1980s) from about 2.8 in the early 1950s to about 1.2 in the first half of 2000s (fig. 3.15). When the USSR disintegrated, Russia had about 140 million people; by 2000 that total had declined by 5 million, and by 2005 the population shrank by another 3 million. The medium variant of the United Nation's (2005) long-term forecast puts the number at about 129 million by 2025 and less than 112 million by 2050, when the high and low variant totals are expected to be, respectively, 134 million and 92 million. No other modern industrialized country is expected to experience such a population decline: in absolute terms, ~35 million people, or a Canada-sized population loss, in just two generations; in relative terms, a decline of ~25% between 2000 and 2050. And none share a major reason for it, Russia's falling life expectancy.

This reality runs against a long-standing trend of rising life expectancies in the richest nations to more than 75 years for men and more than 80 years for women. By the late 1960s, Russia's post–World War II health and nutrition gains brought its combined (male and female) life expectancy to within about 4 years of the EU average, but while the latter's kept improving, Russia's came to a standstill, rose briefly during the late 1980s, and then plunged. By the year 2000 the gap between

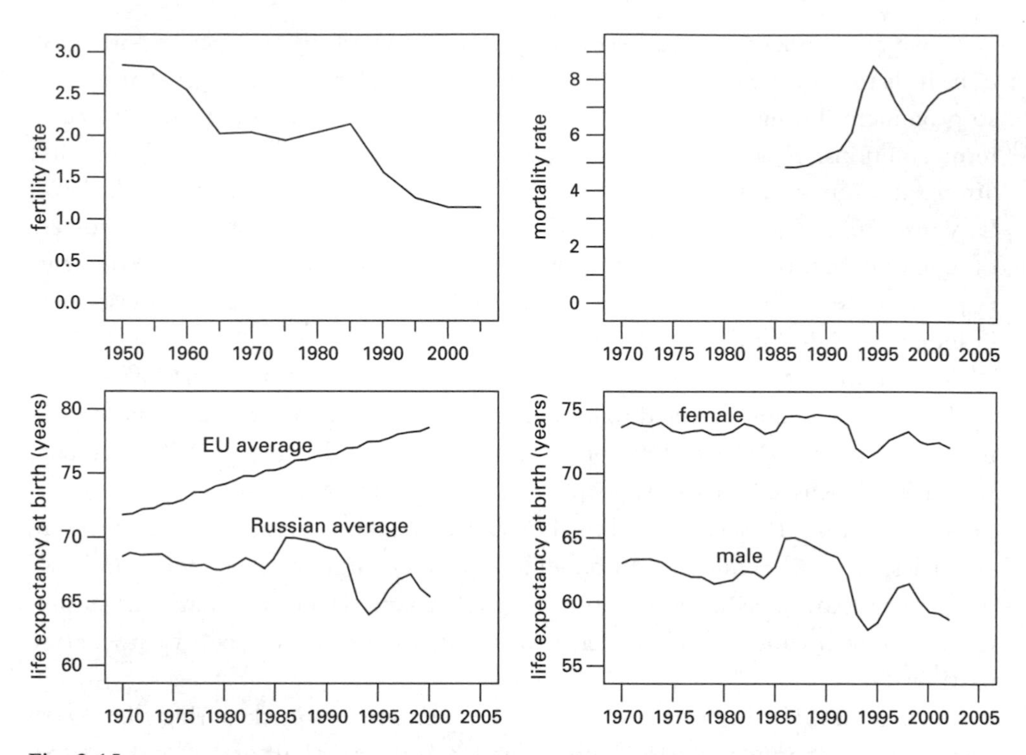

Fig. 3.15
Russia's demographic miseries: fertility, mortality, life expectancy. Based on World Bank (2005) and United Nations (2005).

Russia and the EU was more than 12 years: less than 66 years versus slightly more than 78 years (United Nations 2005).

Several factors account for this large gap. Russia's infant mortality has been falling, but by 2000 it was still 15 deaths/1,000 live births, roughly twice the mean of Western nations. At about 40 per 100,000 live births, Russia's maternal mortality is more than four times as high as in the West. But the main reason for the gap is an extraordinarily high adult mortality, in particular, male mortality. A life expectancy gender gap is normal in every society (~5 years in UK to ~7 years in Japan), but in Russia it was 13 years by 2002: 72 years for women, 59 years for men (see fig. 3.15). This means that by the early 2000s Russian male life expectancy at birth was lower than the Indian mean of about 63 years and far behind the Chinese average of about 70 years (United Nations 2005).

A number of factors contribute to this deficit. Russian cardiovascular disease mortality is about three times higher than in the West; premature cancer mortality is also considerably higher. In the early 2000s the Russian death rate due to road accidents was nearly 0.4/1,000, compared to about 0.15/1,000 in the United States (WHO 2004a). During the late 1990s the Russian suicide rate peaked at more than 50/100,000, and since 2000 it has been about 40/100,000, four times the rate in the EU and the United States (World Bank 2005). Russia is unequaled in its toll due to alcohol poisoning, over 30/100,000.

Poor nutrition (saturated fat), high rates of smoking (60% of adult Russian men smoke, more than twice as many as in the United States), spreading drug use (a total addict population estimated at ~4 million in 2005), inadequate preventive health care, and mental stress contribute to Russia's high premature mortality due to noncommunicable diseases. All these factors have one common denominator, the chronic epidemic of alcoholism. The overall level of officially reported average adult per capita alcohol intake in Russia has been only 10%–25% above the Western mean. This difference alone cannot explain the extraordinary impact of drinking on Russian society, but are explanation can be found in the actual consumption of alcohol and the kind of drinks consumed. The unrecorded consumption of *samogon* (home brew) and bootleg liquor may as much as double the official total, which means that an average annual intake per adult Russian amounts to at least 15 liters of pure alcohol, roughly 50% more than in Western countries (McKee 1999; WHO 2004). A 2002 survey found that about 70% of Russian men and 47% of Russian women were drinkers, and whereas 50%–70% of Western alcohol intake is in the form of beer or wine, in Russia spirits, especially vodka, account for 80% of all alcohol consumed. The history of Russia's attempts to deal with its epidemic of vodka alcoholism shows how intractable it is (Herlihy 2002; Tartakovsky 2006; Vodka Museum 2006).

The first temporary prohibition was introduced in 1904 during the time of Russo-Japanese war, the second one at the beginning of World War I, on August 2, 1914. Because these cut state revenues by more than one-quarter, they helped to speed up the demise of the Czarist empire, but they did nothing to change the population's attitude toward "little water." Bolsheviks reintroduced the ban in 1917, but it was reversed by 1925. Stalin, Khrushchev, and Brezhnev enforced the state monopoly on vodka but did not dare to proscribe the drink. That was done again only by Mikhail Gorbachev in May 1985, but his Proclamation ("On the Improved Measures Against Drunkenness and Alcoholism"), which once again cut state revenues, increased the consumption of dubious (and often deadly) substitutes, and it was reversed five years later.

Yeltsin's new market economy included the abolition of the state monopoly on vodka (on June 7, 1992), a decision that led to a large increase in hazardous drink and alcohol poisoning as well as to significant revenue loss. The latter was the main reason why the monopoly was reestablished just a year later. Not surprisingly, with the average Russian adult drinking some 20 liters of vodka per year (~10 liters of pure alcohol), a large share of adult nonaccident mortality (male cardiovascular disease, liver cirrhosis), at least half of all automobile accidents, and many suicides, homicides, and fatal workplace and home injuries are related to this intractable vodka abuse. And the latest concern aggravated by alcohol is the rapidly worsening HIV/AIDS problem. This addiction to vodka has no effective short-term solutions. The resulting population trends cannot be readily reversed, a fact well understood by Russian leaders.

In his first States of the Nation address in July 2000, Putin acknowledged that Russia is "facing the serious threat of turning into a decaying nation." Six years later he called the nation's demographics—a loss of almost 700,000 people per year—"the most acute problem facing our country today." He singled out new programs for improving road safety, preventing the import and production of bootleg alcohol (in his 2005 address he noted that some 40,000 people died each year from alcohol poisoning), and the early detection, prevention, and treatment of cardiovascular diseases. He also called for the encouraging immigration by attracting skilled compatriots from abroad. Even if these measures are successful, the positive impacts will not be seen for many years.

That is perhaps why Putin emphasized ways to increase the birth rate by a introducing a number of pro-natal policies, including doubling the benefits for a first child (to 12,500 rubles/month), doubling that amount for a second child, giving maternal leave of 18 months at 40% of pay, and more generously subsidizing preschool child care. Altogether, these financial incentives should total at least 250,000 rubles (about ~$9,300) per family, a very large sum in a country where the average 2005 gross national income (in terms of PPP) amounted to about $10,000 and the average monthly wage was 8,530 rubles (CBRF 2006). But pro-natal programs have a decidedly unimpressive record, and it is unlikely that Russia's population decline can be stopped or reversed.

Russia's strengths have improved the country's domestic situation and international influence far above the levels to which it sank during the post-Soviet convulsions, but its current weaknesses are most likely to keep it from reoccupying its once undisputed superpower position.

If, as is highly probable, Russia's superpower resurgence fails to materialize, there is already an established trend that alone is powerful enough to accelerate the relative decline of Western economic and military power during the first quarter of the twenty-first century: China's rise to superpower status.

China's Rise

Historians of dynastic China would say that *return* would be a more accurate description than *rise*. For about two millennia China was the preindustrial world's largest economy. Maddison (2001) credited it with some three-quarters of the global economic output at the beginning of the Common Era, two-thirds by the year 1000, and still nearly 60% by 1820. There is little doubt that under Qianlong (1736–1795), the longest reigning of all Qing dynasty emperors, it was on average more prosperous in per capita terms than England or France (Pomeranz 2001). That was surely the emperor's perception as he tersely dismissed the British offer to trade and ordered George III to "tremblingly obey" his warnings (Qianlong 1793). Half a century later, Britain inflicted the first Western defeat on China, and soon afterwards the ancient empire was fatally weakened by the protracted Taiping rebellion, one of the key transformational mega-wars of the past two centuries (see chapter 2). The empire staggered on until 1911; its dissolution was followed by four decades of internal and external conflicts.

The establishment of Maoist China in 1949 did not end violence and suffering. Collectivization campaigns and anti-intellectual drives of the 1950s were followed by the world's worst famine, overwhelmingly Mao-engineered, which claimed at least 30 million lives between 1959 and 1961 (Ashton et al. 1984; Chang and Halliday 2005). Then came the decade of the incongruously named Cultural Revolution (1966–1976), which ended only with Mao's death. By the end of 1979, Deng Xiaoping began to steer the country toward economic pragmatism and reintegration with the world economy. This process actually intensified after the 1989 Tian'anmen killings as the ruling party kept its priorities clear: maintain firm political control by buying people's acquiescence through any means that strengthen the age-old quest for wealth. China is thus finally reclaiming what its leaders feel is its rightful place at the center of the world: ReOrient, Frank's (1998) apt label.

China's rapid rates of economic growth (though not as rapid as indicated by the country's notoriously unreliable statistics) have been the result of several interacting factors. Since the mid-1980s China has been receiving a rising influx of foreign direct investment, with the 2005 total surpassing $60 billion, compared to less than $4 billion invested by foreigners in India's economy. China's one-child policy cut the population growth rate, with the total rising from 999 million in 1980 to 1,308 million by the end of 2005 (NBS 2006).

The country thus has a huge pool of cheap and disciplined labor (about 760 million people in 2005, compared to 215 million in the EU and 150 million in the United States), which has been moving from the impoverished interior to coastal cities, where most of the new manufacturing plants are located. It is the largest and most rapid urbanization in history. China has followed the Japanese and South Korean example by promoting export-oriented, labor-intensive manufacturing, but rapid economic growth has also created a new huge domestic market for costly consumer items. Finally, China has shown a readiness to innovate, unfortunately not only by setting up well-supported research facilities but also through widespread infringement of intellectual property rights and massive commercial and industrial espionage.

Some 200 million workers have been deployed in China's new manufacturing enterprises since 1980. The unprecedented conquest of global markets has helped to decimate U.S. and European manufacturing. China produces more than 90% of Wal-Mart's merchandise and contributed just over $200 billion, or 26%, to the U.S. trade deficit of $767.5 billion in 2005 (USBC 2006). What a remarkable symbiosis: a Communist government guaranteeing a docile work force that labors without rights and often in military camp conditions in Western-financed factories so that multinational companies can expand their profits, increase Western trade deficits, and shrink non-Asian manufacturing (ICFTU 2005). More of the same is yet to come, and before this wave is exhausted, the country's manufacturing will dominate the global market for common consumer items as completely as it now dominates Wal-Mart's selection.

This economic surge has attracted a great deal of attention (Démurger 2000; Fishman 2005; Shenkar 2005; Sull and Wang 2005; Prestowitz 2005; Kynge 2006; Hutton 2006). Unfortunately, many of these writings fail to acknowledge some fundamental difficulties with Chinese statistics (inflated estimates and fabricated data are common), provide no long-term context of these necessarily temporary developments, and contain "abundant sycophancy, though nowadays the kow-tow is to the managed market system rather than to a single individual like Mao" (E. L. Jones 2001, 3).

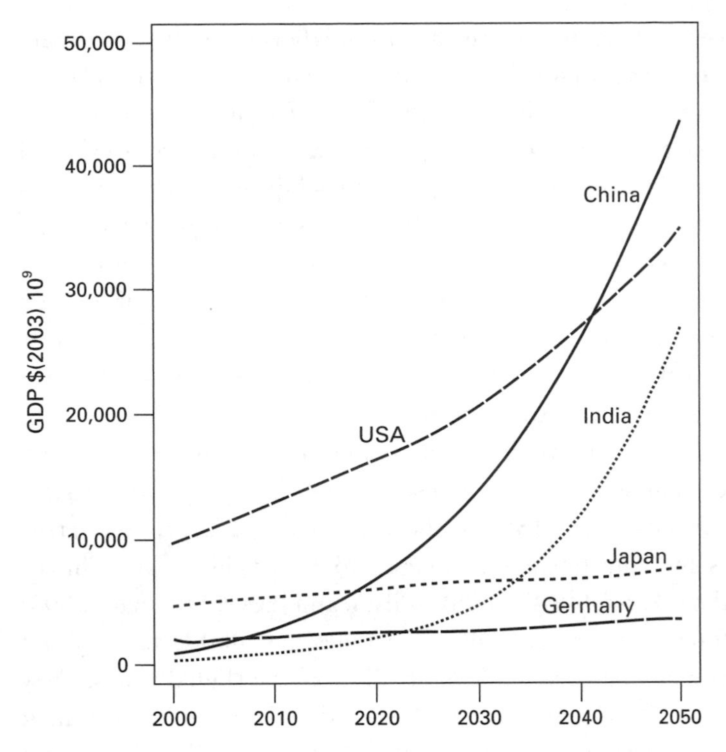

Fig. 3.16
China's rapid GDP growth would make it the world's largest economy perhaps by the 2020s. Plotted from data in Wilson and Purushothaman (2003).

Continuation of high growth rates would make China the world's largest economy at some time between 2025 and 2040. Wilson and Purushothaman (2003) project China's GDP (in exchange-rate terms) to surpass that of Germany by 2007 and that of Japan in 2016 and to reach the U.S. level by 2041 (fig. 3.16). But in per capita terms China would still be far behind the United States. By 2040 the two countries will have, respectively, about 1,430 and 380 million people, so identical GDPs would leave China's per capita level at roughly one-fourth of the U.S. rate. A projected per capita GDP rate of about $19,000 per year (in 2003 monies) would make China as rich as today's Greece. In PPP terms, China is already the world's second largest economy, but in 2005 it was still less than half the United States' size.

What will China do with this new power? During most of the 1990s, China's external actions were seen as overwhelmingly mercantile as the country appeared to be preoccupied with employing its huge surplus rural labor force. But since then, China has become aggressive in a global quest for raw materials, in particular oil (Zweig and Bi 2005). Many commentators see the flood of China's manufactured products as an entirely welcome trend (they keep Western inflation rates low), and many CEOs speak favorably about the strategic partnership of the Untied States with China. But some Chinese strategists and policymakers think differently. Their arithmetic is made clear by the following calculation, which I have heard most frankly expressed by a senior Chinese governmental adviser on strategic affairs. By 2020, China's continuing high economic growth rate will allow it to spend on its military as much as the United States spends today, and this will make it a real superpower impervious to any threat or pressure.

This may not be wishful thinking. All these calculations depend on the conversion rates used to compare Chinese and U.S. GDPs. Official exchange rates pegged China's GDP in 2005 at only about 18% of the U.S. total ($2.25 trillion versus $12.5 trillion), whereas the adjustment for PPP (up to $9.4 trillion) put China's GDP at about 75% of the U.S. total (IMF 2006). Wilson and Purushothaman (2003) projected China's 2020 (exchange-rate) GDP at 6.5 times the 2000 level, or about $7.1 trillion (in 2003 monies) compared to $16.5 trillion for the United States. They also estimated that the value of Chinese currency could double in ten years' time if growth continued and the exchange rate were allowed to float. This adjustment would lift China's 2020 real GDP close to $15 trillion, near the U.S. level at that time. With a higher share of it going to the military, China could indeed match U.S. defense spending by 2020. The Pentagon estimates that by 2025, Chinese defense spending will be as high as $200 billion (USDOD 2004). Again, multiplied by 2, this gives a level above the U.S. FY 2004 defense budget.

Moreover, in contrast to the decades of the Cold War when the United States was in no way economically dependent on the Soviet Union, China is already helping to prop up the U.S. economy by supplying it with essential goods at cut-rate prices and by buying up part of ballooning U.S. debt. Not everyone sees this as threatening: some see peaceful economic cooperation (Zheng 2005; Zhu 2005); others, an unpeaceful expansion (Mearsheimer 2006), China's becoming a strategic adversary of the United States, and a Sino-U.S. military contest in the Pacific (Kaplan 2005). Both views have a great deal of validity.

For example, a complete ban on China's imports (if such a thing were practical) would bring a surge of products from other Asian exporters (South Korea, Taiwan, Vietnam, Indonesia, Thailand) but hardly a resurgence of U.S. manufacturing (many of its sectors are simply defunct). And anticipations of high military spending by China are based on its continuing a defensive and offensive military buildup that has been under way for years and on bellicose statements of some of its policymakers. But such views ignore a multitude of internal and external weaknesses that militate against the China's becoming a superpower during the next two generations.

All large, populous countries face limits and challenges, but in China's case these checks are uncommonly numerous, and ignoring them is to repeat the mistakes made before 1990, when the West saw the USSR as a formidable superpower to be feared or Japan as inevitably becoming a global economic leader. I note here key trends concerning China's population, economic progress, environmental degradation, and power of ideas. These trends, rather than the endlessly discussed possible outcomes of the China-Taiwan dispute (Bush 2005; Tucker 2005), will determine the reach and limits of China's rise during the first half of the twenty-first century.

China's population rests on an uncommonly uneven foundation. Since 1979 the traditional preference for sons has been exacerbated by the compulsory one-child policy, and this has left China with a shockingly aberrant sex ratio at birth. A 2001 national family planning and reproductive health survey showed total fertility below the replacement level, the ratio of males to females at birth at 123 (Ding and Hesketh 2006). The normal ratio of boys to girls is 106, the global mean (affected by Asia's preference for boys) is 107, but some Chinese provinces are above 115, and a recent study found 20 rural townships in Anhui province with a ratio of 152 (Walker 2006; Wu, Viisainen, and Hemminki 2006).

This reality will disrupt China's social fabric in several worrisome ways. Leaving aside the fundamental moral question (what happened to all those missing girls?), it condemns tens of millions of men to spouseless (shorter, less satisfying) lives. It has already led to waves of rural abductions of women by criminal gangs that supply brides to urban bachelors willing to pay a price. And a surplus of young unattached men, who are responsible for most of the crime in any society, could be a factor in contemplating foreign aggression with more equanimity (Hudson and den Boer 2003). A single war could go a long way toward returning China's skewed gender ratio closer to normal.

China's birth-planning policy will also result in a rapid aging of the country's population (England 2005; Jackson and Howe 2004; Frazier 2006). In 2005 about 11% of China's population was 60 years and older, compared to some 17% in the United States, but by 2030 the two shares will be equal, and China's age structure

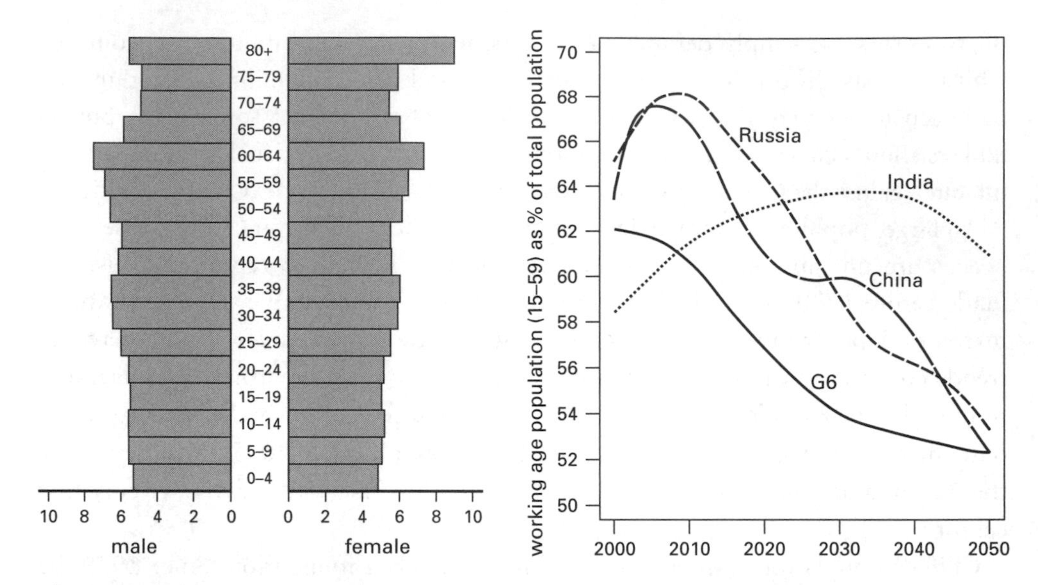

Fig. 3.17
Left, China's rectangular age and gender population structure in 2050; *right*, countries' declining shares of working population. Based on Kaneda (2006) and Wilson and Purushothaman (2003).

will resemble that of Japan in 2005 but with only about one-fifth of its per capita income. By 2050, China will have more old people (about 30%) and a higher dependence ratio than the United States (fig. 3.17). The proportion of persons of working age in the total population, currently about 68%, will fall to 53% by mid-century, equaling the corresponding figure for the G-6 countries. China's pension reforms already require transfers of large subsidies from the central government to local levels, and the possibility of a government default on a substantial part of the rising pension debt poses a threat to the country's future social stability (Frazier 2006), particularly given the strongly held expectations regarding the state's role in the provision of security.

This burden will be aggravated by the fact that some three-quarters of all Chinese have no pension plans, leaving tens of millions of young men responsible for two parents and four grandparents. The problem will be particularly acute in China's villages (Zhang and Goza 2006). Unlike the cities, rural areas usually do not have any institutions to care for the elderly, but they have proportionately larger numbers of elderly people than urban areas, which attract a young labor force. As Dali Yang (2005) notes, the aging process will be felt during the next 15 years as the number of entry-level, low-skilled workers keeps on shrinking, making it more difficult to recruit migrant labor at depressed wages.

Economic reforms have employed tens of millions in new industries, transformed villages to large cities in less than a single generation, attracted enormous inflows of foreign investment, conquered global markets in many industrial categories, and elicited worldwide admiration of Chinese economic progress. But they have been also responsible for one of the world's fastest increases of income inequality (Khan and Riskin 2001). They have brought poverty and marginalization, and created a massive urban underclass of uncounted destitute migrants and unemployed city workers (Solinger 2004). Economic reforms have created an elite enjoying excessive levels of private consumption and interested in maintaining the marriage of convenience between the unchecked power of the ruling party and illicit wealth (Pei 2006a; 2006b).

Tens of millions of peasants lead a precarious existence subject to the arbitrary actions of party leaders, state officials, and ambitious businessmen, including violent (and uncompensated) expropriation of their land and punishing taxation (Chen and Wu 2004; Friedman, Pickowicz, and Selden 2005). Poverty also keeps rural China unhealthy (Dong et al. 2005), and corruption is severe and endemic (Manion 2004; Wedeman 2004; Ying 2004). Transparency International (2006) puts China into the same class as the family-run Saudi Arabia, hardly a sign of a progressive society aspiring to a global leadership.

The degradation of the environment has been exceptional in China in its extent and intensity. Pre-1949 China was extensively deforested, suffering from heavy erosion and regional shortages of water: Maoist policies made all problems much worse and added enormous burdens of industrial air and water pollution even as its propaganda was brainwashing Western admirers with tales of exemplary environmental achievements. I will never forget the disbelief and doubts with which my first survey of China's environment (Smil 1984) was met in the United States and Europe: things could not possibly be that bad. Eventually China opened up most of its territory to visitors, Earth observation satellites provided more detailed coverage of many environmental phenomena, and the results of pollution monitoring of China became publicly available. During the 1990s there could be no doubt that the country had few rivals in the extent and intensity of its air and water pollution and chronic water shortages (Smil 1993; World Bank 1997).

But when people asked me how soon China would reach an environmental breaking point, I could not give unequivocal answers. As Deng Xiaoping's reforms led

China to quadruple its GDP within a generation, the country's environment got better in some respects and worse in others, and this dynamic situation made it difficult to assess the net outcome. The post-Mao leadership adopted a number of measures that abolished the most irrational Maoist policies, including the conversion of orchards, wetlands, and slopelands to grain fields and a ban on private wood fuel lots. The quality of afforestation efforts improved quite impressively, major cities acquired at least primary waste water treatment, particulate air pollution from large stationary sources was controlled by electrostatic precipitators, and higher energy efficiency of modernized industries reduced the waste streams per unit of products.

Three decades after Mao's death China is definitely greener and more efficient, but a close look makes it clear that we should be worried about the state of its environment more today than we were a generation ago. China's food production and energy demand are the two key reasons. In 2005, affluent Western nations had about 800 million people, or some 12% of global population, but they cultivated nearly one-quarter of the world's farmland, averaging about 0.5 ha per capita (FAO 2006). Western populations will grow by less than 5% by 2025, but because about three-quarters of their staple grain harvests are fed to animals to support high average meat and dairy intakes, North America, the EU, and Australia could forgo the cultivation of a large share of their farmlands by simply eating less animal protein. For decades China's statistics greatly underestimated the country's cultivated land, putting it at just 95 million hectares in 2000 (Smil 2004). Then the total was revised to 130 million hectares, about a 37% increase, but by 2005 it was reduced to 122.4 million hectares largely because of conversions to nonagricultural uses (NBS 2006).

This means that China, with 20% of the world's population in 2005, had only 9% of the world's farmland, or just a little over 0.1 ha per capita. The only two poor populous countries with less farmland per capita are Egypt and Bangladesh, but nearly 300 million Chinese already live in provinces where the per capita availability of arable land is lower than in Bangladesh. Moreover, as China undergoes the biggest construction boom in its history, the conversion of farmland to other uses as well as excessive soil erosion (over at least one-third of China's fields), salinization, and desertification have steadily reduced the country's arable land. Even without further acceleration of recent trends, the average per capita availability of farmland will be no more than 0.08 ha/person by the year 2025 or not much more than the Bangladeshi mean. And the loss of usable farmland will continue. For example, these are plans to build a national trunk highway system whose length, 85,000 km, will surpass that of the U.S. interstate highways by more than 10%.

Deng Xiaoping's agricultural reforms made China basically self-sufficient in food, and at a higher level than at any time in the country's long history. It basically matches Japan in average daily per capita food availability, although not in nutritional quality (Smil 2004). But because of large regional disparities China still has several tens of millions of malnourished people. In order to eliminate this deficit, to produce adequate food for the additional 135 million people to be added between 2005 and 2025 (United Nations 2004), and to improve the quality of average intakes, China would have to expand its food output by at least 20% by 2025.

This would require an incremental food supply roughly equivalent to the total current food consumption in Brazil. Yet this food production would have to come from a steadily diminishing area of farmland because the country will not have the Japanese or South Korean option. Those two countries import most of their food in exchange for value-added industrial products. China could certainly produce enough manufactures to buy most of its food, but that amount of grain and meat is simply not available on the global market. In 2005, China produced about 428 Mt of food and feed cereals, whereas in 2004 the global exports of all grains amounted to 275 Mt (FAO 2006).

China could thus absorb all of the world's grain exports and still satisfy less than two-thirds of its demand. The meat situation makes for an even wider gap: about 28 Mt of all meat varieties and processed meat products entered the global market in 2004, but China's meat output was put officially at 77 Mt (NBS 2006). Even if all of the world's traded meat were shipped to China, it would meet less than 40% of domestic demand. The official meat output total is almost certainly exaggerated, but this in no way changes the basic conclusion. China can never rely on imports to satisfy most of its enormous food demand. Cropping intensification is the only way to produce more food from less arable land, and irrigation is key. In absolute terms China already irrigates more land than any other country, and in relative terms it ranks only behind Egypt and Israel, but its water supply situation is already very precarious.

China has only 7% of the world's freshwater resources. The provinces north of the Yangzi, with some 40% of population and GDP, have only about 20% of the southern average, or just over 500 m^3 per capita. In 2000, China's nationwide mean of annual freshwater availability was about 2000 m^3 per capita, and in 2030 (when China's population is ~1450 million) this will fall to less than 1800 m^3 per capita (and in the northern provinces to barely half of that). By contrast, global availability in 2000 averaged about 7000 m^3 per capita, in the United States nearly 9000 m^3 per capita, and in Russia, 30,000 m^3 per capita (WRI 2000). Even if it were possible to use every drop of northern stream runoff, per capita water supply in China would be less than one-quarter of America's actual U.S. per capita water consumption.

Actual per capita northern supply in China for all uses—agriculture, industry, services, and households—amounts to little more than the amount used in the United States to flush toilets and wash clothes, dishes, and cars. In addition, 90% of water sources in China's urban areas are polluted, and in some northern provinces water tables have been sinking by several meters per year. Such are the northern water shortages that the Huanghe, the region's principal stream, regularly ceases to flow long before it reaches the sea. Between 1985 and 1997 the river dried up in some sections every year. In 1997 the stream did not reach the Bohai Bay for a record 226 days, and the dry bed extended for more than 700 km from the river's mouth (Liu 1998). Massive, costly, and environmentally risky south-north water transfer from the Yangzi to the Huanghe basin is a controversial (and hardly a lasting) solution (fig. 3.18) (Smil 2004; Stone and Jia 2006). The need to feed at least 100 million additional people, to satisfy rising urban demand, and to secure water for growing industries means that the northern China's strained resources will be, even with the south-north transfer, under more pressure during the next two decades.

Nitrogen fertilizer is the other key ingredient of cropping intensification, but China is already the world's largest user of synthetic fertilizers, with average per hectare rates four times as high as U.S. rates and with some double-cropped rice fields receiving annually in excess of 400 kg N/ha (Smil 2001). Higher rates of average applications may make little difference because the crop response to high applications of nitrogen has been declining, overall efficiency of nitrogen use in China's cropping is very low—below 50%, and for rice often less than 30% (Cassman et al. 2002)—and nitrogen leaching causes widespread contamination and eutrophication of streams, reservoirs, and coastal waters.

Future energy demand will impose a tremendous pollution burden on China. In 2005, China's consumption of primary commercial energy amounted to about 9% of the global total, again much less than the country's population share. And less than 15% of the low per capita rate, equivalent to about half a tonne of crude oil per year, is used by households, compared to about 40% in the West. In order to

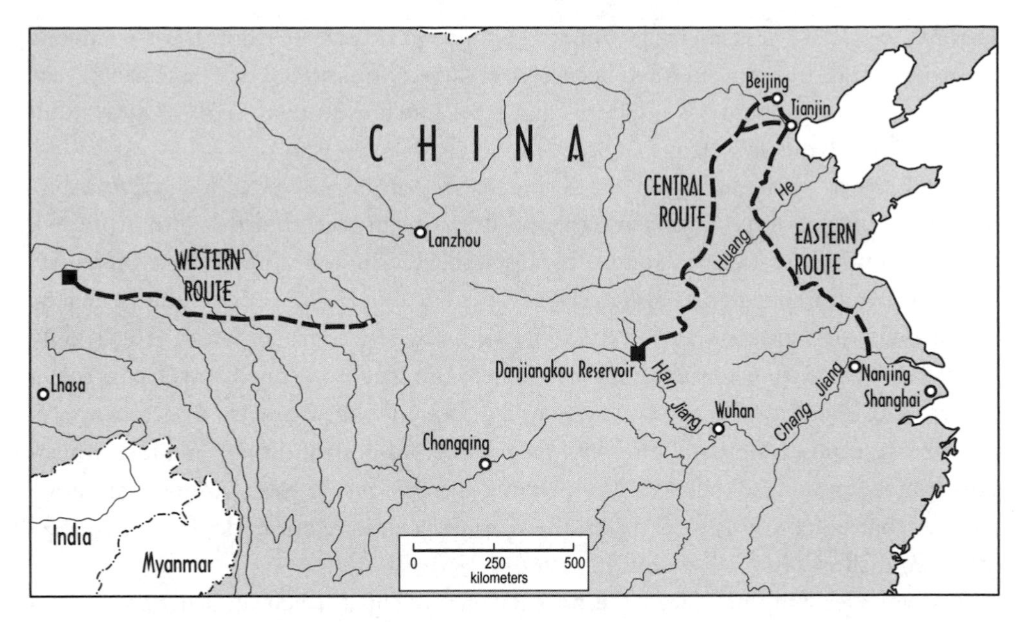

Fig. 3.18
Three routes for China's south-north water transfer. The eastern and central routes are under construction; the planned western route would be expensive and difficult to construct. Based on Smil (2004) and Stone and Jia (2006).

join the ranks of developed nations China's per capita energy consumption would have to be at least twice the current mean. China has been the world's largest source of sulfur emitted from combustion since 1987 (Stern 2005), and only extensive desulfurization of flue gases could prevent it from producing an even larger share. Nitrogen oxides from large power plants and rising combustion of refined fuels, in particular from vehicular traffic (crude oil demand tripled between 1990 and 2005) will increase the levels of semipermanent photochemical smog that aggravates respiratory illnesses and reduces crop yields (Aunan, Berntsen, and Seip 2000).

In addition, by 2007 China has already become the world's largest emitter of greenhouse gases, and it will take a crucial and contentious role in any effort to stabilize and reduce their generation. The most appealing nonfossil alternative has its own environmental problems. Accelerated development of hydroelectric generation, exemplified by the controversial Sanxia mega-project (Smil 2004), has caused extensive flooding of high-yielding farmland, mass population resettlement, and rapid reservoir silting. Even a partial quantification of China's environmental burdens reveals a considerable impact. Economic losses attributable to China's environmental degradation have been conservatively quantified as equal every year to 6%–8% of the country's GDP, or almost as much as annual GDP growth (Smil 1996; World Bank 1997).

Finally, there is the intangible but critically important power of ideas. No aspiring superpower can do without them, and it can be argued that they are as important as economic or military might. In this respect, China has no stirring offerings; the power of the ruling party still derives from stale and rigid Marxist-Maoist tenets. As for Beijing's "socialist market economy with Chinese characteristics," this is only a label for a mixture of relatively free enterprise and continued party control, a rather unoriginal idea with elements copied from Japan, Taiwan, and Singapore. Even more fundamentally, China has yet to face old deep internal wounds; official government policy still silences any probing discussions of the two greatest catastrophes that befell China after 1949, the world's largest, Mao-made, famine (1959–1961) and the Cultural Revolution (1966–1976).

Postwar Germany has faced the horrors of the Third Reich, and it has worked in many ways to atone for its transgressions. Russia began to face its terrible Stalinist past when Khrushchev first denounced his former master (in 1956), opened the gates of the gulag, and had the dictator's corpse removed from the Red Square mausoleum. But the portrait of Mao still presides over the Tian'anmen, hundreds of his statues still dot China's cities, and Maoism remains the paramount ideology of the ruling party. This amnesia is hardly a solid foundation for preaching moral superiority. And as for serving as a social and behavioral model, China—despite (or perhaps because of) its ancient culture, and in a sharp contrast with the United States—has little soft-power appeal to be a modern superpower of expressions, fashions, and ideas.

Its language can be mastered only with long-term devotion, and even then there are very few foreigners (and fewer and fewer Chinese) who are equally at ease with the classical idiom and spoken contemporary dialects. Its contemporary popular music is not eagerly downloaded by millions of teenagers around the world, and how many Westerners have sat through complete performances of classical Beijing operas? China's sartorial innovations are not instantly copied by all those who wish to be hip. Westerners, Muslims, or Africans cannot name a single Chinese celebrity. And who wants to move, given a chance, to Wuhan or Shenyang? Who would line up, if such an option were available, for the Chinese equivalent of a green card?

In the realm of pure ideas, there is (to chose a single iconic example) no Chinese Steven Jobs, an entrepreneur epitomizing boldness, risk taking, arrogance, prescience, creativity, and flexibility, a combination emblematic of what is best about the U.S. innovative drive. And it is simply unimaginable that the turgid text of the country's Communist constitution would be ever read and admired as widely as is that hope-inspiring 1787 document, the U.S. Constitution, whose stirring opening, I assume, you know by heart. Here is the first article of China's 1982 constitution:

> People's Republic of China is a socialist state under the people's democratic dictatorship led by the working class and based on the alliance of workers and peasants. The socialist system is the basic system of the People's Republic of China. Sabotage of the socialist system by any organization or individual is prohibited.

Those who laud the new China might re-read this a few times. And anybody familiar with today's China knows how eagerly the Chinese people themselves imitate U.S. ways even as they profess nationalistic, anti-American fervor.

This brings me to the hardest task of all national assessments—a look at prospects for the United States. I tip my hand before I begin. Even a problem-ridden China, a self-absorbed Europe, a faltering Russia, and a relatively nonconfrontational Islam would not guarantee the continued global primacy of the United States. External forces are important, but as with all great states, a great danger is the rot that works from within.

The United States' Retreat

The United States is a superpower in gradual retreat. Its slide from global dominance has been under way for some time, but in the first years of the twenty-first century many components of this complex process have become much more prominent, coalescing into a new amalgam of worrisome indicators that point unequivocally toward a gradual decline.

I emphasize *gradual*. Unless the country sustains a massive unanswered nuclear attack (an event of negligibly low probability; its nuclear triad guarantees devastating retaliation), there is no way its power can vanish instantly. Nor is it easy to come up with rational scenarios (large-scale urban terrorist attacks, collapse of the country's economy that would leave other major players unaffected) that would see its power drained away in a matter of months or years. Its hegemony will devolve at varying speeds and in a nonlinear fashion across decades.

Using an irresistible analogy with Rome, and assuming the end of that Western empire at 467 C.E., we cannot know if the United States is already at a point comparable to when the Roman legions began withdrawing from England (383 C.E.) or when Diocletian first divided the empire (284 C.E.), but it is surely not at a point comparable to the time before the Roman empire's extensive conquests in the Middle East (90 B.C.E.). Signs of ennui, of unmistakable overstretch, classic markers of an ebbing capacity to dominate, have been noted by historians and political commentators (Nye 2002; Ferguson 2002; Johnson 2004; Cohen 2005; Merry 2005). The title of Wallerstein's (2002) article, "The Eagle Has Crash Landed," referring to the 1969 moon landing at the pinnacle of U.S. power, perhaps most evocatively summarizes these analyses. They say the Pax Americana, the period of relative peace in the West since the end of World War II coinciding with the dominant military and economic position of the Untied States in the world, is over.

The limits of U.S. military power have already been demonstrated, for instance, by the wars in Korea and Vietnam and an ill-fated 1992–1993 Marine mission in Somalia. But the attacks of 9/11; the occupation of Iraq; Iran's bid for the Middle Eastern leadership; a defiant North Korea; a newly confident Russia, economically powerful and assertive; a militarizing China and Europe, more reluctant than ever to follow almost any U.S. initiative—all of this has left the United States in such a bind that no Houdini-like contortions can release it with its power and global influence intact.

Other foreign challenges demonstrate the limits and essential irrelevance of U.S. military power, and these realities leave policymakers with no appealing options. Foremost among them is the inability to prevent millions of illegal immigrants, especially Mexicans, from crossing the country's borders, an influx that divides populace along many split lines (Hanson 2005). The current immigration challenge is not comparable with the task of absorbing massive waves of immigrants from Europe between 1880 and 1914. These inflows were controlled (the names of virtually all arriving immigrants can be checked in passenger lists and immigration records), and it was not dominated by any single national or religious group, a reality conducive to relatively rapid acculturation and integration.

By contrast, the United States has no control of the latest immigration wave. The Border Patrol will not even speculate how many border crossers evade capture (the official total of those captured crossing from Mexico was 1,241,089 in 2004), and the total number of illegal immigrants is not known with an accuracy better than ±5 million. More important, Most Mexican immigrants actually do not feel that they are emigrating. In a 2002 survey, 58% of Mexicans maintained that the U.S.

Southwest "rightfully belongs to Mexico" (Ward 2002). *Reconquista* of the U.S. Southwest (Aztlan, in the parlance of radical groups) has been in high and accelerating gear since the mid-1980s, and the fences that have been built or planned along parts of the U.S.-Mexican border are nothing but useless, pathetic attempts that do not (and will not) prevent anybody's entry into the country (Skerry 2006).

Some fence designs actually facilitate ingress: horizontally laid corrugated steel panels unintentionally provide ladder-like supports for easier scaling, and in order not to "offend" Mexico, the flange on top of the fence points toward the U.S. side, making it easier to roll over. Huntington (2004) argued that the new immigration is fundamentally different from previous (transoceanic) inflows because it taps into a large pool of poorer, less well-educated migrants who come from a culturally resilient place and who thus may not be amenable to swift integration with U.S. culture. Illegal immigrants waving Mexican, not U.S., flags during their protest marches in U.S. cities only reinforce this observation.

Even if such views only partly identify the true situation, the political and economic consequences of such a change would be profound. After all, if everyday U.S. realities came to resemble Mexico's, the change would be in a very undesirable direction: toward a *narcotraficante* economy governed almost entirely by corrupt arrangements. This would further alienate U.S. citizens from participating in the political process (the percentage of U.S. population voting has dropped steadily and is now lowest among all Western countries). A key comparison hints at the shift that could occur: Mexico ranks 70 on the 2006 Transparency International scale of corruption perceptions (the same as China), whereas the United States ranks 20, ahead of France and Japan. Everything from politics to policing, would change if U.S. conditions cause to resemble Mexico's in this respect—a clear reason to worry about a rapid, nonassimilating Hispanicization, and indeed balkanization (as other immigrant groups are assimilating more slowly), of the country.

Even if there were no challenges from abroad, there is no shortage of homegrown trends whose continuation will undermine U.S. global primacy. A prominent economic concern is the country's rising budget deficit and deteriorating current account balance. Despite a brief spike of budget surpluses between 1998 and 2001, the cumulative gross federal debt had doubled between 1992 and 2005 to about $8 trillion, or 65% of the country's GDP in those years (fig. 3.19) (USDT 2006). The country's net international investment position was positive (in market value terms) from the end of World War I until 1988 (USBC 1975; Mann 2002). A subsequent slide brought it to about –$300 billion by 1995; more than –$1 trillion in 1998;

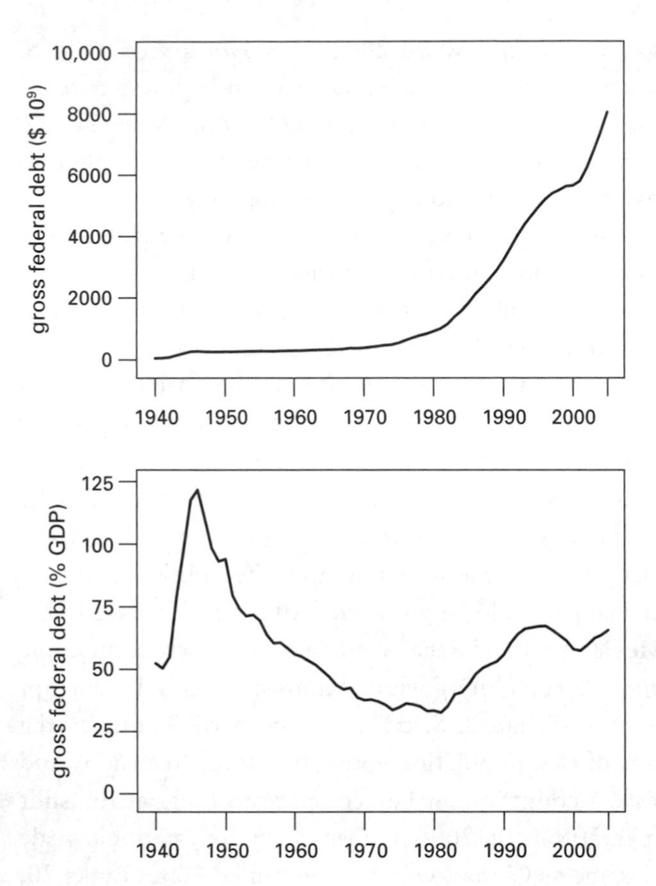

Fig. 3.19
U.S. gross federal debt, 1940–2005. Plotted from data in BEA (2006).

–$1.58 trillion in 2000; –$2.33 trillion in 2001; and –$2.5 trillion in 2005, equal to about 20% of the country's GDP in that year (BEA 2006).

By 2005 the annual current account deficit reached about $800 billion, or 6.5% of GDP, and it has been absorbing some two-thirds of the aggregate worldwide current account surplus, both unprecedented levels. Some critics see these deficits inevitably ending in a severe crisis. Cline (2005) asked how long the world's largest debtor nation can continue to lead. Obstfeld and Rogoff (2004) concluded that after taking into account the global equilibrium consequences of an unwinding of this deficit, the likelihood of the collapse of the dollar appears more than 50% greater than their previous estimates. And Fallows (2005) used the term *meltdown*.

In contrast, others see this deficit as a sign of economic strength. Dooley et al. (2004) believe that a large current account deficit is an integral and sustainable feature of a successful international monetary system. R. W. Ferguson (2005) and Bernanke (2005) concluded that the deficit is primarily driven not by U.S. extravagance but by the slump in foreign domestic demand, which has created excessive savings, resulting in the large current account surpluses in Asia, Latin America, and the Middle East being invested in the United States. But Summers (2004, 8) asked, "How long should the world's greatest debtor remain the world's largest borrower?" and suggested the term "balance of financial terror" to describe "a situation where we rely on the costs to others of not financing our current account deficit as assurance that financing will continue."

The countries that finance the U.S. current account deficit will not do so indefinitely. Even now, as Summers (2004, 9) stressed, "A great deal of money is being invested at what is almost certainly a very low rate of return. To repeat, the interest earned in dollar terms on U.S. short-term securities is negative." In addition, the investor countries with fixed exchange rate regimes (most notably, China; its revaluations of yuan have been, so far, an inadequate adjustment) are losing domestic monetary control. Summers also noted that a similar kind of behavior was behind Japan's excesses of the 1980s as "much of the speculative bubble . . . that had such a catastrophic long-run impact on the Japanese economy was driven by liquidity produced by a desire to avoid excessive yen appreciation." A house of cards comes to mind when thinking about these arrangements.

Are these the fiscal foundations of a superpower? By October 2007 the two largest holders of U.S. Treasury securities were Japan, with about $592 billion, and China, with about $388 billion (USDT 2007). Hence one can think of China as financing the U.S. Department of Defense bill for forces and operations in Iraq and Afghanistan for almost three years, and of Japan covering the U.S. budget deficit for almost two years. One can—but is such a thinking correct? Not according to Hausmann and Sturzenegger (2006), who deduce that the United States must have more foreign wealth than is apparent ("black matter") and conclude that it is actually still a net creditor nation. This is their best way to explain how despite its enormous accumulated deficit, the country still earns more on its foreign assets than it pays out to service its foreign debt and has no current account deficit. U.S. Treasuries still have the highest rating, and its currency is still the global standard.

Given the disagreement among economists regarding even the near-term outlook (ranging from the dollar's imminent collapse to a comfortable continuation of rising deficits that might actually be "black matter" surpluses), I think it is more useful to point out the two most worrisome features of the U.S. trade deficit, namely, the degree to which it has become embedded and hence impervious to any near-term reversal, and its consequences for the country's position as a global leader in manufacturing and scientific and technical innovation. Both of these are perilous trends; both steadily diminish the country's great power status, and neither of them can be denied, argued away, or eliminated by creative accounting.

After decades of trade deficits, caused by the demands of a rapidly developing post–Civil War economy with its huge infrastructural needs, the United States began to run a surplus on its foreign trade in 1896 and maintained it throughout economic cycles and wars for the next 75 years. An uninterrupted but fluctuating sequence of deficits followed between 1974 and the late 1990s, and then came a free fall. The 1997 deficit more than doubled in just two years, and the record 1999 deficit nearly doubled by 2003 and then grew by another 24% in 2004 and 17% in 2005, when it stood nearly four times above its 1996 level (fig. 3.20).

This transformation has not been merely a matter of high consumer spending or, as some economists maintain, a benign unfolding of a new global system. Continued enlargement of the trade deficit is unsustainable (it would have the country importing every year more than its current GDP in less than 25 years). This long-term trend will result in structural deficiencies and strategic vulnerabilities. A resolute administration can drastically cut, even eliminate, the budget deficit, but the elimination of the trade deficit is highly unlikely.

The U.S. trade deficit was $494.9 billion in 2004, $611.3 billion in 2005, and $716.6 billion in 2006 (USBC 2006). Almost exactly 40% of the 2005 deficit was due to imports of industrial supplies and materials. The country has become steadily more dependent on foreign energy, metals, chemicals, and other raw materials, and in 2005 it bought 2.2 times as much of these goods as it sold abroad. In 2005 crude oil imports accounted for one-quarter of the total trade deficit. In January 2006, President George W. Bush, a Texas oil man, acknowledged the country's addiction to oil purchased from a Venezuelan Castroist (11% of U.S. oil imports in 2005), Nigerian kleptocrats (the country ranks next-to-last on the global corruption perception index) who supplied more than 8% of U.S. oil in 2005, and a country run by a family of princes in Riyadh, Saudi Arabia, that supplied more than 11% of U.S. oil in 2005 (EIA 2005).

After adding other fossil fuels and electricity, the total deficit in the U.S. energy trade amounted to $265 billion in 2005. In order to satisfy its excessive energy use,

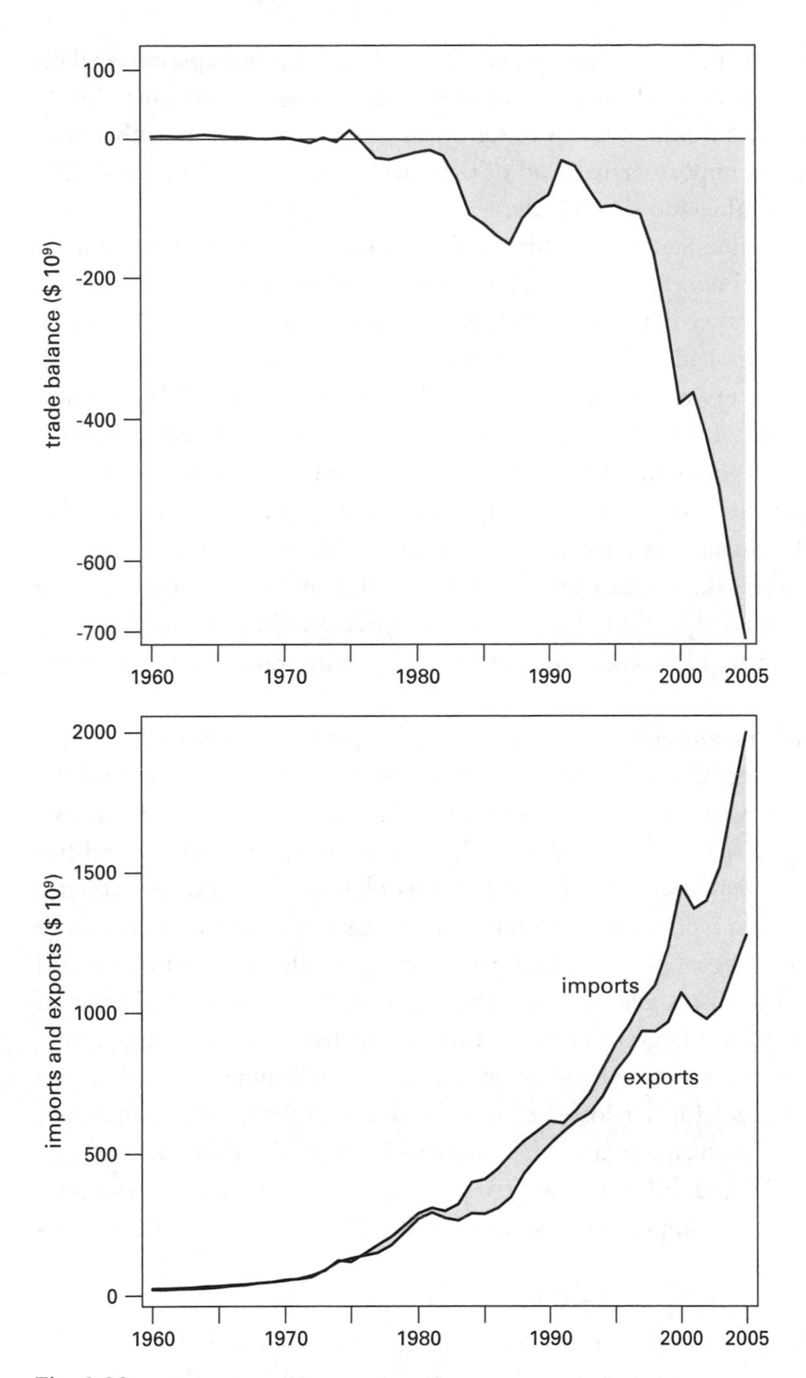

Fig. 3.20
U.S. trade balance, imports, and exports, 1960–2005. Plotted from data in USCB (2006).

the country thus had an embedded annual deficit of about one-quarter trillion dollars, and with higher world prices this has increased substantially since 2005. Still, these large imports would not have to signify any large overall trade imbalance. Japan or South Korea import virtually all of these basic supplies and materials but then turn them into value-added products.

By contrast, the United States has a huge deficit in trading manufactured goods, mostly because of its imports of automobiles, parts, and engines, which in 2005 was about 2.4 times larger than the country's automotive exports. In addition, by 2005 the United States had a deficit in 17 out of 32 major categories of capital goods (led by computers and their accessories: $45.5 billion sold, $93.2 billion bought) and in 26 out of 30 categories of consumer goods. The aggregate deficit in the trade of consumer goods was $291 billion; the only (small) surpluses were from toiletries and cosmetics, books, records, tapes and disks, and numismatic coins (USBC 2006). All affluent countries have seen a relative decline in manufacturing, but in the United States these losses have gone farther than in Europe or Japan. The sector employed about 30% of the labor force during World War II, but by 2005 the share was less than 12%, compared to 18% in Japan and 22% in Germany (USDL 2006).

Many economists, unconcerned about this decline, repeat the mantra of new jobs in services providing greater prosperity. But this ignores both the large embedded trade deficits and increasing strategic vulnerability of a country that has already lost entire manufacturing sectors. Virtually every kind of mundane manufacturing has either completely or nearly vanished (cotton and wool apparel, cookware, cutlery, china, TVs) or has been reduced to a small fraction of its former capacity (furniture making, writing and arts supplies, toys, games, sporting goods). In consumer goods categories alone, these losses add up to an embedded deficit of about $200 billion per year. Car making, the largest manufacturing sector, has been in an apparently unstoppable slide as the adjustments made by the country's leading automakers are never enough to prevent further losses of market share. In 1970, U.S. companies produced 85% of all vehicles sold in the country; in 1998 the share was still at 70%, but by 2005 it had fallen below 60%, and it may drop below 50% even before 2010. Automotive imports add almost $150 billion per year to the chronically embedded deficit.

Dominance in high-quality steel production, still the cornerstone of all modern economies, is a thing of the past in the United States. In 2005, U.S. Steel was the seventh-largest steel company in the world, outranked not only by European and

Japanese producers but by giants whose names are unknown even to well-educated Americans: Lakshmi Mittal's company, South Korea's POSCO, and Shanghai Baosteel (IISI 2006). And the United States has been losing the leadership of one of its last great high-tech manufacturing assets, its aerospace industry. Wrested from Europe during the 1930s with such iconic designs as DC-3 and Boeing 314 (Clipper), it was strengthened during World War II with superior fighters and bombers and extended after the war with remarkable military designs and passenger jets.

The latter category included the pioneering Boeing 707, introduced commercially in 1957; the Boeing 737, the most successful commercial jet in history; and the revolutionary Jumbo, Boeing 747 (Smil 2000). By the mid-1970s the U.S. aviation industry dominated the jetliner market, and its military planes were at least a generation ahead of other countries' designs (Hallion 2004). The subsequent decline of the U.S. aerospace industry forced the commission examining its future to conclude that the country had come dangerously close to squandering the unique advantage bequeathed to it by prior generations (Walker and Peters 2002).

By 2000 only five major airplane makers remained in the United States; the labor force declined by nearly half during the 1990s; Boeing was steadily yielding to Airbus in the global market (between 2003 and 2005 it sold fewer than 900 planes, compared to about 970 for Airbus); most top military planes (F-16,F/A-18, A-10) had been flying for more twenty-five years; and U.S. rocket engines were outclassed by refurbished Russian designs. In 2003, Airbus began intensive work on the A380 superjumbo jet (Airbus 2004). Boeing went ahead with a smaller but superefficient 7E7 (787, Dreamliner) but only after it subcontracted much of its construction to foreign makers (Boeing 2006). By 2005 the U.S. surplus in trading civilian aircraft, engines, and parts was less than $19 billion, considerably less than the imports of furniture or toys. So much for high-tech manufacturing making up for the losses in traditional sectors.

A rapid reduction of the country's traditionally large surplus in trading agricultural commodities is the most recent, and perhaps most surprising, component of the U.S. decline. The country has been a net exporter of agricultural commodities since 1959, and the export/import ratio of the value of this trade was 1.83 as recently as 1995. By 2000 it was 1.30; in 2004, 1.18, with the surplus falling below $10 billion a year; and in 2005, 1.08. This trend points to deficits before 2010 (Jerardo 2004; USDA 2006). As a result, by 2005 the net export gain from the huge agricultural sector (just $4.74 billion) paid for less than half of the country's rising imports of fish and shellfish in that year (worth $11.9 billion).

When the accounting is done in terms of a larger, actually more meaningful, category that comprises all traded crops (food and feed), animal products, processed foods, and beverages (including wine, beer, distillates, and liqueurs), the United States already had a substantial trade deficit in 2005 (about $9.1 billion), and it is almost certain that the country will become increasingly dependent on imported food. In 2005 imports (by weight) claimed nearly 15% of all food consumed, with individual shares surpassing 80% for fish and shellfish, 30% for fruits, juices, nuts, sweeteners, and candy, and 10% for meat and processed meat products. This has been an inexplicably unreported shift with enormous long-term impacts. After borrowing abroad to support its need for crude oil, gadgets, and toys, the United States will be borrowing just to feed itself. Autarky in food is an impractical and unattainable goal for many small nations, but were the United States to become a permanent net food importer, who would be left to do any net exporting—Brazilians and heavily subsidized French farmers?

There is no possibility that the trade in services can ever make up for the massive deficits in traded goods. Although the annual surplus in service trade rose from $52 billion in 2003 to $66 billion in 2005, all that the latter sum could do was to lower the overall trade deficit by about 8% (from $783 billion to $717 billion). In a world where the theft of intellectual property is a ubiquitous activity that is not only rarely prosecuted but that is tacitly condoned by many governments (particularly in Asia, with China being the most blatant transgressor), U.S. trade salvation will not come from royalties and licensing fees. Nor it will come from tourism, unless the U.S. dollar becomes incredibly cheap. Given an embedded trade deficit of about $0.5 trillion per year, which cannot be swiftly eliminated by policy changes, and a growing dependence on energy imports, it is hard to imagine how this dismal trading trend could be reversed short of a substantial devaluation of the dollar and a precipitous decline in the standard of living.

Even if some form of "black matter" assets existed, one would still have to ask, How long will the Japanese, Chinese, Middle Eastern and other OPEC creditors, and Caribbean banking centers holding recycled drug trade profits be willing to extend credit to the superpower at a cost to themselves? For pessimists, the only remaining uncertainty is how the end will come. Will it be a controlled, drawn-out fading or a precipitous fall? In any case, a United States forced to live within its means would be a very different place. It has already turned into a very different place because of a multitude of interconnected demographic, social, and behavioral trends that have weakened it from the inside.

The aging of the U.S. population, although far less pronounced than in Europe or Japan, and a multitude of social ills will only accelerate the inevitable transformation of the country. The aging of the population will have similar effects on health budgets, pensions, and the labor market as in Europe or Japan, but given much more widespread stock ownership, its principal undesirable effect may be on the value of long-term investments. There will be too few well-off people in the considerably smaller post-boomer generations to buy the stocks (and real estate) of aging affluent baby boomers at levels anywhere near peak valuations. That is why Siegel (2006) expects that stock prices in rich countries could fall by up to 50% during the coming decades, unless newly rich investors from Asia, the Middle East, and Latin America step in. But that intervention would have to be on a truly massive scale. Siegel's calculations indicate that for rich countries' stocks to perform at their long-run historic rate, most multinational corporations would have to be owned by non-Western investors by 2050.

Underperforming U.S. education leads to well-documented dismal scores in international assessments in math and science. For example, the latest international comparison of mathematical skills in 29 OECD countries puts U.S. 15-year-old students only above those of Portugal, Greece, Turkey, and Mexico (OECD 2003). Despite the decades-long war on drugs, street prices have remained low and stable (or falling), and distribution (particularly of highly addictive methamphetamine), is more widespread. Cross-border seizures and potency of marijuana are at unprecedented levels, and the overall trend in drug use has not shown a decline since 1990 (fig. 3.21) (White House 2006). Drugs are also a major reason for the country's extraordinarily high rate of incarceration, and for what some commentators call America's prisons nightmare (DeParle 2007). Troubled health care and pension systems may be headed toward bankruptcy, yet endless congressional debates cannot offer effective solutions (Kotlikoff and Burns 2004; Béland 2005; Derickson 1998).

A visible deterioration of the nation's physical fitness makes the United States the most obese and physically unfit nation in Western history (CDCP 1998). No wonder, when the nation is the undisputed leader in supersized restaurant meals and highly popular, exceedingly disgusting, contests in competitive eating where the top competitors consume in a matter of minutes more than 20 grilled cheese sandwiches or hot dogs (Fagone 2006). Recent attempts to portray obesity as a fairly innocuous condition and grossly overweight people as victims are ludicrous. The links between obesity and greater morbidity are too well established, and in most cases obesity is a matter of choice, not of somatic inevitability. In 1991 only 4 of

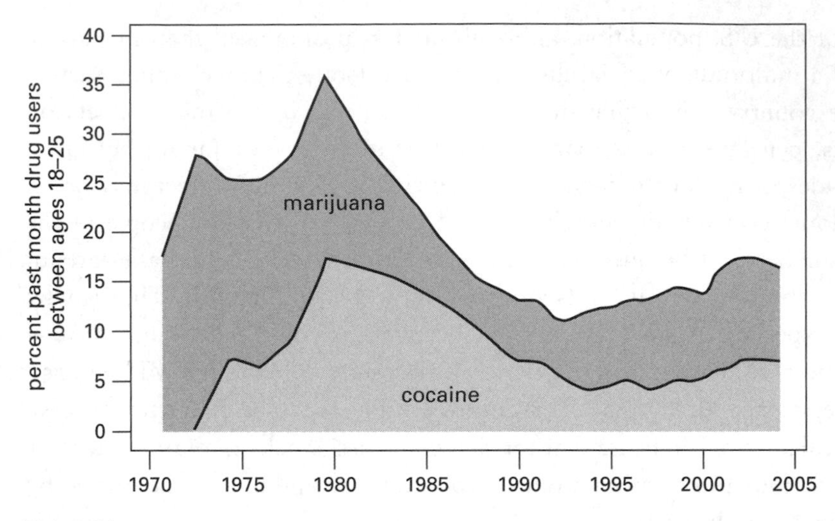

Fig. 3.21
Cocaine and marijuana use among U.S. users aged 18–25, 1990–2004. From White House (2006).

45 surveyed states had an obesity prevalence rate of 15%–19%; none was over 20%. But by 2001 the nationwide mean was 20.9%, with 29 states having rates of 20%–24% (Mokdad et al. 2003).

Nader (2003) included displays of gluttony among his four signs of societal decay, together with electoral gerrymandering, ubiquitous corporate crime, and corporate excess. The flaunting of possessions has become a new norm, exemplified by the size of new homes and vehicles. By 2005 the national average size of new houses reached 220 m^2 (12% larger than a tennis court and 12% above the 1995 mean). The average for custom-built houses rose to 450 m^2, houses in excess of 600 m^2 were not uncommon, and the mega-structures of billionaires claimed as much as 3000 m^2 or even more than 4000 m^2 (Gates's house cluster measures to 4320 m^2). The sizes of the largest vehicles used as passenger cars climbed past 2 and 3 tonnes to the most massive brands, weighing nearly 4.7 t (Hummer H1) and almost 6.6 t (CXT, designed to be just 1 pound lighter than the weight requiring a trucking license).

These displays of private excess have been accompanied by spreading public squalor (abandoned housing, derelict areas of former manufacturing compounds) and the unraveling of essential social supports, including one that people actually paid for, their pension plans. Many corporations (led by airlines, car parts companies, and steel companies) have defaulted on their pension responsibilities, transferring them to the Pension Benefit Guaranty Corporation, whose deficit reached about $23 billion by 2005.

One does not have to watch inane TV shows or read supermarket tabloids in order to feel that public mores and tastes are driven toward the lowest common denominator. This process entails a pathetically emotive Oprah-ization of America, an eager discarding of privacy, proudly vulgar displays of immature behavior, an endless obsession with celebrities (and a mass yearning to become one of them, fueled by so-called "reality" shows), rampant legalized gambling, and frenzied purchases of lottery tickets.

All of these symptoms have been discussed at length in vibrant and cacophonous electronic and printed media; scathing self-examination and self-criticism show no signs of decline (God bless America!). But this will make no difference as long as there is no commitment among the policy-making elites to address at least some of these matters in a practical, effective fashion, and as long as that commitment is not combined with a mature willingness among the country's population to live at least closer to (if not entirely within) their means, to curb the worst excesses, and to think and act as if the coming generations mattered. Regrettably, the latter possibility is even less likely than the first proposition, and both could become real (perhaps) only if the country were thrown into a truly deep financial and existential crisis. But by that time it might be too late for the United States to regain its great power status. It is very clear that it is living on borrowed time and yet has no imminent intentions to do otherwise.

Place on Top

As with every change of global leadership, the retreat of the United States from global leadership will be widely felt, but because of the country's pervasive impact (no matter if positively or negatively perceived), it will have many consequences whose definite contours we cannot foresee. For some four generations the country has been the world's dominant agent of change, an unselfish savior, a reluctant arbiter, and a brash trendsetter. At the same time, it has never been averse to realpolitik, as attested by détente with the USSR, support of dictatorial regimes in Latin America, Africa, the Middle East, and Asia, and rapprochement with Communist China. But in its fundamentals, and very often in its execution, U.S. foreign policy has been imbued with moral (and moralistic) convictions about the duty to act as a global promoter of freedom, a call to action that unites John F. Kennedy and George W. Bush.

The dominant call would be very different if it came from the mullahs speaking for a resurgent Islam, from the *énarques* running a new United States of Europe, from the councils of a rejuvenated Russian military, or from confident politbureau strategists in Beijing. Before any of these alternatives might happen, role of the United States must weaken sufficiently, perhaps even to the point where it would cease to be the first among equals.

As I said earlier, I do not intend to offer any forecasts or suggest firm dates. There seem to be only two things that could be done. The first is to look at the fates of past powers and see if there are, despite the obvious singularities of every case, any helpful conclusions regarding their longevity in general, and the rate of their decline in particular. The second is to look at the most likely new configuration of power on top, appraising the chances for today's other key international players to become the new trendsetters during the first half of the twenty-first century.

Dominance and Decline

Falling from a position of power, dominance, and affluence (in absolute terms or relative to other contenders at a given time in history) is always a painful process, but the rate of decline makes a great difference. Think of Germany's accelerated rise and demise, as the Thousand Year Reich was compressed between 1933 and 1945; or of the USSR's demise and Russia's ensuing pitiful economic and social position during most of the 1990s. As already noted, barring an unanswered (and hence extremely improbable) nuclear attack, such a precipitous fall is not in the cards for the United States. Under normal circumstances large and powerful states take time to unravel.

How much time? The Roman experience, even when limited to the Western Empire, is unlikely ever to be repeated. Taking Gibbon's span of decline and fall (180–476 C.E.) it was nearly 300 years. But this is dubious, given the fact that the empire's cohesion was a shadow of its former self and its dependence on foreign legions became critical long before its formal demise. The durabilities of other major premodern empires do not offer any better guidance because the common attributions of their average duration, such those used by Ferguson (2006), are similarly questionable. The Holy Roman Empire was legally dissolved by Napoleon in 1806, but to date it between 800 and 1806 is misleading; for most of its existence it was not a coherent entity. And the Ottoman empire actually spent more than half of its duration (1453–1918) in deep decay.

As for modern hegemonies, the British one, with the peak in 1902 (end of the Second Boer War) and the end in 1947 (quitting of India) took only 45 years to unravel. (The violent Malay and Kenyan conflicts after the British left India just wrapped up Britain's retreat.) The Soviet retreat took almost exactly the same amount of time, from the WW II victory in 1945 to Yeltsin's disbanding of the union in 1991. But the duration of less than half a century of the British and Soviet retreats provides no directly applicable insight into the likely extension of the United States' place on top. Adding 45 to 1945 gives us 1990, yet by 1991, after its victory in the Gulf War (and with the USSR falling apart), the United States appeared to be at the apex of its military powers rather than on its way out.

There is no doubt that the closing months of 1945, after the defeat of Germany and Japan and after the country became the sole nuclear power, marked the peak of U.S. military and economic power (fig. 3.22). At that time the country was responsible for 35% of the world's economic product; this share fell (adjusted for inflation) to a bit above 30% in 1950 (as the USSR, Europe, and Japan began to

Fig. 3.22
When the United States was at the peak of its power: moments before the Japanese delegation signed the documents of Japan's surrender on board the USS *Missouri* on September 2, 1945. Photo from Naval Historical Center, Washington, D.C.

recover from war damages), and to just below 25% in 1970. During the following two decades it remained stable, but by 2005, after a generation of China's rapid economic growth, it slipped to about 20% (IMF 2006) as China surpassed Japan to claim the second place with about 15% of the total.

But the Gulf War could be seen as just one last delayed victory, a blip briefly interrupting the declining trend, with a number of defensible dates for its onset. An economist might opt for August 15, 1971, when the Nixon administration ended more than a quarter century of the Bretton Woods global monetary regime by stopping the convertibility of dollars to gold. A geostrategist might date the beginning of the U.S. decline to April 30, 1975, when the North Vietnamese tanks drove into the former U.S. embassy in Saigon, ending a decade-long war in the Southeast Asia with a U.S. defeat, the country's first in its 200-year history. (The Korean war was a draw, a return to *status quo ante*, and its eventual outcome was the creation of the world's seventh-largest economy, a success by any measure.) But arguments can be made that in the long run the Vietnamese victory was not a strategic defeat because Southeast Asia did not become a Communist stronghold and Vietnam is now following China by integrating its trade-oriented economy into the global market (and it welcomed the U.S. president on a state visit in 2006).

That is why many would argue for 9/11 ("the date everything changed") as the beginning of the end of U.S. supremacy. Admittedly, "everything" is a hyperbolic statement, but the notion that nothing much changed on that day (Dobson 2006) is indefensible. At the same times, I am not sure if the date fits into the lineup of events marking the decline of U.S. power. A generation from now, 9/11 may not seem so much a marker of an unstoppable retreat as the beginning of a temporary reassertion of U.S. power.

I note only that downward trends are no less susceptible to sudden shifts (which may lead to temporary changes and readjustments) than are upward trajectories. Moreover, the magnitude and complexity of interrelated financial, political, and strategic arrangements that bind the United States with the rest of the world provide a significant degree of buffering that makes gradual moves much more likely than any sudden catastrophic shifts.

Setting the arguable duration of U.S. dominance aside, the most important conclusion from reviewing the likely national trends is that U.S. global leadership is in its twilight phase, approaching the latest of the infrequent power transitions taking place on this scale. Indeed, an argument can be made that the coming transition will be unprecedented because the United States was history's first true global power. None of the great powers of antiquity, not even the Roman Empire or the Han dynasty of China, were global powers; their reach was far too restricted for that. The history of global power began only with Europe's grand maritime expansion, which gradually brought all continents into interaction.

Before the United States assumed the leading position, there were states with far-flung possessions and commercial interests, but given the relatively weak level of global economic integration—and in earlier eras the absence of instant communication and the impossibility of rapid long-distance projection of power—their place on top was not of the same import as the U.S. primacy. This quasi-global role has been played by a single state only twice since the beginning of the early modern period and the U.S. ascent: by Spain from 1492 to 1588 (from the conquest of Granada and the Atlantic crossing to the defeat of the Armada) and, on a much grander scale, by Britain from 1814 to 1914 (from Waterloo to the trenches of WW I). During the seventeenth and eighteenth century centuries there was no clear hegemon. Powerful Qing China (particularly under emperor Kangxi, 1661–1722) dominated in the East and pulled in bulk of the world's silver, but it did not engage the rest of the world either politically or militarily, while the West experienced a prolonged competition among waning Spain and expanding France and England.

But which power will fill the gradually expanding vacuum left by a strategically retiring, economically much weakened, technically less competent, more corrupt (looking more like *Estados Unidos Mexicanos*) United States, with its receding memories of superpower glory? Reviewing the likely national trends, I conclude that none of the possible candidates is likely to do that in ways even remotely similar to the U.S. post-1945 dominance. By 2050 the Muslim world of some 2 billion people will almost certainly wield more influence than today, but three major factors will prevent its emergence as a coherent, trendsetting actor on the global stage.

First, its ancient religious discord, the more than 1,300-year-old *sunnī–shī'ī* rift, is not going to disappear in a generation or two: that feud and its passions cannot be easily set aside.

Second, the diverse national interests composing the Muslim world have repeatedly overridden the characteristically flowery language of pan-Arabism, and they would be even harder to set aside to achieve a new Spain-to-Pakistan caliphate.

Third, Muslim countries have a multitude of internal troubles and economic challenges. In a few decades Iran's oil production will be in steep decline; Indonesia's output has already fallen by one-third between 1991 and 2005; and it is very likely that the output of Nigeria (partly Muslim) will also fall. The overall economic progress of the Muslim world will be very uneven, further accentuating the current divide between de facto modern, relatively affluent states (Kuwait, Qatar, UAE, Oman) and overcrowded, unruly, and relatively very poor countries (Pakistan, Bangladesh). Such realities are not at all conducive to setting up a cohesive faith-based caliphate (or even fashioning an EU-like economically based entity).

Only a wishful thinking can conjure the transmutation that would be required to remake aging Europe into a new consensual hegemon, and a resurgent (but still depopulating) Russia could not fill the multifaceted superpower niche. After all, even the much mightier USSR was never a true peer of the United States. Leaving its enormous natural resources and militarized economy aside, its industrial and agricultural mismanagement (the country could not feed itself and had to rely on imparted U.S. grain), rigid political system, and lack of basic freedoms did not allow it to rise to that level. China will become the world's largest economy (in absolute terms), but, as explained, its further rise will be checked by a internal limits and external complications.

This leaves India as the only remaining plausible contender. Barring major (and highly unlikely) shifts in fertility, its global population primacy is only a matter of time. The United Nations (2005) medium forecast has India surpassing China in population by 2030 (1.449 billion vs. 1.447 billion) and reaching a total of 1.59 billion people (vs. 1.39 billion for China) in 2050. This primacy and India's belated economic takeoff (following the relaxation of autarkic and nationalistic policies that the long-ruling Congress Party imposed for decades on India's businesses) have led commentators to extol India's long-term advantages. Several expect India to be right behind (and eventually ahead of) China in the global race to the top (Huang and Khanna 2003; Srinivasan and Suresh 2002; Rahman and Andreu 2006; Winters and Yusuf 2007). These are understandable but unrealistic expectations.

I have mustered a variety of reasons, ranging from economic to environmental to cultural, that militate against China's becoming the world's leading superpower. Analogous reasons weigh even more heavily against India's occupying that place. A long array of indicators quantifies India's relative weaknesses vis-à-vis China. Even after one discounts the official exaggerated claims, China is well ahead in terms of economic performance. In 2005 the country's PPP GDP reached nearly $5.5 trillion (almost 10% of the global level), or more than $4,000 per capita. By contrast, India's PPP GDP was about $2.3 trillion (roughly 40% behind Japan and about 4% of the global total), or about $2,100 per capita (World Bank 2007). And while

China's manufactures have captured more than 6% of the global market, India's have less than 1%.

China's rapid economic growth spurt put nearly 85% of the country's population above the poverty line, but in 2005 one-third of all people in India remained below it. And I have noted a 1 OM gap in foreign direct investment. India's one notable positive indicator is lower income inequality (Gini index below 35 compared to China's 45), but to a large extent this reflects, as was the case in Mao's China, only more widely shared poverty. By most other measures India is a deeply unequal country because of its legacy of social stratification and exclusions (a de facto caste system persists), and as its economic growth accelerates, this inequality is almost certain to grow. More important, China is well ahead of India in the quality of human capital.

For some key indicators the differences are not just twofold (infant mortality) or threefold (in China ~15% of children younger than five years are stunted; in India ~45%) but even fivefold and sixfold (fig. 3.23). Low birth weight is an important predictor of future health; in China its share is 6%, in India 30%. China's adult illiteracy is less than 7%, India's more than 35%. In 2005, China's share of people engaged in R&D was more than five times that of India (WHO 2006; UNDP 2006). And India's rate of immunization against childhood diseases has actually been falling; for measles, it is well below the rate in Bangladesh and on a par with such modernization laggards as Haiti or Chad (IMF 2006).

Similar disparities are true insofar as modern infrastructures are concerned (see fig. 3.23) (World Bank 2006; Long 2006). Marginal differences include such important measures as access to improved urban sanitation (~70% in China, ~60% in India; India is a bit ahead in assured rural access to clean water). China has built more expressways in a year than India has since 1947; China's per capita electricity-generating capacity is nearly three times that of India; and China's modern subways, airports, and container ports are superior (e.g., in 2005 China's container ports handled about 17 times the volume of India's) (IMF 2006).

India will find it hard to catch up with China in average per capita terms because its population still grows much faster (1.5% vs. 0.7% in early 2000s); its total fertility rate (3.0 vs. China's 1.7 in 2005) is projected to equal the Chinese level only after 2040. To be sure, India has advantages. On a 1–7 scale (1 best) India outranks China in political freedom (2 vs. 7) and civil liberties (3 vs. 6). (World Audit 2006). And India has eagerly embraced just about every facet of the new electronics economy, from call centers and accounts-processing facilities serving

Fig. 3.23
Quality-of-life and economic indicators show a large gap in performance between China and India at the beginning of the twenty-first century. Plotted from data in UNDP (2006), IMF (2006), and BP (2006).

numerous Western businesses (Bhagat 2005) to the world-renowned software designers of Bangalore who, at least in the eyes of Indian experts and an adoring public, are far better than Silicon Valley's.

At the same time, India's electoral process has been violent and corrupt, and in general India has few equals in the level of corruption and bribery permeating its personal, governmental, and business spheres. On Transparency International's (2006) corruption perception's index (0–10 scale, with 10 best) China scores 3.4 (71st of 146 countries), whereas India scores 2.8 (90th), and Indian companies have the highest propensity to offer bribes when operating abroad.

India's progress will necessarily be complicated by the country's heterogeneity; its diverse cultures and religions and its often extreme political groups (including some of the world's last Marxists, Communists, and Maoists) do not make for natural consensus politics.

There are potentially serious external factors that can slow down or derail India's progress. Any substantial change of the monsoonal flow caused by global warming (delayed onset, higher variability, more violent rains) would have enormous consequences, as would the disappearance of a large chunk of Bangladesh under repeated storm surges of an elevated Bay of Bengal and the migration of tens of millions of people into India. Long-term water supply problems and the contamination of streams are as acute in parts of India as they are in China. The future course and outcome of the 60-year-old (now nuclear-armed) conflict with Pakistan cannot be foreseen, nor can the state of India's Muslims in a world of radicalized Islam.

But even if everything goes rather well, the India of coming decades will be too preoccupied with its economic and social modernization, its quest to move from large-scale poverty and meager subsistence to a modicum of widely shared income and lifestyle security, to become a globally dominant superpower. By 2050, India will have the world's largest population. It might have the world's second-largest economy, but in per capita terms it will still most likely be no richer than Mexico is today, and its rich native cultural heritage of Hinduism and Sikhism will not become any more transferable abroad than it is today. That India will be a major global power in 2050 is obvious; that it will be a globally dominant superpower is wishful thinking.

Who is on top matters—be it as savior, hegemon, pacesetter, model, irresistible attractor, or brutal enforcer. The United States may have been one or the other of these to different nations at different times, but its retreat from such roles will not create a more stable world, particularly if there is no clearly dominant power or grand alliance. This conclusion is perhaps the easiest to defend: the demise of U.S. global dominance will not bring any multilateral balance of power. Conditions in the absence of a global leader in a world swept by the forces of globalization would resemble those following the retreat of Roman power that underpinned centuries of coherent civilization: chaotic, long-lasting fragmentation, inimical to economic progress, which would greatly exacerbate many of today's worrisome social and environmental trends.

About 2 billion people already live in countries that are in danger of collapse (fig. 3.24), and there are no convincing signs that the number of failing states will diminish in the future. A century ago a failure or chronic dysfunction of a small

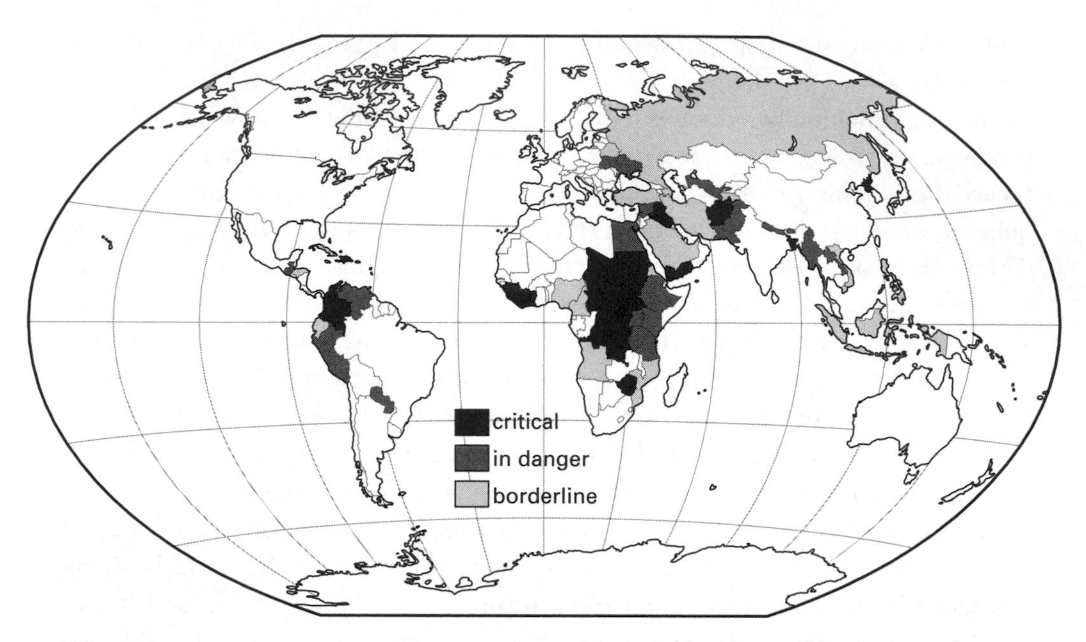

Fig. 3.24
Failed states. According to the Failed States Index (2005) of *Foreign Policy*/Fund for Peace, the Republic of South Africa and Djibouti are the only two stable countries in Africa, and Mongolia, South Korea, Japan, and Oman the only ones in Asia.

(particularly a landlocked) state would have been a relatively inconsequential matter in global terms. In today's interconnected world such developments command universal attention and prompt costly military and humanitarian intervention. Prominent recent examples include Afghanistan, Bosnia and Herzegovina, Congo, East Timor, Iraq, Liberia, Sierra Leone, Somalia, and Sudan. Were a number of such state failures to take place simultaneously in a world without a dominant power, who would step in to defuse at least the most threatening ones? As Ferguson (2004, 39) has warned, "Be careful what you wish for. The alternative to unipolarity would not be multipolarity at all. It would be apolarity—a global vacuum of power. And far more dangerous forces than rival great powers would benefit from such a not-so-new world disorder."

Many Western strategic planners, forced to make contingency plans for a massive launch of thermonuclear weapons, wished for the end of the Cold War and yearned for a world without superpower confrontation. They got their wish sooner than any think tank could have imagined, and now they look back wistfully at a world with an identifiable, rational enemy: the Soviet politburo had no wish to immolate the country by launching the first strike. Today's young European leftists may get their wish of a severely hobbled and introverted America even before they need glasses to read their copies of *The New Statesman, Il Manifesto*, or *Junge Welt*. But how much will they then enjoy the ambience of Eurabia with edicts defining permissible readings, clothes, and investments?

Much has been made of unfavorable sentiments toward the United States shown by public opinion surveys around the world. But do these sentiments indicate a U.S. retreat would be widely welcomed? The numbers, insofar as published surveys go, are real. Even in Australia the share of respondents who worried about radical Islam matched the share of people who perceived U.S. foreign policies as a potential threat. Commentators ascribe this trend to such mythical causes as an evolving national character (made up of values that "look increasingly ugly to many foreigners") that is a liability to foreign relations (Starobin 2006), or to such predictably partisan reasons as anti-Bush sentiments that have mutated into broader anti-Americanism (Kurlantzick 2005).

Realities are not that simple. As Applebaum (2005) points out, the world is not uniformly awash in anti-American sentiment; the United States continues to exert a powerful inspirational and aspirational pull for hundreds of millions of people. A great deal of anti-Americanism has always mingled outrage and envy (and among European intellectuals, disdain) with grudging admiration and carefully suppressed affection, a stance summed up by "Yankees go home—but take me with you!" As Mandelbaum (2006, 50) noted, the world complains about U.S. recklessness, arrogance, and insensitivity, but we should "not expect them to do anything about it. The world's guilty secret is that it enjoys the security and stability the United States provides."

Globalization and Inequality

In addition to this complex assortment of national factors, we must consider globalization, a supranational trend that makes a single nation's claiming an undisputed place on top even less likely. Globalization has profound personal implications for each of us because it affects our place along the ever-shifting continuum of individual and familial well-being. Perhaps its most personally relevant impact is the increasing frequency and range of income and social inequality.

Globalization was discovered by the media during the 1990s, and it has become one of the most prominent, loaded terms of the new century. The process is condemned by activists and enthusiastically greeted by free-marketers. What is new about it is its intensity and pace, but globalization has been under way since the beginning of the early modern era. Initially it was restricted to certain segments of the economy, and its benefits were enjoyed only by the richer strata in a small number of countries.

On a personal level the process manifested itself as an increasing accumulation of possessions made in different parts of the world. By the sixteenth century the homes of wealthy Dutch and English merchants commonly displayed paintings, maps, globes, rugs, tea services, musical instruments, and upholstered furniture of wide provenance (Mukherji 1983). By the eighteenth century such private collections had items from China, Japan, India, the Muslim countries, and the Americas. During the nineteenth century the single most important driver of large-scale commercial globalization was the accelerated expansion and maturation of the British Empire (Cain and Hopkins 2002).

At the end of the nineteenth century food grains (from the United States and Canada) and frozen meat (from the United States, Australia, and Argentina) joined textiles as major consumer items of new intercontinental trade, which was made possible by inexpensive steam-driven shipping. By the middle of the twentieth century the highly uneven distribution of crude oil resources elevated tanker shipments to the most global of all commercial exchanges. But only after 1950 did the trend embrace all economic activities. Inexpensive manufacturing was the first productive segment affected by it, beginning with the stitching of shirts and the stuffing of toy animals; later came the assembly of intricate electronic devices.

During the 1990s specialization and product concentration reached unprecedented levels as increasing shares of particular items originated in highly mechanized or fully automated facilities owned worldwide by just a few companies or located in just a few countries in East Asia. At the global level, two companies (Airbus, Boeing) now make all large jetliners; two others (Bombardier, Embraer) make all large commuter jets; and three companies (GE, Pratt and Whitney, Rolls-Royce) supply all their engines. Four chip makers (Intel, Advanced Micro Devices, NEC, Motorola) make about 95% of all microprocessors; three companies (Bridgestone, Goodyear, Michelin) sell 60% of all tires; two producers (Owens-Illinois, Saint Gobain) press two-thirds of the world's glass bottles; and four carmakers (Toyota, GM, Ford, and Volkswagen) assemble nearly 50% of the world's automobiles.

This concentration trend did not bypass highly specialized industries or services. Japan's Jamco makes virtually all lavatories and custom-built galleys and inserts in commercial jetliners; and no matter where a burning oil or gas well is located, one will call International Well Control of Houston or Safety Boss of Calgary. These examples make it clear that in the concentration of many productive sectors globalization has already gone about as far as possible. For many services a similar level of saturation is likely before 2020, and the process, although it may experience temporary setbacks, appears to be generally unstoppable. Its critics use this increasing concentration of economic activities as a worrisome example of dangerous usurpation of power by a handful of corporate entities, entirely missing the fact that the most long-term consequence of this trend is its impact on national aspirations for strategic dominance.

Perhaps the most helpful way to think about globalization would be to get rid of the term, and not just because that noun has become so emotionally charged. The term *interdependence* describes much more accurately the realities of modern economies. Once they left behind the limited autarkies of the preindustrial era, states have come to rely on more distant and more diverse sources of energy, raw materials, food, and manufactured products and on increasingly universal systems of communication and information processing. No country can now escape this imperative, and as this process advances, it will become impossible for any nation—no matter how technically adept or how militarily strong—to claim a commanding place on top.

Analogies with ecosystems are always useful when thinking about economic organization and modern states. The most complex ecosystems abound with many fierce forms of specialization, competition, and aggression, but they are fundamentally founded on enormous webs of symbiosis and cooperation. They have top (carnivorous) predators, many omnivores, and a very large number of herbivores, but there is no dominant species able to claim an excessively large share of resources, and all macroorganisms depend critically on environmental services provided by microorganisms, mostly bacteria and fungi. An outstanding example of this is the fact that African termites may consume annually as much biomass per unit of savanna as do elephants (Smil 2002).

Today's global economy still has its dominant top carnivore. In 2005 the United States, with less than 5% of the world's population, accounted for 20% of the world's economic output and claimed about 23% of the world's commercial primary energy. But its influence has been declining, and its relative importance will further diminish by 2050. As its complexities and interdependencies increase, the modern world thus begins to resemble a coral reef rather than a tundra, and there is actually no alternative to this shift, short of the system's collapse and a return to premodern existence with all that it implies for the quality of life.

There is enough evidence to conclude that in natural ecosystems greater complexity promotes system stability. Analogously, greater interdependence of national economies is (in the long run) a stabilizing factor. But true stability also requires that the benefits of globalization be reasonably well distributed, both internationally and within individual countries. Globalization has been good for complexity and interdependence, but has it been good for most of the people whose lives it affects? This is, of course, a loaded question because the quality of life is notoriously difficult to measure, and if it were possible to demonstrate clear, defensible, measurable benefits of the process, then all ideologically driven debates could become irrelevant.

A telling and intuitively impressive way to gauge this performance (and one that also happens to be favored by economists, so they cannot complain about a biased selection of "soft" variables) is to ask, Has globalization, after a few centuries of slow progress and two generations of rapid advances, helped to narrow huge income gaps that exist on a global scale, that is, has it had tangible global rewards? And, fortunately, there is a revealing comparison of national macroeconomic achievements that can be used to measure the extent of economic inequality: its decline would demonstrate greater benefits of globalization; its increase would indicate more limited benefits.

The simplest choice is to use national averages of GDP per capita (in terms of PPP) derived from standard national accounts. A better choice is to weigh the national averages by population totals. And the most revealing choice is to use average disposable incomes (from household surveys) and their distribution within a country in order to assign a level of income to every person in the world. The best available studies (fig. 3.25) show that unadjusted global inequality changed little between 1950 and 1975 but subsequently increased (Milanovic 2002). Population-weighted calculations show a significant convergence of incomes across countries since the late 1960s, but a closer look reveals that this desirable shift was mostly due to China's post-1980 gains. A weighted analysis for the world without China shows little change between 1950 and 2000. Similarly, comparisons of post-1970 studies that use household incomes and within-country distributions indicate only minor improvement, slight deterioration, or basically no change (Sala-i-Martin 2002).

Since 1945 inequality of per capita GDPs has been declining among the Western nations, and North America and Western Europe, the principal architects and ben-

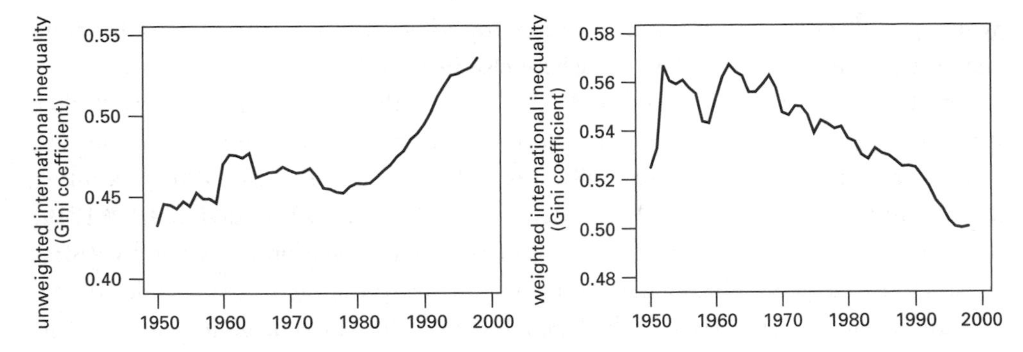

Fig. 3.25
International inequality, 1950–2000. *Left*, unweighted inequality of per capita GDPs shows considerable divergence; *right*, population-weighted assessment indicates gradual convergence, but that trend was due mostly to China's post-1980 progress. Based on Milanovic (2002).

eficiaries of globalization, have been pulling ahead of the rest of the world. The U.S. GDP per capita was 3.5 times the global mean in 1913, 4.5 times in 1950, and nearly 5 times in 2000 (Maddison 2001). Even China's spectacular post-1980 growth did not narrow the income gap with the Western world very much (but China pulled ahead of most of the low-income countries). The obverse of this trend has been the growing number of downwardly mobile countries. In 1960 there were 25 countries whose GDP per capita was less than one-third of the poorest Western nation; in 2000 there were nearly 80. Africa's post-1960 economic (and social) decline accounted for most of this negative shift (by the late 1990s more than 80% of that continent's countries were in the poorest category), but (China's and India's progress aside) the Asian situation also deteriorated. Income-based calculations confirm this trend.

Perhaps the most noticeable consequence of these inequalities is what Milanovic (2002) called the emptiness in the middle, the emergence of a world without a middle class. By 1998 fewer than 4% of world population lived in countries with PPP GDPs between $8,000 and $20,000, and 80% were below the threshold. Everything we know about social stability tells us that this is a most undesirable situation. Unfortunately, national statistics reveal that this trend is at work even in all the world's four largest economies. Income inequality has been on the rise not only in the United States, where riches have always been rather unequally distributed, but also in Japan, for decades a paragon of well-distributed riches, in China, where globalization of the country's economy lifted all boats but at the same time made them more unequal than anywhere else in East Asia, and in Russia.

The U.S. trend between 1966 and 2001 was analyzed in depth by Dew-Becker and Gordon (2005). During that period the median income rose by 11%, but the rise for people in the 90th percentile was 58%; for those in the 99th percentile, 121%; and for those in the 99.99th percentile (about 13,000 individuals), 617%. This trend severed the previously well-documented development whereby rising productivity helped to lift nearly all incomes. During the last three decades of the twentieth century, only 10% of the U.S. labor force had income gains that matched the growth of the country's productivity, whereas between 1997 and 2001 the top 1% of earners had a higher income gain than the bottom half. There is little doubt that this redistribution of income upward has many undesirable, even poisonous (Lardner and Smith 2005) consequences. The fraying of the middle class, the bastion of U.S. stability and future hopes, has been a very worrisome trend.

Soviet Russia never had a strong middle class but, contrary to the official goal of creating a classless society, had substantial disparities between the ordinary populace and the ruling elite, including top Communist party members, secret police, and military-industrial bureaucrats. Given the fundamentally criminal nature of post-Soviet privatization and an unprecedented concentration of fabulous riches by assorted "oligarchs," it is no surprise that Russia's income inequality rose substantially during the 1990s, with the highest quintile claiming about 47% of the national income by 2000 and only a negligibly lower share by 2005 (Goldman 2006).

Japan's income inequality has been on the rise since the early 1980s. According to the comprehensive survey of living conditions, done regularly by the Japanese Ministry of Health, the country's Gini coefficient (perfect equality of income 0, perfect inequality 1), which used to be on a par with that of the Nordic countries (around 0.25), rose to 0.37, well above the German or French level (Ota 2005). The rise has been most pronounced among young workers because a rising share of young adults do not have regular employment (even Toyota, the world's most profitable company, has increased its hiring of short-term contract workers). The actual Gini coefficient among these young cohorts is even higher than the official rate because the available wage data capture only a small part of nonregular earnings.

Before Deng Xiaoping's reform began to dismantle the egalitarian edifice of Maoism, China did not have a broad middle class. Although its rising economic tide brought impressive reductions of extreme poverty and improved average incomes in all provinces, the country (in many ways the greatest beneficiary of globalization) has seen perhaps the fastest increase of economic inequality in history. The Gini index of rural income inequality rose from 0.21 in 1978 (still basically a Maoist system) to 0.416 by 1995 and then declined slightly to 0.375 by 2002 (SSB 2002; Khan and Riskin 2005). The 2002 coefficient is from the latest available nationwide household survey by China's Academy of Social Sciences, which showed slight declines in income inequality in both rural and urban areas, with the urban Gini index falling marginally from 0.332 in 1995 to 0.318 in 2002.

But the overall nationwide Gini coefficient remained unchanged, at 0.45, because of a further increase in the income gap between villages and cities. The urban/rural income ratio rose by more than 20%, from 2.47 in 1995 to 3.01 in 2002 (Khan and Riskin 2005), and it increased further to 3.22 by 2005 (NBS 2006). Rural incomes only one-third of urban earnings represent an extremely high difference in international comparisons (ICFTU 2005). China's nationwide Gini coefficient is also considerably above that of other major economies in Asia. Recent values are about 0.32 in South Korea and India and about 0.34 in Indonesia. Moreover, China's real urban inequality is definitely much higher because the index excludes the massive transient urban labor force of rural migrants as well as all foreign employees, and as it also underestimates the highest incomes of China's newly rich. When these realities are taken into account, China's urban income inequality may be among the highest in the world.

Manifestations of this new reality are ubiquitous. On one hand, China's *nouveau arrive* are pushing the sales of luxury products by 50%–60% per year, so China's share of that market will rise from less than 2% in 2005 to 10% by 2010 (*China Daily* 2005). On the other hand, tens of millions of displaced peasants are exploited, often in demeaning conditions, to produce goods for the global supermarket. While China's bureaucrats, managers, and middlemen are buying Prada perfumes and Bulgari brooches and installing Italian marble floors in their oversize suburban villas, the urban poor, who now can see everything but can afford little, are left with disappointment and anger (Davis 2005). Excessive banqueting leads to enormous waste of food, but China still has tens of millions of malnourished people.

Perhaps no other indicator of China's growing inequality is as shocking as the disparity in medical care coverage. In 2003, 38.5% of people surveyed in large cities had no medical care coverage, but that share was 88.6% in those rural areas that are officially classified as places where the necessities of daily life have reached an adequate level (Zhao 2006). Consequently, it is hardly surprising that a ranking of the fairness of financial contributions to health systems makes China 188th of 191 countries, only marginally above Brazil, Myanmar, and Sierra Leone (WHO 2000). Surely these are not the solid foundations of a new superpower.

Intensifying globalization has had an equivocal effects on the overall stability of the global economic system. It is very difficult to decide if the positive (system-stabilizing) effects of increasing interdependence matter more than the negative (system-destabilizing) consequences of unchanged international inequality and increases in intranational inequality. But these two contradictory trends have similar long-term effects on the global distribution of power. The first weakens national dominance because it weaves individual economies tighter into the global web of interdependencies; the second weakens national dominance because it makes the greatest beneficiaries of globalization (United States, China, the EU) less stable in the long run.

But all of this jostling for a place on top may matter much less during the next 50 years than it did during the past half century. All human affairs unfold on the irreplaceable stage of the Earth's biosphere, whose considerable resilience, elaborate integrity, and amazing complexity are being seriously endangered by human actions. Many consequences of this relentless impact are amenable to inexpensive social and economic fixes, whereas others have technical solutions but require considerable investment. But for the first time since the emergence of our species, some of these changes result in undesirable transformations on a global scale, and their pace and intensity may determine humankind's fortunes much more decisively than any economic policies or strategic shifts.

4

Environmental Change

Mihi contuenti semper suasit rerum natura nihil incredibile existimare de ea.
(When I have observed nature, she has always induced me to deem no statement about her incredible.)

Gaius Plinius Secundus (Pliny the Elder), *Naturalis Historia*, XI.ii, 6

There is nothing new about large-scale impacts of human activities on the biosphere. Conversion of forests, grasslands, and wetlands to crop fields, and deforestation driven by the need for wood and charcoal to heat homes and smelt metals, and for lumber to construct cities and ships, changed natural ecosystems on a grand scale in preindustrial Europe and Asia. Even the pre-1492 American societies had a greater impact on their environment than previously surmised. An assumption has been that these changes transformed the environment only on a local or regional scale, deforestation in the Mediterranean countries or in North China being perhaps the best-known examples.

But Ruddiman (2005) argues that human beings began to influence global climate as soon as Neolithic populations adopted shifting agriculture. Biomass burning released CO_2, and later settled agricultures, through flood irrigation and animal husbandry, became a major source of methane, a gas whose atmospheric warming potential is 1 OM greater than that of CO_2. Despite their relatively small numbers, humans thus hijacked the entire global climate system, and they have remained in control ever since. A more conventional interpretation dates the onset of significant human interference in global climate to the latter half of nineteenth century, when rapidly rising combustion of fossil fuels and massive conversion of natural ecosystems to farmlands in the Americas, Asia, and Australia led to large emissions of CO_2. In any case, by the 1990s global warming driven by anthropogenic emissions of greenhouse gases had become the most prominent environmental concern because of its historically unprecedented nature and its likely impacts.

There are many other worrisome large-scale environmental changes. Excessive soil erosion is increasing even as most countries keep losing prime farmland to nonagricultural uses. The global water cycle will always be dominated by massive evaporation from the oceans, but human actions have altered its local, regional, and national flows to such an extent that water supplies are near or below adequate levels in many densely populated countries. The most uncertain impact of human activities on the ocean is the extent to which global warming will reduce the polar ice cover and the great ice sheets of Greenland and Antarctica.

The biogeochemical carbon cycle has received the most attention because of its role in global warming, but human activities have changed the biospheric cycle of nitrogen much more. Losses of biodiversity could be aggravated by global warming, but they would proceed even with a stable climate, as would the spread of invasive species.

Finally, I note an invisible but fatal environmental change. Pandemic-carrying viruses (see chapter 2) are not the only constantly mutating microorganisms. Bacteria, their more complex allies, are nearly as adept at defeating our controls, and they, too, add to a growing list of emerging infections (Morens, Folkers, and Fauci 2004). The most worrisome result of bacterial mutations is the emergence and diffusion of antibiotic-resistant strains of potentially lethal bacteria. In most hospitals there are now only one or two antibiotics that stand between us and an incurable attack of antibiotic-resistant microbes.

Global Warming and Its Consequences

The greenhouse gas effect is indispensable for life on the Earth; it is the weakness or excessive strength of the effect that is a matter of concern. The effective radiative (blackbody) temperature of a planet without an atmosphere is simply a function of its albedo (the share of incoming radiation that is directly reflected to space) and its orbital distance. The Earth (albedo 30%) would radiate at –18°C, compared to –57°C for Mars and –44°C for Venus, and all these planets would have permanently frozen surfaces. A planet ceases to be a perfect radiator as soon as it has an atmosphere some of whose gases—above all, water vapor, CO_2, CH_4, and nitrous oxide (N_2O)—can selectively absorb part of the outgoing infrared radiation and reradiate it both downward and upward. Such an atmosphere is highly (though not perfectly) transparent to incoming (shortwave) solar radiation, but it is a strong absorber of certain wavelengths of the outgoing (longwave) infrared radiation (IR) that is produced by the reradiation of absorbed sunlight. This absorption of IR is known as the greenhouse effect.

In the absence of water vapor on the Earth's two neighbors, it is the presence of CO_2 that generates their greenhouse effect, a very strong one on Venus (average surface temperature 477°C) and a very weak one on Mars (average −53°C). The Earth's actual average surface temperature of 15°C is 33°C higher than its black-body temperature, with the absorption by water vapor accounting for almost two-thirds of the difference (fig. 4.1). CO_2 accounts for nearly one-quarter of the

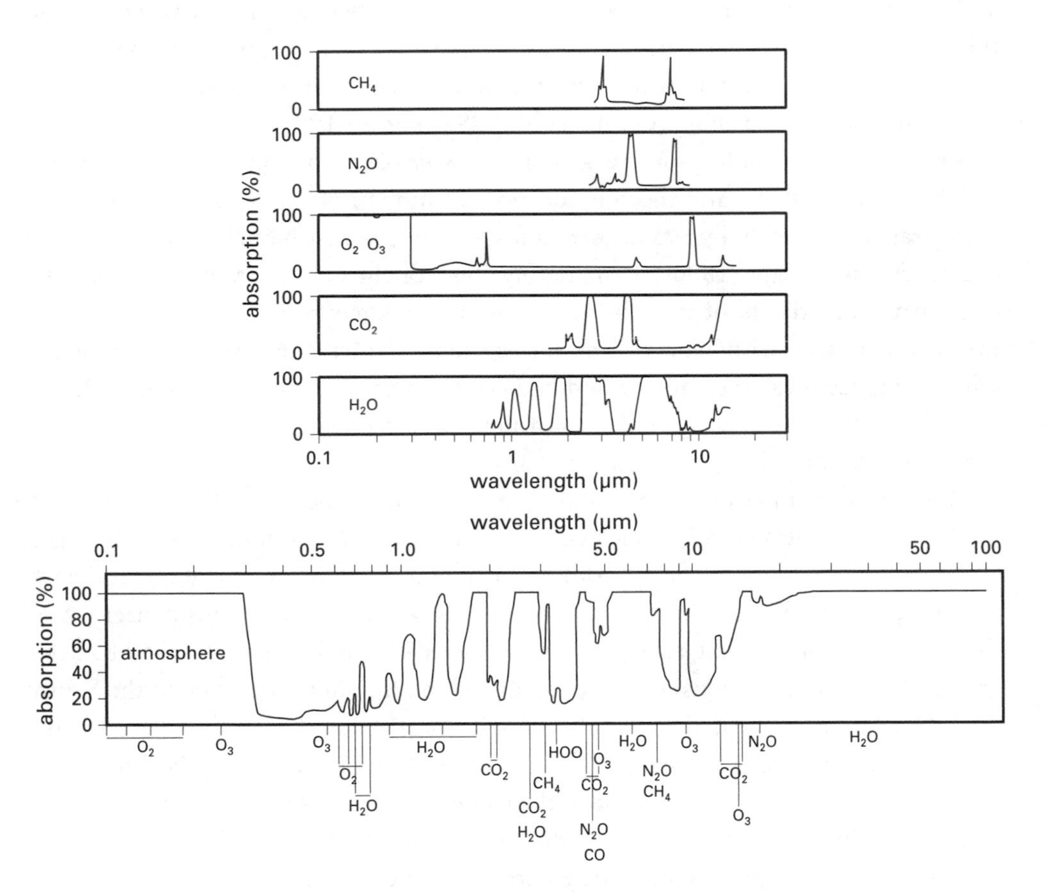

Fig. 4.1
Absorption bands of six major greenhouse gases and their aggregate effect on the ultraviolet, visible, and infrared wavelengths reaching the Earth's surface. From Smil (2002).

temperature difference; CH_4 and N_2O come next, with minor natural contributions from NH_3, NO_2, HNO_3, and SO_2 (Ramanathan 1998). A greenhouse effect amounting to at least 25°C–30°C had to operate on the Earth for nearly 4 billion years for life to evolve.

But water vapor, the Earth's most important greenhouse gas, could not have been responsible for maintaining relatively stable climate because its changing atmospheric concentrations amplify rather than counteract departures of the surface temperatures. Water evaporation declines as the atmosphere cools, and rises as the climate warms. In addition, changes in soil moisture do little to chemical weathering. Only long-term feedbacks between fluctuating CO_2 levels, surface temperature, and the weathering of silicate minerals explain a surprisingly limited variability of the mean tropospheric temperature. Lower tropospheric temperatures and decreased rates of silicate weathering will result in gradual accumulation of the emitted CO_2 and in subsequent warming (Gregor et al. 1988; Berner 1998).

Scientific understanding of the greenhouse gas phenomenon goes back to the 1820s (Fourier 1822), and the consequences of human interference in the process were grasped correctly by Svante Arrhenius, one of the first Nobel prize winners in chemistry, during the 1890s. Remarkably, his conclusions contained all the key qualitative ingredients of modern understanding: geometric increases of CO_2 will produce a nearly arithmetic rise in surface temperatures; the resulting warming will be smallest near the equator and highest in polar regions; the Southern Hemisphere will be less affected; and the warming will reduce temperature differences between night and day (Arrhenius 1896).

Modern interest in this fundamental biospheric process began with a classic paper by Revelle and Suess (1957), which characterized fossil fuel combustion as "a large-scale geophysical experiment of a kind that could not have happened in the past nor be reproduced in the future" (18). An important response to this concern was the setting up of the first two permanent stations for the measurement of background CO_2 concentrations, near the top of Mauna Loa in Hawaii and at the South Pole. Measurements from these baseline monitoring stations began showing a steady rise of atmospheric CO_2, but a global cooling trend continued during the 1960s and early 1970s. The possibility of rapid anthropogenic global warming became a matter of considerable public attention only during the late 1980s, and it is now undoubtedly the most prominent worldwide environmental concern.

This attention has not been able to eliminate many fundamental uncertainties. The phenomenon itself is complex (the relation between the atmospheric concentration of greenhouse gases and the mean tropospheric temperature is nonlinear, and it is subject to many interferences and feedbacks), and any appraisals of future impacts are immensely complicated by two key uncertainties: because the future rate of greenhouse gas emissions is a function of many economic, social, and political variables, we can do no better than posit an uncomfortably wide range of plausible outcomes. And because the biospheric and economic impacts of higher temperatures will be both counteracted and potentiated by numerous natural and anthropogenic feedbacks, we cannot reliably quantify either the extent or the intensity of likely consequences.

Rising Temperatures

Arrhenius (1896) predicted that once the atmospheric levels of CO_2 (at that time at about 290 ppm) doubled, to nearly 600 ppm, the average annual temperature increase would be 4.95°C in the tropics and just over 6°C in the Arctic. His findings applied to natural fluctuations of atmospheric CO_2. He concluded (correctly) that future anthropogenic carbon emissions would be largely absorbed by the ocean and (incorrectly; he grossly underestimated the rate of future fossil fuel combustion) that the accumulation would amount to only about 3 ppm in half a century. Just before World War II, Callendar (1938) calculated a 1.5°C rise with the doubling of pre-industrial CO_2 levels and documented a slight global warming trend of 0.25°C for the preceding half a century. The first computerized calculation of the radiation flux in the main infrared region of CO_2 absorption indicated an average surface temperature rise of 3.6°C with the doubled atmospheric CO_2 (Plass 1956).

Atmospheric CO_2 levels are now known for the past 650,000 years thanks to the ingenious analyses of air bubbles from ice cores retrieved in Antarctica and in Greenland. During that period CO_2 levels never dipped below 180 ppm and never rose above 300 (fig. 4.2) (Raynaud et al. 1993; Petit et al. 1999; Siegenthaler et al. 2005). More important, during the time between the rise of the first advanced civilizations (5,000–6,000 years ago) and the beginning of the fossil fuel era, atmospheric CO_2 levels fluctuated within an even narrower range of 250–290 ppm. Continuous measurements at Mauna Loa began with an annual mean of about 315 ppm in 1958, and by 2005 the atmospheric CO_2 level surpassed 380 ppm, nearly 20% above the 1958 level (see fig. 4.2) (Blasing and Smith 2006).

By contrast, worldwide instrumental temperature measurements (with large spatial gaps) are available only since the 1850s (more extensive coverage began only after WW II), and hence all pre-1850 reconstructions of global temperature means

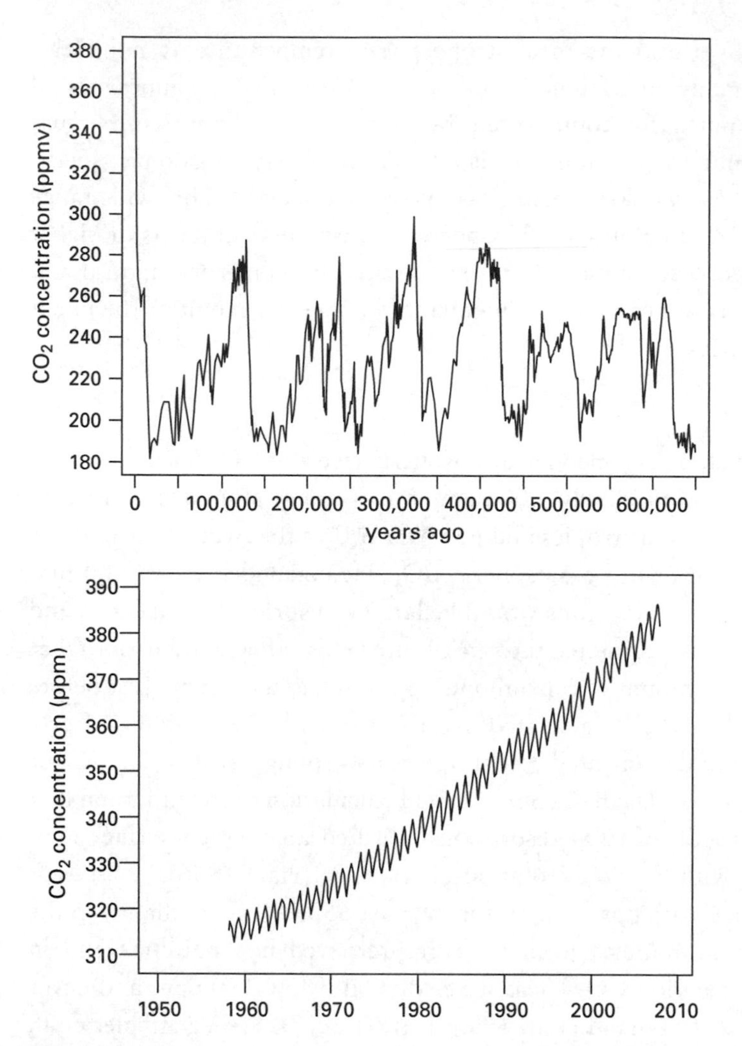

Fig. 4.2
Atmospheric CO_2 concentrations during the past 650,000 years (top), and measured at Hawaii's Mauna Loa volcano, 1958–2005 (bottom). From Siegenthaler (2005) and Blasing and Smith (2006).

have to rely on a variety of proxy variables such as tree rings, ice cores, bore hole temperatures, glacier lengths, and historical documents. The best available reconstructions indicate broadly similar anomalies during the past 1,000 years. Warming during the early Middle Ages (900–1100 C.E.) was followed by noticeable temperature declines culminating in a relatively cool period (Little Ice Age), 1500–1850. Subsequent instrumental records confirm a relatively rapid global warming; the gain was nearly 0.8°C (0.57°C–0.95°C) between 1850 and 2000 (IPCC 2007). The temperature increase has been accelerating. The updated 100-year trend for 1906–2005 is 0.74°C, and during the past 30 years the gain has averaged 0.2°C/decade (Hansen, Sato et al. 2006).

Consequently, it can be stated with a high degree of confidence that the mean temperatures during the closing decades of the twentieth century were higher than at any time during the preceding four centuries, and it is very likely that they were the highest in at least the past 13 centuries (NRC 2006b; IPCC 2007). However, the NRC (2006b) review of temperature reconstructions also concludes that little confidence can be placed in the conclusion by Mann et al. (1999) that the 1990s were the warmest decade and that 1998 was the warmest year in at least a millennium. The unreliability of all pre-1600 temperature reconstructions does not allow such a claim. This conclusion is supported by the best available reconstruction of temperature variations from China's rich proxy indicators, which shows that the closing decades of the twentieth century were no warmer than the warmest decades of the medieval warm period during the eleventh century (B. Yang et al. 2002).

Looking ahead, recent emission scenarios (assuming various rates of fossil fuel combustion) result in a wide range of CO_2 concentrations, 540–970 ppm by the year 2100. These would not be the highest-ever concentrations of CO_2. Boron-isotope ratios of planktonic foraminifer shells point to levels above 2000 ppm 60–50 Ma ago (with peaks above 4000 ppm), followed by an erratic decline to less than 1000 ppm by 40 Ma ago. But since the early Miocene era of 24 Ma ago atmospheric levels of CO_2 have been relatively stable and low, remaining below 500 ppm (Pearson and Palmer 2000). Consequently, continued large-scale combustion of fossil fuels could increase atmospheric CO_2 to levels unseen since large herds of horses and camels grazed on grassy plains of America.

This preoccupation with CO_2 misses nearly half of the problem. In 2005 slightly over half (1.66 W/m^2) of the post-1880 anthropogenic radiative forcing, averaging globally about 3 W/m^2, has been due to CO_2, with CH_4, chlorofluorocarbons (CFCs), O_3, and N_2O (all less common but more potent greenhouse gases) contributing the

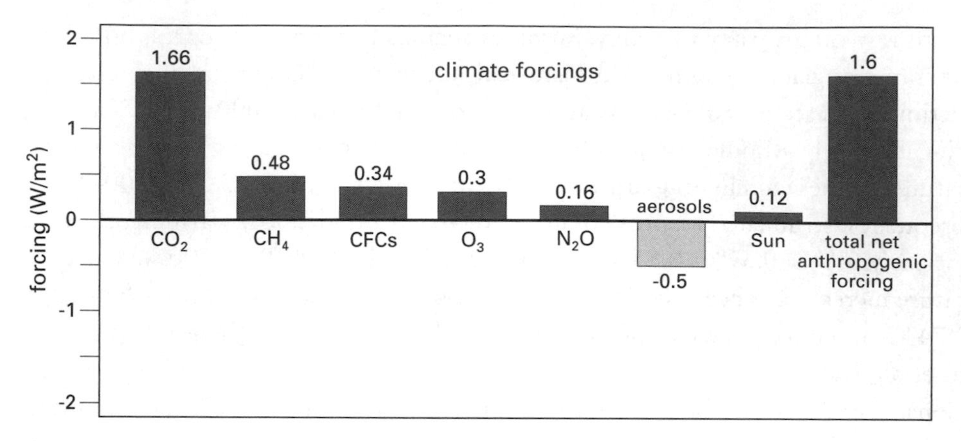

Fig. 4.3
Cumulative radiation forcings by greenhouse gases and other anthropogenic and natural factors. Based on Hansen, Sato et al. (2000).

rest (fig. 4.3) (Hansen, Nazarenko et al. 2005; IPCC 2007). Leaving CFCs aside (their production was outlawed by the 1967 Montreal Protocol), the other greenhouse gases accounted for at least one-third of all positive forcing, so we cannot talk only about CO_2 doubling but must express the overall warming potential in CO_2 equivalents.

The anthropogenic forcing process is also heavily influenced by nongaseous drivers of climate change as well as by intricate feedbacks. During the past 150 or so years the opposite effects of land use changes (mainly deforestation, which increases albedo) and reduced snow cover (which reduces it) have nearly canceled each other. But higher solar irradiance added 0.12 (0.06–0.30) W/m^2, black carbon from combustion contributed roughly twice as much, and reflective tropospheric aerosols lowered the overall forcing (directly and by changing cloud albedo) by about –1.2 W/m^2. Since 1750 the net anthropogenic radiative forcing has thus amounted to about 1.6 W/m^2, and it raised average global temperature by about 0.8°C compared to the preindustrial level (see fig. 4.3).

In order to forecast the additional warming that might take place by the year 2050 we must rely on a set of highly uncertain assumptions. We do not know the positive forcing, the future rates of fossil fuel combustion, land use changes, fertilizer use, and meat production. They will depend on the continuing increases of energy use, the extent of discoveries of new hydrocarbon deposits, the rates of penetration of nonfossil energy conversions, national land use policies, disposable incomes, and the overall vitality of the global economy. A multitude of possible outcomes based on these variables opens an unhelpfully wide range of possibilities.

Whatever the outcome, it must first be adjusted for the effects of future negative forcing caused by anthropogenic aerosols. The latest satellite measurements put them at $-1.9\ W/m^2$ in 2005 (Bellouin et al. 2005). This rate, higher than assumed by most of the climate change models, implies future warming greater than predicted as worldwide particulate emissions continue to decline (as expected). A complete elimination of today's aerosol effect would boost the positive forcing by about 60%, and hence even partial controls will have a major effect. Thus we are reduced largely to speculation when thinking about the net effects of numerous feedbacks unleashed by rising temperatures.

To what extent will higher temperatures (accompanied by an intensified water cycle) boost global photosynthesis, and to what extent will this additional production be sequestered in long-lived carbon compounds (tree trunks, decay-resistant soil organic matter)? How much additional carbon will be released from soils whose organic matter will be subject to faster rates of decomposition in a warmer world? Data from two National Soil Inventories (1978 and 2003) in England and Wales have already found that annual carbon losses (relative to the existing soil carbon) average 0.6% regardless of land uses, with the maximum rate of more than 2% in carbon-rich soils (>100 g C/kg) (Bellamy et al. 2005). Perhaps even more critically (given the fact that CH_4 is a much more powerful greenhouse gas than CO_2), how much methane will be released from warmer sub-Arctic and Arctic wetlands and lakes? In one Siberian region these rates may already be five times higher than previously estimated (K. M. Walter et al. 2006).

We cannot do better than offer the most plausible ranges of outcomes based on the best available global climate models and increasingly complex simulations that incorporate interactions between the atmosphere, hydrosphere, and biosphere. Their results are very similar to those simple calculations offered by Arrhenius during the 1890s, by Callendar in the late 1930s, and by Plass in the mid-1950s. The first report by the Intergovernmental Panel on Climatic Change put the increase of average surface temperature at 1°C–5°C by 2100 (IPCC 1990), the second narrowed the range to 1°C–3.5°C (IPCC 1995), the third widened it to 1.4°C–4.8°C (IPCC 2001), and the latest narrows it to 2°C–4.5°C with a best estimate of about 3°C, temperature increases less than 1.5°C very unlikely, and values substantially above 4.5°C impossible to exclude (IPCC 2007).

Some studies conclude that climate sensitivity, defined as the response to a doubling of CO_2 (or equivalent) may be substantially higher than the IPCC's upper bound, perhaps even higher than 11°C (Stainforth et al. 2005). Considering that these are global means, the differences between the ecosystemic, health, and economic impacts of 1.5°C and 10°C warming would be enormous. In the first case, warming of ~0.15°C/decade would actually be a bit lower than the mean since the 1970s (~0.2°C/decade), and most societies could cope well with such a relatively moderate change through gradual adaptation. In the second case, the mean decadal warming would surpass the warming experienced during the entire twentieth century (0.74°C), and most societies would find it impossible to adapt to changes precipitated by such a rapid and pronounced temperature rise.

Recent studies relying on several types of evidence (better reconstructions of past mean annual temperature, density of tree rings, temperature changes following volcanic eruptions) were able to constrain the most likely range of climate sensitivity. Hegerl et al. (2006) reduced the 5%–95% range of climate sensitivity to 1.5°C–6.2°C and the 20%–80% range to about ~1.6°C–3.8°C with the median value of 2.6°C. Annan and Hargreaves (2006) suggested an even tighter range, 1.5°C–4.5°C for the 5%–95% range and found it impossible to assign a significant probability to any warming exceeding 6°C. If these findings hold, the prospects appear much more clearly defined. There would be a very low probability that twenty-first-century warming would be comparable to the twentieth-century temperature increase (<1.5°C), but highly worrisome increases in excess of 5°C would be almost as unlikely, and the range of 2.5°C–3°C would be most probable (fig. 4.4).

The qualitative consequences are clear. This climate change would cool the stratosphere and raise the tropospheric temperatures, with the warming more pronounced on land (and at night), and increases about two to three times the global mean in higher latitudes in winter than in the tropics, and greater in the Arctic than in the Antarctic. But even a perfect knowledge of climate sensitivity would not make it possible to quantify many potential effects of global warming because it could not eliminate a multitude of uncertainties concerning the eventual climatic, ecosystemic, health, and economic impacts of the change. There are many other variables and complex feedbacks that will codetermine the eventual consequences of global warming. Consequently, even our most complex models are only elaborate speculations. We may get some particulars right, but it is beyond us to have any realistic appreciation of what a world with average temperature 2.5°C warmer than today's would be like.

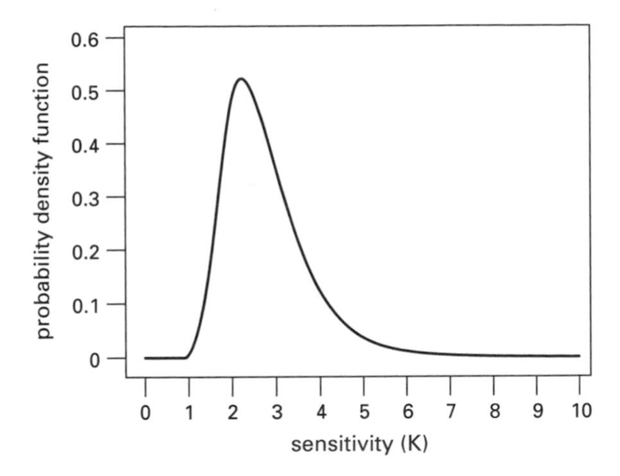

Fig. 4.4
Estimated probability density functions for equilibrium climate sensitivity to CO_2 doubling based on a range of paleoreconstructions to 1850 and on instrumental data for the twentieth century. Based on Hegerl et al. (2006).

If one were to assess the future of global warming based on recent headlines—"Be worried, Be very worried," "Earth at the tipping point," "Climate nears point of no return"—the outlook would be dismal indeed. Unfortunately, many scientists have gone along with this sensationalizing wave and have not stressed enough the limitations of our knowledge. Some researchers have gone even farther by presenting extreme possibilities as unavoidable futures. I do not wish to add to this dubious genre and thus present only the best available evidence and point out important uncertainties regarding the major impacts of future global warming.

Ocean's Rise, Dynamics and Composition

Ocean covers roughly 70% of the Earth's surface. It is the key regulator of global climate and the dominant source of the planet's precipitation. It also provides a convenient medium for inexpensive, long-distance transportation and supplies nearly 10% of the world's food protein. An increasing share of humanity lives along or near its shores. Clearly, any significant changes to ocean's properties and to its mean level are bound to have a major impact on the fortunes of modern civilization. A warmer atmosphere will inevitably produce a warmer ocean, and this transfer is already measurable.

Barnett et al. (2005) found that since 1960 about 84% of the total human-induced heating of the Earth (oceans, atmosphere, cryosphere, uppermost crust) has gone into warming the oceans. The strength of this warming signal varies by ocean and depth. North and South Atlantic warming, by as much as 0.3°C, reaches as deep as 700 m, whereas Pacific and Indian Ocean warming is mostly limited to the top 100 m. This warming may strengthen the ocean's natural thermal stratification and hence weaken the global overturning circulation, but its most immediate consequence is the thermal expansion of water, the most important contributor to mean sea level rise. A higher sea level would not only encroach directly on existing infrastructures but also accelerate the rates of coastal erosion, increase the damage due to storm surges, and contaminate coastal aquifers with salt water.

The mean sea level has been rising ever since the last maximum glaciation 21,000 years ago, when it was 130 m lower than today (Alley et al. 2005). Tide gauge measurements indicate a total twentieth-century rise of 17 (12–22) cm, an increase in the average annual rate compared to the nineteenth-century. The rate was about 1.8 mm/year between 1961 and 2003, and it accelerated to 3.1 mm/year between 1993 and 2003 (IPCC 2007). Thermal expansion of the ocean currently accounts for a rise of at least 0.5 mm/year, and ocean mass change contributes 1.4 ± 0.5 mm/year (Lombard et al. 2005). IPCC models have a range of 18–59 cm of additional sea level rise during the twenty-first-century, with most of increase coming from thermal expansion and the rest from the melting of mountain glaciers and ice sheet. Narrowing the estimates of future sea level rise is extremely difficult. The entire process exemplifies the many uncertainties facing assessments of environmental change produced by global warming.

Sea level rise during the coming generations will be determined by the uncertain rate at which polar ice sheets retreat or parts of them actually grow. Among a few things about the consequences of global warming that we know with absolute certainty is the fact that higher tropospheric temperatures will increase evaporation and hence intensify the global water cycle and result in an overall (but unevenly distributed) increase in precipitation. Polar regions should be major beneficiaries of this trend because they will receive more snow.

Some recent observations and simulations show that this is exactly what is taking place, whereas others obtain exactly opposite results. While Greenland's ice sheet margins have been thinning and its coastal glaciers receding, altimeter data for the decade 1992–2003 confirm an average ice increase of 6.4 cm/year in Greenland's vast interior areas above 1500 m (fig. 4.5) (Johannessen et al. 2005). Satellite altim-

Fig. 4.5
Satellite image of southern Greenland, July 7, 2002, shows thick ice in the interior and considerable ice loss along the island's margins, especially along the western fjords. From <http://rapidfire.sci.gsfc.nasa.gov/gallery/>.

etry indicates that between 1992 and 2003 increased precipitation added 45 billion t/year to the East Antarctic ice sheet interior, enough to slow sea level rise (currently now ~1.8 mm/year) by 0.12 mm/year (Davis et al. 2005).

By contrast, Doran et al. (2002) found that between 1966 and 2000 the Antarctic continent experienced net cooling of 0.7°C/decade, and Monaghan et al. (2006) concluded (based on ice core observations and simulations) that there has been no significant change in Antarctic snowfall during the past 50 years, which means that there is no mitigation of global sea level rise. But these findings are hard to reconcile with a comprehensive evaluation by Zwally et al. (2005), which shows that between 1992 and 2002 the thinning of Greenland's marginal ice sheet (–42 Gt/year) and the growth of inland ice (+53 Gt/year) resulted in an overall mass gain of about 16 Gt/year and that western Antarctica was losing mass at –47 Gt/year while eastern Antarctica was gaining 16 Gt/year, for a net loss of 31 Gt/year.

Velicogna and Wahr (2006b) used scaled and adjusted satellite gravity surveys to show the Antarctic ice sheet losing 138 ± 72 Gt/year, an annual equivalent of 0.4 ± 0.2 mm of global sea level rise. In another study, Velicogna and Wahr (2006a) used satellite gravity surveys to conclude that Greenland is losing ice mass at a rapidly increasing rate (up to 250% when comparing the 2002–2004 period with 2004–2006) and that the 2006 rate of 224 ± 33 Gt/year is equivalent to a global annual sea level rise of 0.5 mm. Even assuming that the complex corrections needed to convert the gravity data (whose spatial resolution is no better than a few hundred kilometers) into mass equivalents do not introduce major errors, these results, showing much higher losses than comparable surveys, cannot be interpreted as a definite trend. Many more years of unassailable observations are needed for that.

And to make matters even more confusing, Zwally et al. (2005) also found that while the western Antarctic shelf was losing 95 Gt/year, the eastern Antarctic shelf was gaining 142 Gt/year, for an annual net gain of 32 Gt. This mass is virtually identical to the annual net loss from the continent's ice sheets and hence an overall neutral balance for Antarctica's ice sheets and shelves. Clearly, some of these contradictory conclusions must be wrong. All of them require careful interpretations and adjustments (a tricky example is correcting the radar altimetry data for changes in temperature-driven firn compaction), and all of them have large margins of error.

According to Zwally et al. (2005), Antarctic ice sheets may be shedding as little as 19 Gt/year and as much as 43 Gt/year, and they put the net gain for the Greenland ice sheet at 11 Gt/year. Box, Bromwich, and Bai (2004) have the Greenland ice sheet (for the nearly identical period 1991–2000) losing 70 Gt/year. Luthcke et al. (2006) found an overall 113 Gt/year ice loss for Greenland. In addition, these processes have large interannual to decadal variability, and spans of 10 or 11 years are insufficient to confirm any significant long-term trends. simulations by Box, Bromwich, and Bai (2004) put the interannual variability for Greenland's entire ice sheet at ±168 Gt, a rate much larger than the best estimates of recent net flows.

I review these complex, uncertain, and contradictory results in such detail in order to make a strong point: in this case (and in many other similar instances) it is irresponsible to draw any definite long-term conclusions from these conflicting findings. Predicting the future of ice sheets is inherently difficult, and the uncertainty about their state may persist for some time (Vaughan and Arthern 2007). Not surprisingly, the IPCC offers a rather broad range for the mean sea level increase during the twenty-first-century; its most likely estimate (+50 cm) is bracketed by a range that is 80% as large as the mean (±40 cm). About 10 cm of this rise would come from the melting of mountain glaciers and a limited retreat of ice caps. However, Raper and Braithwaite (2006) reexamined the evidence and concluded that mountain glaciers and ice caps might contribute only about 5 cm, or about half the previously projected rate. Given these manifold uncertainties, a cautious conclusion would be for a mean sea level rise of ~15 cm (maximum 20 cm) by 2050, clearly a non-catastrophic change.

And here is just one more detail on the sea level rise. The Maldives, a group of small islands comprising some 20 atolls in the central Indian Ocean, have been used as a prominent example of the imminent danger of submergence. Because their elevation is only 1–2 m above sea level it was expected that they would disappear in 50, or at most 100, years. However, a detailed on-site examination found no sea level rise, and the conclusion was, "The people of the Maldives are not condemned to become flooded in the near future" (Mörner 2004, 149). Just the opposite is true because the Maldives have experienced a sea level fall since the early 1970s, most likely due to increased evaporation and intensified monsoon flow across the central Indian Ocean. But White and Hunter (2006) found no evidence of the sea level fall at the Maldives and confirmed that at 2 mm/year the mean sea level rise for tropical Pacific and Indian Ocean islands is close to the best estimate for the global mean.

By far the ocean's most important compositional change is increasing acidity, caused by rising emissions of CO_2 and falling carbonate ion concentrations (Royal Society 2005). Since the onset of industrialization the oceans have absorbed about half of the CO_2 emitted from fossil fuel combustion, and the dissolution of this gas and formation of carbonic acid lowered their average pH (8.2 ± 0.3 due to local and seasonal variations) by about 0.1 unit (Caldeira and Wickett 2003). As the surface ocean absorbs additional CO_2, its pH will keep falling, and it may decline by as much as 0.4 units by 2100 (Orr et al. 2006). Acidity is measured on a logarithmic scale; the decline so far equals a 30% rise in the concentration of hydrogen ions, and the eventual maximum increase would correspond to a 150% increase and a simultaneous decline of carbonate ion concentrations.

Carbonate ion concentrations have already dropped by about 10% compared to preindustrial levels, and their continuing decline would eventually undersaturate the surface water, first with respect to aragonite, later with respect to calcite in some cold seas. These shifts would slow down calcification, the biogenic formation of $CaCO_3$, and they would have serious consequences for biomineralizers ranging from low-latitude corals (which form reefs out of aragonite) to high-latitude pteropods (phytoplankton that forms its shells out of calcite) as well as for such ubiquitous coccolithophorids as bloom-forming *Emiliania huxleyi* (fig. 4.6) and *Gephyrocapsa oceanica*, which dominate the transfer of $CaCO_3$ to the sediments. A negative impact on coral growth would have cascading effects on one of the Earth's richest ecosystems, and reduced productivity of phytoplankton would affect the ocean's entire trophic chain. But because calcification releases CO_2 to the surrounding water, and hence to the atmosphere, its reduced rate may act as a negative feedback in the future high-CO_2 world (Riebesell et al. 2000; Gattuso and Buddemeier 2000).

Warmer oceans will also affect the dynamics of the atmosphere. Climate models indicate that weaker surface winds have already changed the thermal structure and circulation of the tropical Pacific Ocean (Vecchi et al. 2006). Perhaps most important, the twentieth-century warming has increased the west to east temperature gradient in the Pacific Ocean, a change that might increase the probability of strong El Niños (Hansen, Sato et al. 2006). These periodic westward expansions of warm surface waters begin off the coast of South America as the westward trade winds relax, and by early winter they often extend all along the equator to join warm water off Australasia. Their consequences include high rainfall and destructive flooding, not only in coastal Peru but also in the southern United States, while the western United States and Australia experience prolonged droughts. More intensive and more common El Niños could have a serious economic impact on water availability, forest and shrub fires, and crop yields in all regions affected by these climatic teleconnections.

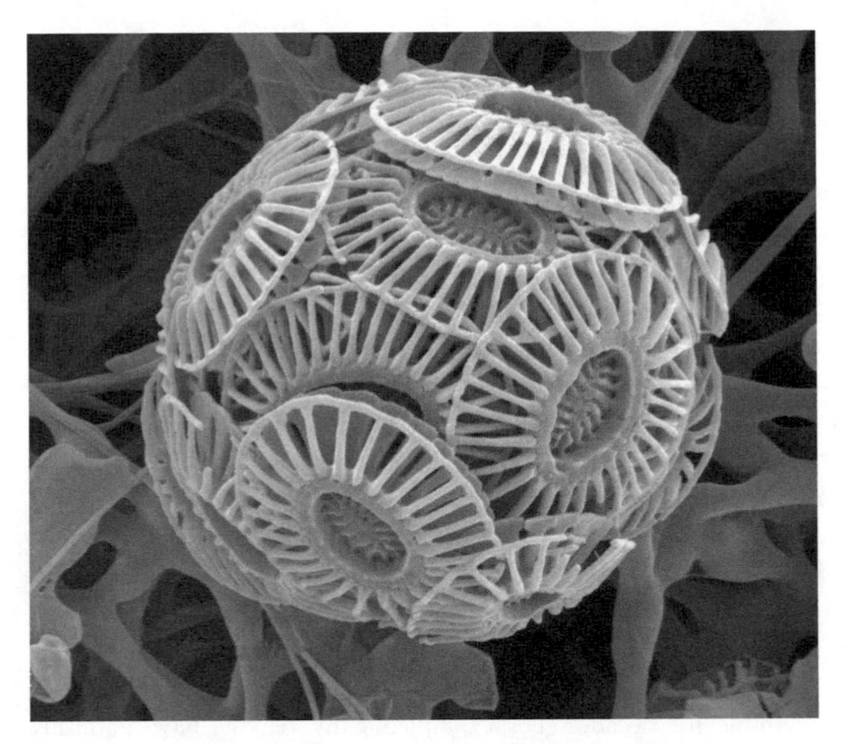

Fig. 4.6
Emiliania huxleyi, a ubiquitous marine coccolithophore, would have increasing difficulty forming protective armor in an acidifying ocean. Photo from the National Oceanography Center, Southampton.

Ecosystems and Economies

Temperature is a key determinant of the extent of life on the Earth, and of its intensity, because temperature highly correlates with the metabolic rates of decomposers, photosynthetic rates and zonal ranges of plants, and stocks, growth rates, habitat ranges, and migration patterns of heterotrophs ranging from zooplankton to top predators as well as with the diffusion rates of pathogens and disease resistance of hosts (Stenseth et al. 2002; Harvell et al. 2002). Biota will react to rising temperature in so many ways that a mere listing of plausible changes would fill many pages. Many direct, first-order ecosystemic responses to recent climate change are already obvious (Walther et al. 2002). Global meta-analyses of nearly 1,500 (Root et al. 2003) and 1,700 (Parmesan and Yohe 2003) responses of wild animal and plant species to warmer temperatures demonstrate that a significant impact of the process is already discernible among species ranging from marine zooplankton to woody plants and birds, with poleward range shifts averaging 6.1 km per decade and advancement of spring events surpassing two days per decade.

I give just three examples of these responses, one clearly positive, one neutral, and one clearly undesirable. The warming of the Eurasian land mass has intensified summer monsoon winds, and these in turn have enhanced upwelling and boosted the average summertime phytoplankton mass in the Arabian Sea by over 300% offshore and more than 350% in coastal regions. Warming has made the Arabian Sea more productive (Goes et al. 2005). Warming has a predictable effect on plant flowering, growth, and maturation. In Britain the first flowering dates of plants (very sensitive to the mean temperature of the previous month) advanced by an average of 4.5 days for 385 species during the 1990s (Fitter and Fitter 2002).

One of the most undesirable consequences of earlier melting of mountain snowpacks, of warmer and drier springs, rain-deficient summers, and reduced soil moisture will be a higher frequency and longer duration of wildfires. Westerling et al. (2006) found such a trend for the western United States. By 2003 (compared to the 1970s) the length of the active wildfire season increased by 78 days, and the average burn duration quintupled from 7.5 to 37.1 days. But many second-order changes may not be discernible for decades. For example, many regions have naturally variable precipitation, not just on annual but on decadal scales, and hence it is not easy to discern a warming-induced trend among the natural fluctuations.

And not all suspected effects can be tested experimentally. For example, a common assumption that the effects resulting from exposures of experimental ecosystems to a large single-step rise in CO_2 can mimic well the responses to a gradual increase unfolding over many decades may often be wrong because some biota are clearly sensitive to abrupt shifts but can cope quite well with incremental changes (Klironomos et al. 2005). The inherent complexities of these effects are even more difficult to unravel than the dynamics of future ice sheet melting. I illustrate these complexities by focusing on two key determinants of carbon flow in the biosphere: Is global vegetation a net sink or source of carbon? and Will future warming turn soils into a major carbon source? In both cases, the answers make a critical difference: Will soils and biota help to counteract future global warming, or will they intensify it?

Because we do not have an accurate carbon budget for the biosphere and are not certain about the long-term consequences of CO_2 enrichment for plants and soils, we cannot make any reliable appraisals of future interactions between the carbon cycle and other biogeochemical and climatic processes (Falkowski et al. 2000; Smil 2002). We know with a high degree of accuracy (because the emissions are known with an error smaller than 10% and atmospheric concentrations are measured fairly precisely) that during the 1990s roughly 6.3 ± 0.4 Gt C/year were emitted from the combustion of fossil fuels and 3.2 Gt ± 0.2 Gt C were retained in the atmosphere. Absorption by the ocean cannot be quantified as accurately (–2.4 ± 0.7 Gt C), and emissions from land use changes (2.2 ± 0.8 Gt C, mainly deforestation) were even more uncertain (Houghton 2003).

The two sources and the three sinks leave a residual of 2.9 ± 1.1 Gt C/year. Carbon transport by rivers may account for 0.6 Gt C/year, and the only plausible sink for the remainder (equal to one-third of all carbon emissions from fossil fuel combustion) is the Earth's vegetation. However, numerous studies of national and continental plant carbon balances (usually using satellite observations as input into several models of net primary productivity) offer annual fluxes that not only differ by as much as 1 OM but that cannot be interpreted often to conclude whether a particular ecosystem is a significant sink or a major source of carbon. On the global level the calculated net terrestrial fluxes are substantially smaller than the just noted residual of at least 2 Gt C/year, typically about 0.7–1.5 Gt C/year. Moreover, all of these fluxes have a significant temporal variability.

North American vegetation appears to be a consistent carbon sink (~0.2–0.3 Gt C/year), but the highest estimate of carbon sink for the continent (mostly south of 51°) has been as much as 1.7 Gt C/year (Fan et al. 1998). Carbon storage in Eurasian forests, about 0.3–0.6 Gt C/year (Potter et al. 2005), has been generally increasing, thanks to forest expansion and better management in Europe (fig. 4.7) (Nabuurs et al. 2003), net uptake of the Russian boreal forests (Beer et al. 2006), extensive replanting in China (Pan et al. 2004), and significant phytomass accumulation in Japan (Fang et al. 2005). But it is not certain whether Amazon forests are a large net source of carbon (–3 Gt C/year) or a major sink, sequestering annually perhaps as much as 1.7 Gt C (Ometto et al. 2005). The latter conclusion received new support from the studies of vertical profiles of CO_2, which indicate that northern lands take up carbon at rates lower than previously thought, while the undisturbed tropical ecosystem appears to be major carbon sinks (Stephens et al. 2007).

There is no consensus about net global values. Measurements of atmospheric O_2 concluded that the terrestrial biosphere and the oceans sequestered annually 1.4 (±0.8) and 2 (±0.6) Gt C between mid-1991 and mid-1997 and that this rapid storage of the element contrasts with the 1980s, when the land biota were basically

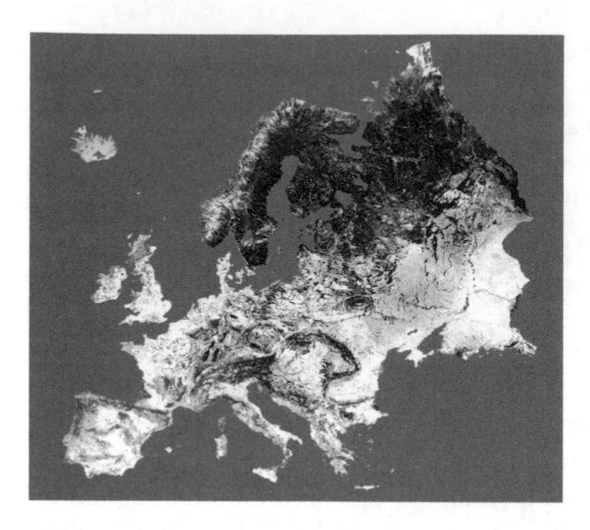

Fig. 4.7
Satellite image of European forests. The darkest shade indicates 88% or more forest, the lightest less than 10%. From VTT Technical Research Centre of Finland, <http://www.vtt.fi/services/cluster1/topic1_5/euro_sat.jsp?lang=en>.

neutral (Battle et al. 2000). Additional photosynthetic sequestration of carbon due to CO_2 fertilizing effect has been estimated at 1.2–2.6 Gt C/year (Wigley and Schimel 2000).

But Potter et al. (2005) concluded that between 1982 and 1998 the carbon flux for the terrestrial biosphere ranged widely between being an annual source of –0.9 Gt C/year to being a large sink of 2.1 Gt C. Houghton (2003) put the net terrestrial flux at –0.7 Gt C during the 1990s. The latest regional flux estimates for the 1990s credit the continents with the net uptake of between 1.1–2.4 Gt C/year and the oceans with being a nearly identical sink of 1.2–2.1 Gt C/year (Baker 2007). Moreover, global terrestrial carbon fluxes appear to be about twice as variable as ocean fluxes (precipitation and surface solar irradiance being the key drivers), and at different times they can be dominated by either tropical or mid- and high-latitude ecosystems (Bousquet et al. 2000).

Several important considerations will influence the future net trend. The carbon sequestration potential of croplands has most likely been overestimated (Smith et al. 2005), whereas the role of old forests, generally thought to be insignificant sinks of carbon, has almost certainly been underestimated (Carey et al. 2001). Because the availability of nitrogen is a key factor for carbon sequestration in many forests and forest soils (De Vries et al. 2006), it is not surprising that in the absence of further specific land management the current European net carbon intake is expected to decline soon (Janssens et al. 2005).

Net fluxes in the Amazon and Congo basins will be driven primarily by future rates of deforestation. The capacity of many forests to act as large sinks of carbon is limited by water and nutrient constraints on photosynthesis, and some ecosystems may experience considerable carbon losses if global warming results in a higher regional frequency of wildfires and longer duration of droughts. At the same time, warming will lengthen the growing seasons everywhere, and because it will intensify global water cycling, many regions will receive higher precipitation. In many ecosystems these changes will result in higher annual productivity, a trend that has been already strongly evident in most of the United States during the latter half of the twentieth century (Nemani et al. 2002).

A warmer world could also produce a worrisome feedback by releasing significant amounts of soil carbon. Globally, soils store more than twice as much carbon as do plants or atmosphere, and because warming will accelerate decomposition and bring additional releases of CO_2 from soils, this process could reinforce warming to an uncomfortable degree. This result was confirmed not only by models and small-scale experiments but also by some large-scale analyses (Bellamy et al. 2005). But it would be premature to extrapolate it worldwide. Few natural processes are as complex and intricately interactive as the stores and pathways of organic carbon in soils (Shaffer, Ma, and Hausen 2001; Davidson and Janssens 2006).

Cycling rates of soil carbon range from minutes to centuries, and different soils have different shares of slow- and fast-cycling carbon stocks. Root respiration and microbial decomposition are, like all chemical and biochemical reactions, temperature-dependent, so warming must be expected to increase their intensity. But no simple relation governs these processes because the compounds involved exhibit an enormous range of kinetic properties. Large, complex, insoluble molecules tend to be resistant; the requisite enzymes may not be available; or environmental constraints (including physical and chemical protection, drought, and flooding that produces anaerobic conditions) may prevent their action. Not surprisingly, there is no agreement regarding the direction of soil carbon change. Will a warming-driven acceleration of soil carbon release create a significant positive feedback, or will a major share of additional photosynthesis be stored in long-lived soil organic matter and thus moderate future atmospheric carbon increases?

As all economies are only subsystems of the biosphere, any significant perturbations of ecosystems must have a multitude of economic impacts. But these are second-order effects. Because we do not have a tightly constrained range of future temperature increases, we cannot quantify with reasonable certainty the potentially most damaging ecosystemic consequences of this rise, and hence we cannot offer good approximations of the ensuing economic impacts. Again, it is safe to outline a number of the most likely qualitative changes that would call for new capital investment (both preventive and after the fact) and higher operating costs. But different assumptions yield economic penalties ranging from globally trivial to locally crippling, and some appraisals actually see major benefits accruing from this challenge.

The latter claims are easy to question. Hawken and colleagues (1999) have suggested that global warming could be controlled for fun and profit by improving overall energy efficiency and material use: "Within one generation nations can achieve a ten-fold increase in the efficiency with which they use energy, natural resources and other materials" (11). This is nonsense, not, as the authors write, "prophetic words." The factor 10 improvement in a single generation would magically erase population size as a factor in economic development and global environmental degradation, and instantly modernize the entire world. Five billion people in Asia, Africa, and Latin America now claim only about one-third of the world's resources, but wringing ten times as many useful services out of their current resource consumption would make their countries into developed nations, by any definition.

North America and the European Union, content with maintaining their present high standard of living, would have no need for 90% of their current resources, a shift resulting in plummeting global commodity prices. Poor countries could then snap up these give-away commodities (the capacity to produce them will not disappear), and using them with ten times greater efficiency, could support another 7 billion people enjoying a high standard of living—all this with no increase in the global use of energy and materials. Who would care if the global population total kept or rising up by another 2 or 3 billion before eventually stabilizing?

Other uncritical appraisals see a profitable solution in renewable conversions. Gelbspan (1998) wrote that reducing reliance on carbon for energy would not require any personal deprivation but usher in a worldwide economic boom. But any such boom is unlikely during the next few decades (see chapter 3). Yet, dismissing claims of cost-free or inexpensive adaptation is not to say that the economic cost of global warming will be cripplingly high. Some of the most realistic assessments of the cost should be expected from the world's leading reinsurance companies, and reinsurance companies have been analyzing closely the rising frequency of weather-related disasters and studying the likelihood of a warmer atmosphere's producing more extreme weather events (Swiss Re 2003).

Munich Reinsurance Company produced an itemized appraisal of the eventual costs of global warming that would result from the doubling of preindustrial greenhouse gas levels (UNEP 2001). This put the worldwide total at just over $300 billion/year, with the largest share due to additional mortality (27%), demands of water management (15%), losses in agriculture (~13%), and damage to coastal wetlands and fishing (~10%). This is a surprisingly small sum even for today's global economy. In 2005 the global world product was (in PPP) about $55 trillion, and hence the burden of global warming would be only about 0.5% of that. There is no need to contrast this burden with the enormous worldwide annual military expenditures in order to make the $300 billion total even less onerous. In 2005 global business spent in excess of $1.2 trillion on advertising, public relations, and communications (WPP 2006).

A different approach obtained a very similar result. Edenhofer et al. (2006) used ten global economy-energy-environment models to explore the implications of stabilizing greenhouse gas concentrations at different levels and found that even for the lowest limit (450 ppm) average discounted abatement costs were no higher than 0.4% of the gross world product. Nordhaus and Boyer (2001) used two integrated models of climate and economy to estimate impacts of a 2.5°C warming at –0.2% and –0.4% of global output for, respectively, output and population weights. However, a reevaluation by Nordhaus (2006) raised these impacts to –0.93% and –1.73%. Even so, the market impacts of a moderate warming appear to be relatively small. The highest costs would arise from catastrophic events precipitated by the process, and predictably the modeling results showed marked regional variation, ranging from slight economic gains in Russia and Canada to losses of about 5% in India and Africa. If a maximum cost of 2% were applied to the 2005 gross world product, it would be about $1.2 trillion, the total slightly larger than that year's Canadian GDP. In per capita terms it would prorate to about $180/year (50 cents/day), a trivial sum in all affluent countries but a substantial burden for hundreds of millions of poor peasants and marginalized urban dwellers in Asia, Africa, and Latin America.

If these estimates turned out to be close to the actual cost, the economic consequences of global warming would not be fundamentally different from many other challenges that require significant capital investment and operating expenses, including the need for the delivery of clean water to hundreds of millions of people who still lack it. But major disparities in wealth and technical capabilities mean that affluent nations should be able to cope with these new outlays relatively easily (and some should even benefit), whereas for many poor countries even moderate warming will add to an already unmanageable burden.

N. Stern (2006) concluded that the cost of climate change can be limited to about 1% of global economic product only if concerted action is taken within the next 10–20 years. Otherwise these will be a perpetual cost equivalent to at least 5% of annual gross world product. After taking a wider range of risks and impacts into account, Stern puts the overall annual damage at 20% of GWP, or even higher. The 2-OM range of the estimated costs of global warming (0.2%–20% of annual GWP) points to outcomes ranging from a negligible obligation to an unprecedented global economic burden. It does not offer any confidence-inspiring foundation for rational policy making.

A major reason for discrepancies in evaluating environmental burdens is the fact that health effects account for such a large part of these impacts and yet their monetization is notoriously difficult. For instance, suppose warmer temperatures doubled the frequency of allergies and environmentally induced asthma attacks plants growing in warmer temperatures produce more pollen (Epstein and Mills 2005). This would substantially increase the occurrence of serious personal discomfort, school and workplace absenteeism, and number of admissions for emergency treatment. But there are no clear valuations for these new realities. The salaries of physicians and nurses and hospital overhead can be used to value the time spent in emergency treatment and lost wages for missed workdays, but what value are we to put on missed school days, on a child's severe discomfort, and on parental anxiety?

Another impact may be more premature deaths. Global climate models indicate that a warmer atmosphere would very likely bring more frequent temperate-latitude heat waves of greater intensity and longer duration, which are associated with semistationary high pressure anomalies that produce air subsidence, clear skies, light winds, and warm air advection (Meehl and Tebaldi 2004). Longer-lasting heat waves, such as the French episode in 2003 (fig. 4.8) (Pirard et al. 2005) or the Chicago spell in 1995 have considerable potential for increased excess mortalities (and for economic disruption), especially given the fact that by 2050 at least two-thirds of humanity will live in cities, compared to 50% in 2005. Again, how should

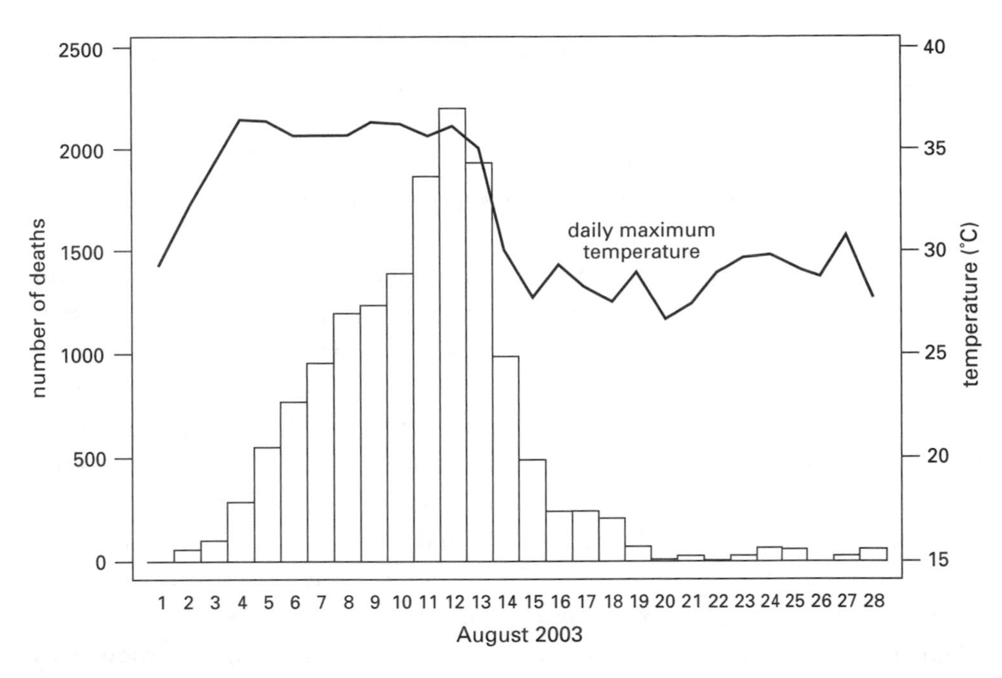

Fig. 4.8
Daily excess deaths and maximum temperatures during the French heat wave of August 2003. Based on Pirard et al. (2005).

we value the truncation of lives, the loss of companionship when a spouse dies prematurely, when a child will never know its grandparent?

Other Global Changes

Concerns about global warming have hijacked most of the attention devoted to environmental problems, but there are other important trends that predate these concerns, and their continuation and intensification can have major (but difficult-to-forecast) global impacts. Loss of arable land to nonagricultural uses and flooding by new large reservoirs are continuing problems, particularly in densely populated Asian countries. An even more common problem is excessive soil erosion on farmland, but we do not have enough solid information to quantify reliably this global trend (Smil 2000).

Many fields around the world are losing their soils at rates two or three times higher than the maximum, site-specific average annual losses compatible with sustainable cropping (mostly 5–15 t/ha) (Smil 2000). The African situation is particularly dire because the nutrients lost in the eroded soil are not replaced by fertilizers. This qualitative soil deterioration is of concern even in places with tolerable erosion. Declining levels of soil organic matter, essential for the maintenance of soil fertility and structure, result from inadequate crop residue recycling, low or no manure applications, and no cultivation of green manures or leguminous cover crops.

But the continuing loss of farmland and its qualitative decline do not imply any near-term global scarcity of arable land or inability to support good yields. New farmland is created by conversions of natural ecosystems (this being, of course, another environmentally destructive process). Significant areas of arable land are held in reserve in many affluent countries (particularly North America), and better management can restore soil quality and bring substantial yield increases everywhere outside the most intensively cultivated areas of the modernizing world (parts of China, Java, Punjab, Nile Delta). Arable land and its quality should not be a major constraint on food production by 2050, but in many places water availability and the use of nitrogen already play this limiting role.

Agriculture is by far the world's largest user of water, and higher productivity from a declining area of farmland will require more irrigation. Similarly, higher yields from smaller areas will have to be supported by more intensive fertilization, and that means, given the inherent inefficiency of the nutrient absorption by plants, greater losses of reactive nitrogen. Another major concern is the continuing reduction of biodiversity, a trend that may eventually have enormous consequences for the maintenance of irreplaceable ecosystemic services. Finally, I address an insidious and ultimately very perilous environmental change, the trend toward widespread bacterial resistance to common antibiotics.

Changing Water and Nitrogen Cycles

Concerns about global warming led to a widespread belief that no other biospheric cycle is subject to so much human interference as the global carbon cycle. This conclusion is doubly wrong. Human interference in the global water cycle is a source of more imminent problems and already a major cause of large-scale premature mortality. And human actions have already changed the global nitrogen cycle much more than carbon cycling, and the ultimate consequences of this multifaceted change may be even more intractable than dealing with excessive CO_2. As a result, if we were a rational society, we would be paying a great deal more attention to these changing water and nitrogen cycles.

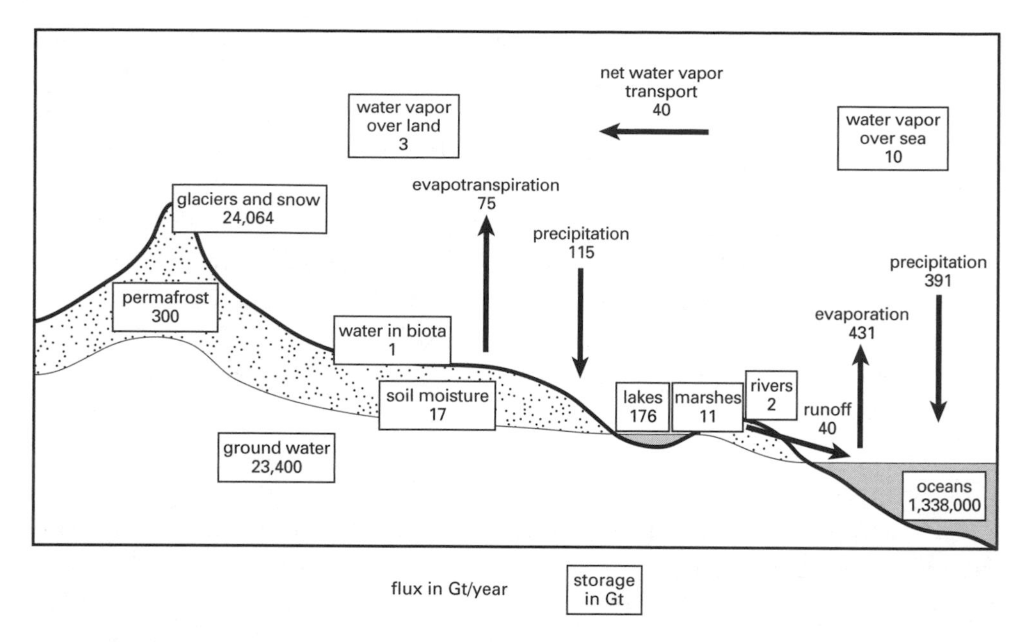

Fig. 4.9
Global water cycle. From Oki and Kanae (2006).

No resource is needed for life in such quantities as water. It makes up most of the living biomass (60%–95%), and its absence limits human survival to just days. The water molecule is too heavy to escape the Earth's gravity, and juvenile water, originating in deeper layers of the crust, adds only a negligible amount to the compound's biospheric cycle. Human activities—mainly withdrawals from ancient aquifers, some chemical syntheses, and combustion of fossil fuels—also add only negligible volumes of water, but they have changed the water cycle (fig. 4.9) in three principal ways, and all of them will intensify during the first half of the twenty-first century (Revenga et al. 2000; WCD 2001; Rosegrant et al. 2002; Shiklomanov and Rodda 2003; United Nations 2003).

Increasing volumes of freshwater are diverted to irrigation, urban and industrial uses (growing a kilogram of wheat needs about as much water as does the production of a kilogram of computer hardware, about 1.5 tons), and electricity generation (there are some 45,000 large dams), and most of this water is released to streams, lakes, and ultimately to the ocean without any or only rudimentary treatment. Moreover, decades of climate change have already intensified the global water cycle, and higher CO_2 levels have affected continental runoff. And yet, global water supplies and their uses and misuses remain a curiously neglected topic of public discourse. Perhaps no investment is as rewarding as spending on clean water, be it in the form of protecting forests in key watersheds or preventing pollution, but the world has failed to make outlays commensurate with the challenge.

This is particularly true for Asia, a continent with 60% of the world's population but only 27% of the Earth's freshwater, which is, moreover, unevenly distributed in time (because of the monsoon) and space (arid Middle East and Central Asia vs. humid South and Southeast). Global warming will produce more but unequally distributed precipitation, and higher CO_2 levels will reduce transpiration (by inducing stomata closure). Hence overall river discharge should increase, but because of continuing population growth the total number of people living with high water stress will increase in the next 50 years (Oki and Kanae 2006).

As Asian and African populations keep growing, the number of countries with stressed water supply (usually taken as the annual rate lower than 1,700 m^3 per capita) and water scarcity (less than 1,000 m^3 per capita) will also increase. In 2005, 17 countries (mostly in the Middle East and North and East Africa) experienced water scarcity, and another 15 countries were in the stressed category. By 2025 (assuming the UN's medium population projections) nearly 30 countries will have water scarcity and another 20 will join the stressed group. The most populous countries in the first category will be Nigeria, Egypt, Ethiopia, and Iran, and in the second category, India and Pakistan (WRI 2006).

Africa's problems will also become serious. A remote sensing analysis concluded that 64% of Africans are already relying on water resources that are both limited and highly variable and that nearly 40% of existing irrigation is unsustainable (Vörösmarty et al. 2005). Moreover, national means hide major regional scarcities that exist in large and populous countries. The most notable example of these regional disparities is China, where acute water shortages in the northern provinces have led to the construction of a massive water transfer project from the Yangzi to the Huanghe basin (see fig. 3.18) (Smil 2004). This is one of the most expensive massive geoengineering tasks ever undertaken and, as with every mega-project, it raises many concerns about its eventual utility and enormous environmental impacts.

Water supply may worsen even in some places with little population pressure. Modeling shows that during the twenty-first century progressively larger areas in Spain, Italy, and the Rhine basin will move into the stressed category (Schröter et al. 2005). A conservative estimate for the year 2050 would put at least 60 countries, with nearly half the world's population, into the water-scarce and water-stressed categories. Only the installation of the most efficient irrigation systems as well as near-complete recycling of urban and industrial water could ease the deficits, but even so there will be a massive new need for desalination (and hence for substantial constant energy inputs).

Two compensating trends may provide relief. Paleoclimatological evidence indicates that in line with expectations, higher tropospheric temperatures have been intensifying the global water cycle and that the twentieth century brought large precipitation gains to regions including the subpolar Arctic, tropical Arabian Sea, and much of temperate Eurasia (Evans 2006). The last instance is illustrated by precipitation in northern Pakistan: the twentieth century was the region's wettest period of the last millennium (Treydte et al. 2006). Moreover, higher atmospheric CO_2 concentrations result in lower evapotranspiration losses from vegetation, and this effect, already detectable in continental runoff records, has increased the amount of water flowing to the ocean (Gedney et al. 2006). Consequently, it is impossible at this time to assess the net outcome of the two countervailing trends (rising demand vs. higher availability of water).

Poor water quality is a much more common problem. In 2005 more than 1 billion people in low-income countries had no access to clean drinking water, and some 2.5 billion lived without water sanitation (United Nations 2003). About half of all beds in the world's hospitals were occupied by patients with water-borne diseases. Diarrhea in its many forms—acute dehydrating (cholera), prolonged with abdominal symptoms (typhoid fever), acute bloody (dysentery), and chronic (caused by waterborne bacteria like *Vibrio*, *Salmonella*, and *Escherichia coli*) is the leading killer (up to 4 billion episodes per year), and dehydration is the principal proximate cause of death. Contaminated water and poor sanitation kill about 4,000 children every day (UNICEF 2005). Deaths among adults raise this to at least 1.7 million fatalities per year. Add other waterborne diseases, and the total surpasses 5 million. In contrast, automobile accidents claim about 1.2 million lives per year (WHO 2004b), roughly equal to the combined total of all homicides and suicides, and armed conflicts kill about 300,000 people per year. The water treatment record of India and China, the world's two most populous countries, is appalling. But even the richest countries have a poor record of water management. Primary treatment is generally in place, but the removal of eutrophication-inducing nitrates and phosphates is still rare.

The natural nitrogen cycle is driven largely by bacteria (fig. 4.10) (Smil 2000). Fixation, the conversion of inert atmospheric N_2 to reactive compounds, is dominated by bacteria. They convert N_2 to NH_3 using nitrogenase, a specialized enzyme that no other organisms carry. Most N-fixing bacteria are symbiotic with leguminous plant roots (some live inside plants stems and leaves) and free living cyanobacteria are present in soils and water. Nitrifying bacteria present in soils and waters transform NH_3 to NO_3^-, a more soluble compound that plants prefer to assimilate. Assimilated nitrogen is embedded mostly in amino acids, which form plant proteins. Animals and people must ingest preformed amino acids in order to synthesize their tissues. Dead tissues undergo enzymatic decomposition (ammonification), which releases NH_3 to be reoxidized by nitrifiers. Denitrification returns the element from NO_3^- via NO_2^- to atmospheric N_2, but incomplete reduction results in emissions of N_2O, a greenhouse gas about 200 times more potent than CO_2.

Cultivation of leguminous crops (done in every traditional agriculture) was the first major human intervention in the nitrogen cycle, and it now fixes annually 30–40 Mt N/year. During the nineteenth century, guano and Chilean nitrate were the first commercial nitrogen fertilizers. The synthesis of ammonia from its elements—demonstrated for the first time by Fritz Haber in 1909 and commercialized soon afterwards by the German chemical company BASF under the leadership of Carl Bosch—opened the way for large-scale, inexpensive supply of reactive nitrogen (Smil 2001). By 2005 global NH_3 synthesis surpassed 100 Mt N/year, with about 80% of it going to produce urea and other nitrogen fertilizers and the rest used in industrial processes ranging from the production of explosives to the syntheses of plastics (Ayres and Febre 1999). The third-largest source of anthropogenic reactive nitrogen is combustion of fossil fuels, which adds almost 25 Mt N/year in nitrogen oxides.

Losses of nitrogen from synthetic fertilizers and manures, nitrogen added through biofixation by leguminous crops, and nitrogen oxides released from combustion of fossil fuels are now adding about as much reactive nitrogen (~150 Mt N/year) to the biosphere as natural biofixation and lighting does (Smil 2000; Galloway and Cowling 2002). This level of interference is unequaled in any other global biogeochemical cycle. Carbon from fossil fuel combustion and land use changes is equal to less than 10% of annual photosynthetic fixation of the element, and sulfur from combustion and metal smelting is equal to only about one-third of the annual flux of sulfurous compounds produced by biota, volcanoes, and sea spray (Smil 2000; D.I. Stern 2005). Not surprisingly, this large anthropogenic fixation of nitrogen has

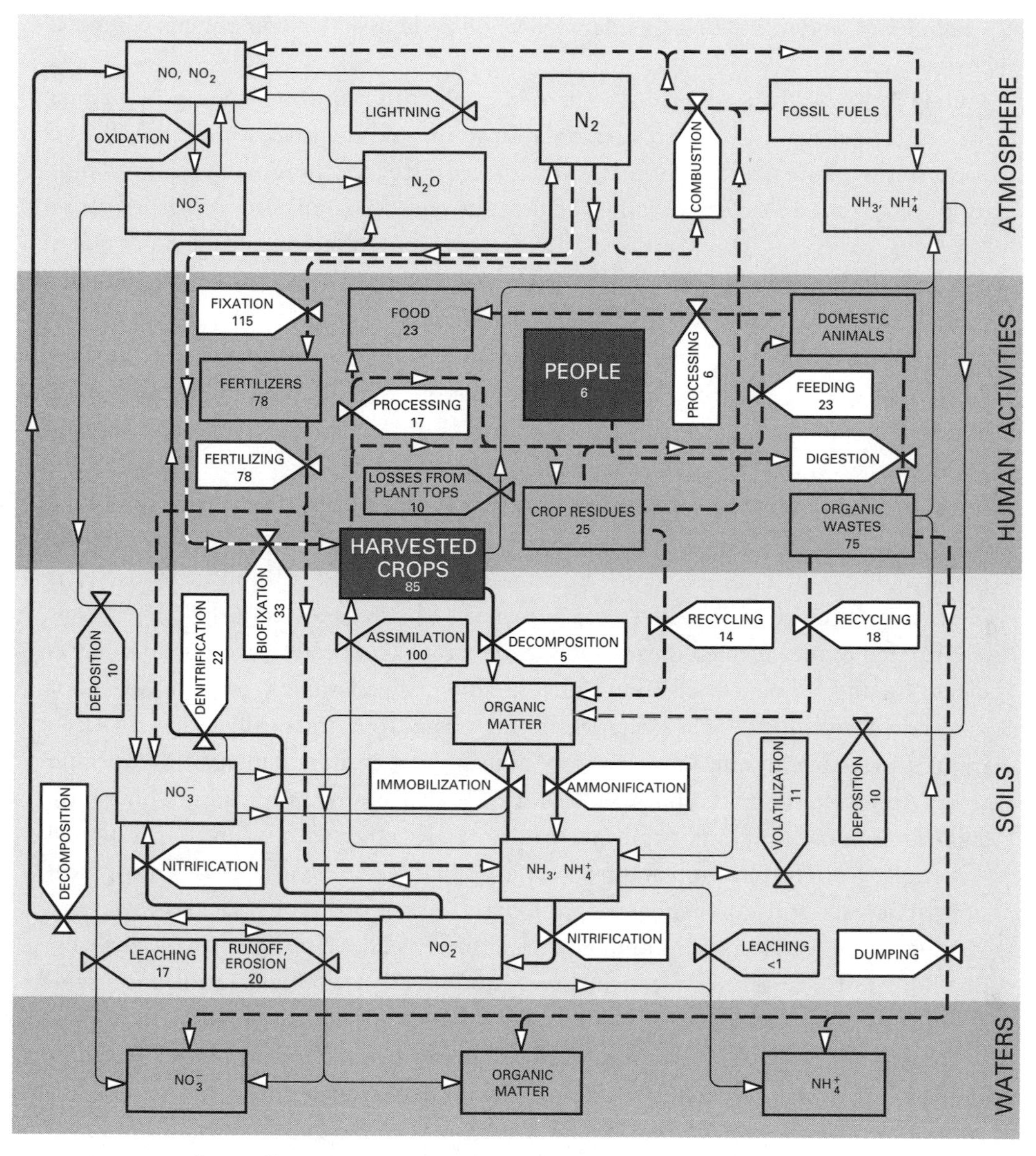

Fig. 4.10
Global nitrogen cycle is governed by bacteria. Human inputs of reactive nitrogen compounds now roughly equal the natural contributions. Adapted from Smil (2002).

a number of undesirable biospheric impacts once the reactive compounds enter the environment.

Only 25%–40% of all fertilizer nitrogen applied to crops is taken up by plants; the rest is lost to leaching, erosion, volatilization, and denitrification (Smil 2001). Because the photosynthesis of many aquatic ecosystems is limited by the availability of nitrogen, an excessive influx of this nutrient (eutrophication) leached from fertilizers promotes abundant growth of algae and phytoplankton. Subsequent decomposition of this phytomass deoxygenates water and reduces or kills aquatic species, particularly the bottom dwellers.

The worst affected offshore waters in North America are in the Gulf of Mexico, where every spring eutrophication creates a large hypoxic zone that kills many bottom-dwelling species and drives away fish (Rabalais 2002; Scavia and Bricker 2006). Other anoxic zones can be found in the lagoon of the Great Barrier Reef, the Baltic Sea, the Black Sea, the Mediterranean, and the North Sea. Algal blooms may also cause problems with water filtration or produce harmful toxins. Escalating worldwide use of urea (besides fertilizer also for animal feed and in industry) is increasing pollution of sensitive coastal waters (Glibert et al. 2006).

Nitrogen oxides formed during high-temperature combustion are essential ingredients for the formation of photochemical smog, a persistent feature of all major urban areas worldwide whose major impacts range from drastically reduced visibility to serious health effects (respiratory ailments) to chronic damage to crops and trees. Atmospheric oxidation of NO and NO_2 also produces nitrates, which with sulfates compose acid rain. Atmospheric nitrates, together with volatilized ammonia (especially from fertilization and from large animal feedlots), also cause eutrophication of forests and grasslands. In parts of eastern North America, northwestern Europe, and East Asia, rains annually bring more reactive nitrogen than fertilizers do. Both nitrification and denitrification produce N_2O, making fertilization contributor to global warming.

Human interference in the global nitrogen cycle is an inherently more intractable challenge than the decarbonization of the world's energy supply. That will not be an easy transition (see chapter 3), but a carbon-free energy system is an eventual inevitability. By contrast, there can be no nitrogen-free organisms, and the larger and more affluent populations of the twenty-first century will demand better nutrition that will have to come largely (given the distribution of future population increments) from higher fertilizer applications, either to secure higher yields of more intensively cultivated land in Asia or to stop the still increasing nutrient mining in agricultural lands of Africa (Henao and Baanante 2006).

Loss of Biodiversity and Invasive Species

The loss of biodiversity usually evokes the demise of such charismatic mega-fauna as Indian tigers, Chinese pandas, and Kenyan cheetahs. All these species are greatly endangered, but in terms of irreplaceable ecosystemic services their loss would not even remotely compare with the loss of economically important invertebrates. Both Europe and North America have seen a gradual decline of pollinators, including domesticated honeybees and wild insects. Pollination by bees is an irreplaceable ecosystemic service responsible for as much as one-third of all food consumed in North America, from almonds and apples to pears and pumpkins (fig. 4.11) (NRC 2006a).

Pollinators are also needed to produce a full yield of seed for such important feed crops as alfalfa and red clover (Proctor et al. 1996). Spreading infestations of varroa mites and tracheal mites (they either kill bees outright or introduce lethal viruses into their bodies) are very difficult to control (NRC 2006a). Introduced African bees, which began their northward expansion in Brazil in 1956, have been destroying native wild colonies in the Americas, and indiscriminate use of pesticides has been the most important human factor in roughly halving the North American honeybee count during the second half of the twentieth century (Watanabe 1994; Kremen et al. 2002). This worrisome trend took a dramatic turn during the winter of 2006–2007, when beekeepers across North America reported widespread collapses of entire bee colonies. The usual suspects included a variety of pathogens, pesticides, and the high-fructose sugar syrup diets used to feed the hives in winter (and their contaminants). The most likely cause has been a virus from Australia (Stokstad 2007).

Loss of biodiversity takes place mostly because of the destruction or substantial alteration of natural habitats (Millennium Ecosystem Assessment 2005). Agricultural land is now the largest category of completely transformed, much less biodiverse land. In 2005 its extent, including permanent tree crops, was about 15 million km^2, and the three largest grain crops—cultivars of wheat (originally from the Middle East), rice (from the Southeast Asia), and corn (from Mesoamerica)—are now grown on every continent and occupy a combined area of about 5 million km^2, more than all remaining tropical forests in Africa. Land under settlement and

Fig. 4.11
European honeybee, *Apis mellifera*, a pollinator without whose toil we would not have many of our fruit and nut crops. Photo from National Human Genome Research Institute, courtesy U.S. Department of Agriculture.

bearing industrial and transportation infrastructures adds up to about 5 million km^2, and water reservoirs occupy about 500,000 km^2. Human activities have thus entirely erased natural plant cover on least 20 million km^2, or 15% of all ice-free land surface.

Areas that still resemble natural ecosystems to some degree but that have been significantly modified by human actions are much larger. Permanent pastures total about 34 million km^2, and at least one-quarter of this area is burned annually in order to prevent the growth of trees and shrubs. A very conservative estimate of the global extent of degraded forests is at least 5 million km^2, and the real extent may be twice as large. Total area strongly or partially imprinted by human activities is thus about 70 million km^2, no less than 55% of nonglaciated land. Another way to appraise the significance of this transformation is to estimate the share of global photosynthesis that is consumed or otherwise affected by human actions. Vitousek et al. (1986) calculated that during the early 1980s humanity appropriated 32%–40% of terrestrial photosynthesis through its field and forest harvests, animal grazing, land clearing, and forest and grassland fires. A more recent recalculation (Rojstaczer, Sterling, and Moore 2001) closely matched the older estimate.

The remaining wilderness has been pushed to subboreal and polar regions. During the late 1980s an inventory of its large (>400,000 ha) contiguous blocks found that more than one-third of the global land surface (nearly 48 million km^2) is still wilderness, but 40% of the total was in Arctic or Antarctic (McCloskey and Spalding 1989). Territorial shares of remaining wilderness ranged from 100% for Antarctica and 65% for Canada to less than 2% for Mexico and Nigeria and, except for Sweden (about 5%), to zero for even the largest European countries. There were also no undisturbed large tracts of land in such previously biodiverse ecosystems as the tropical rain forests of the Guinean Highlands, Madagascar, Java, and Sumatra, and the temperate broadleaf forests of eastern North America, China, and California.

This enormous transformation and degradation will continue (in Asia largely unabated, in Africa much accelerated) during the first half of the twenty-first century, with the tropical deforestation and conversion of wetlands causing the greatest losses of biodiversity (fig. 4.12). Quantifying this demise has been difficult. Nearly 850 species were lost worldwide between 1500 and 2000 (Baillie, Hilton-Taylor, and Stuart 2004), an average rate of one to two per year that is roughly 1 OM higher than the natural rate of extinction. However, for the past century the rate may have been about 100 times faster than indicated by the fossil record, and depending on the rate of habitat destruction, it may become 1,000 times faster during the next 50 years (Millennium Ecosystem Assessment 2005). The best compilation suggests that 12% of known bird species, 23% of mammals, 25% of coniferous trees, and 32% of amphibians are threatened with extinction (IUCN 2001), and unlike recent extinctions, which took place mostly on oceanic islands, continental extinctions will be the norm.

Climate change could become a major cause of extinction in addition to excessive exploitation and extensive habitat loss (to which climate change directly

Fig. 4.12
Deforestation in the Amazon Basin, Bolivia, from logging, ranching, and settlement. Remaining forest shows as dark patches. Satellite image, January 8, 2000, from USGS EROS Data Center Satellite Systems Branch, <http://209.15.138.224/bolivia_maps/s_deforestation_bolivia.htm>.

contributes). The theoretical potential for extinctions driven by climate change is substantial (Lovejoy and Hannah 2005). Thomas et al. (2004) used projections of species distributions (including hundreds of plants, mammals, birds, frogs, reptiles, and invertebrates) to asses extinction risks in a sample region covering about 20% of the continental surface and found that by the year 2050 minimal, medium, and maximum climate change scenarios would condemn, respectively, 18%, 24%, and 35% of species to extinction. Given the many uncertainties, these rates are not to be seen as predictions but as worrisome indicators of the extent of possible change. By 2050 the biosphere is likely to lose thousands of species ranging from Indian tigers to Central American tree frogs to cycad plants.

Further degradation of biodiversity will take place because of the continuing deliberate introductions and unintended invasions of alien plants, animals, and invertebrates that have already changed ecosystems on all continents (Mooney and Hobbs 2000; Baskin 2002; Cox 2004; Sax et al. 2005). Croplands are obviously anthropogenic ecosystems, but many seemingly natural landscapes are as well. For example, 99% of all biomass in parts of the San Francisco Bay is not native (Enserink 1999), and bioinvasions have had particularly devastating effects on islands. Oahu's native vegetation now occupies less than 5% of the island, and even the Galápagos Islands, a national park where 97% of land is protected, have been heavily invaded by rats, pigs, goats, cats, ants, and quinine trees (Kaiser 2001).

The recent rush to cultivate energy crops should be seen "as adding biofuels to the invasive species fire" (Raghu et al. 2006, 1742) because such high-yielding grass species as giant reed, reed canary grass, and switchgrass have been shown to be actually or potentially invasive in U.S. ecosystems. Finally, there are also the largely unexplored consequences of the global spread of microbes in the ballast water of commercial ships (Ruiz et al. 2000). New bioinvasions are largely unpredictable, but their progress is often very rapid, and once they are under way, they are nearly impossible to stop. The rising volume of global trade and travel will only multiply the opportunities for unintended species introductions.

The consequences of bioinvasions often include a profound transformation of affected ecosystems and a high economic cost. Zebra mussels and water hyacinth are two notable examples. The mussels were carried by ships in European ballast water to North America for generations, but they took hold in Ohio and Michigan in 1988, and by the end of 2000 they had penetrated all the Great Lakes and the Mississippi basin. Their massive colonies cloak and clog underwater structure and pipe inlets, reduce the presence of native mussel species, and cause economic damage

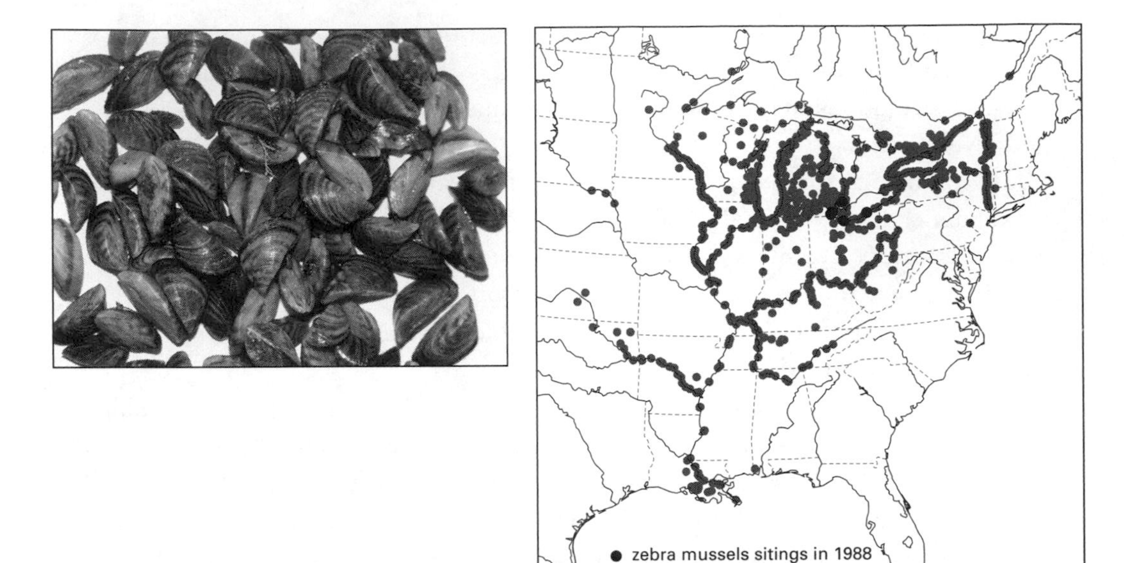

Fig. 4.13
European zebra mussels, *Dreissena polymorpha*, and their distribution in North America. From USGS (2006).

in billions of dollars per decade (fig. 4.13) (USGS 2006). Water hyacinth (*Eichhornia crassipes*, an Amazonian native) has taken over many river, lake, and reservoir surfaces on five continents, with major local and regional economic impacts. It asphyxiates native biota; interferes with fishing, electricity generation, and recreation; increases evaporation from infested surfaces; and serves as a breeding ground for disease vectors (McNeely 1996).

Why does all this matter? It is clear that overexploitation, loss of natural habitats, and species introductions and invasions lead to ecosystemic impoverishment and homogenization. These change previously unique and species-rich communities into homogenized assemblages dominated by a few generalists, be they pests or weeds. Sparrows, crows, pigeons, rats, mice, and feral dogs are the inheritors of the biosphere molded by humans. But there is much more than this esthetic impoverishment and lamentable loss of genetic information that took so many generations to select and perfect. The loss of biodiversity and bioinvasions have major economic consequences, and if severe enough, they can reduce the stability and resilience of ecosystems and seriously compromise the delivery of irreplaceable ecosystemic services.

Antibiotic Resistance

Antibacterial drugs, originally derived from spore-forming microbes and commercially available since the 1940s, were one of the most consequential innovations of the twentieth century. They have been a major cause of extended life expectancy and saved millions of lives that would have been lost to previously untreatable diseases. Less dramatically, they have shortened the course of common infections, lessened patient discomfort, and speeded up recovery. But their spreading use led inevitably to the selection of resistant strains (Levy 1998; WHO 2002). Penicillin-resistant *Staphylococcus aureus* was found for the first time in 1947, just four years after the mass production of this pioneering antibiotic.

As the resistance spread, methicillin became the drug of choice and the first methicillin-resistant strains of *Staphylococcus aureus* (MRSA, causing bacteremia, pneumonia, and surgical wound infections) were encountered in 1961. The MRSA strains are now common worldwide, and in the United States they have been responsible for more than 50% of all infections acquired in hospitals during intensive therapy (fig. 4.14) (Walsh and Howe 2002). Vancomycin was the next choice, but the first vancomycin-resistant enterococci appeared in 1987 in Europe and in 1989 in the United States. The first vancomycin-resistant staphylococci appeared in Japan in 1997 and in 1999 in the United States (Jacoby 1996; Cohen 2000; Leeb 2004). This was a particularly worrisome development because vancomycin has become

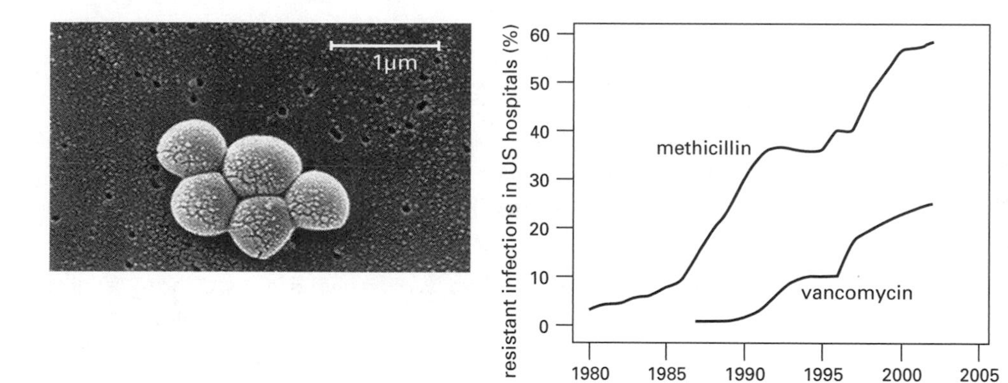

Fig. 4.14
Left, scanning electron micrograph of methicillin-resistant *Staphylococcus aureus*. Photo by Janice Carr/Centers for Disease Control and Prevention. *Right*, methicillin and vancomycin resistance in U.S. hospitals, 1980–2002. From Centers for Disease Control and Prevention.

the drug of last resort after many bacteria acquired resistance to penicillins, erythromycin, neomycin, chloramphenicol, tetracycline, and beta-lactams.

Strains of *Salmonella* with resistance to antimicrobial drugs are now found worldwide, and they are usually transmitted from animals to people through food (Threlfall 2002). A strain of *Salmonella typhimurium* (causing gastroenteritis) is now resistant to as many as six antimicrobial drugs, and multiple resistance associated with treatment failure is also found in *Salmonella typhi* (typhoid fever) in India and Southeast Asia. Multidrug resistance (including to tetracycline and ampicillin) was found among strains of *Vibrio cholerae* in India (Yamamoto et al. 1995). Other common bacteria with multidrug resistance now include *Haemophilus influenzae* (causing pneumonia, meningitis, and ear infections), species of *Mycobacterium* (tuberculosis), *Neisseria gonorrhoeae*, *Shigella dysenteriae* (severe diarrhea), *Streptococcus pneumoniae*, *Clostridium difficile* (severe diarrhea and fever), *Escherichia coli*, and a soil-dwelling bacterium *Burkholderia pseudomallei*, which causes melioidosis (joint inflammation, internal abscesses, and impaired breathing), whose progression may be rapid and whose mortality rate is high (Aldhous 2005).

Resistance to antibiotics would have developed naturally even under the most stringent conditions. It is acquired not only through inevitable spontaneous mutations but also through horizontal gene transfer among neighboring bacteria, even if they happen to be only distantly related species. Moreover, while the antibiotic-resistant strains were assumed to be weaker than their susceptible precursors, recent research found that resistant strains of *Mycobacterium tuberculosis* can be just as aggressive (Gagneux et al. 2006). But the rapid diffusion of drug-resistant strains has been greatly assisted by unnecessary self-medication with antimicrobials and by their prescribed overuse. Overprescribing and improper use of antibiotics in affluent Western countries look like a minor problem compared to the abuse of the drugs in the rest of the world. Amyes (2001) noted that in India more than 80 companies made ciprofloxacin without any license from the patent holder (Bayer) and that almost 100% of the country's healthy population carries bacteria resistant to several common antibiotics. (European rates of resistance range from less than 5% to 40%.)

Other key factors that have promoted the diffusion of resistance are poor sanitation in hospitals and a massive use of prophylactic antibiotics in animal husbandry. Because of this common overuse antibiotics resistance is now present not only in humans and domestic animals but also in wild animals that have never been exposed to those drugs (Gilliver et al. 1999). Gowan (2001) called attention to a significant (but unquantified) economic impact of antimicrobial resistance: premature deaths due to ineffective drugs and extended hospital stays needed to combat the infections.

If antimicrobial drugs were to lose their efficacy completely, the cost would be truly catastrophic. Even with their use, infectious diseases remain the second-highest cause of deaths worldwide. Without them, every annual influenza epidemic might bring many more deaths due to bacterial pneumonia, and tuberculosis and typhoid fever might become very difficult or impossible to treat. Some appraisals of the current situation are very pessimistic. Amyes (2001) concluded that we are slipping into an abyss of uncontrollable infection. Amábile-Cuevas (2003) thinks that, in many ways, the fight against antibiotic resistance is already lost.

The situation has been made worse by the fact that most major pharmaceutical companies have largely withdrawn from the development of new antibiotics. In 2004 only 6 out of 506 drugs in late-stage clinical trials were antibiotics, and all of these were derivatives of existing drugs (Leeb 2004). Their reasoning, as profit-maximizing entities responsible to shareholders, was obvious: antibiotics are drugs of sporadic and time-limited utility, unlike the cholesterol-lowering statins or hypertension-lowering beta blockers that many people take daily for life and that can become multibillion-dollar items on corporate balance sheets. So we simply do not know how close we are, or might become, to a world that resembles the pre-penicillin era of bacterial infections with unpredictable and massively fatal outcomes.

As noted, more than four decades have passed since the introduction of the last new anti-tuberculosis drugs, and neither the existing drug pipeline nor the level of research funding indicates that a new compound could become available by 2010 (Glickman et al. 2006). At the same time, nearly every third person in low-income countries is infected with *Mycobacterium tuberculosis*, and while under normal circumstances fever than 10% of infected individuals develop the disease, the activation rate of latent TB rises once the immune system is weakened, a condition increasingly common with spreading HIV. The worst possible combination is that of HIV-infected patients stricken by extremely drug-resistant (XDR) TB. And while there have been many isolated XDR TB cases, the first virtually untreatable outbreak of the infection, which affected more than 100 patients in rural KwaZulu Natal in South Africa, occurred in 2006 (Marris 2006).

Bacteria have a great evolutionary advantage: some 3.5 billion years of continuous existence that has endowed them with superior mutation and survival capacities. And so a discovery by D'Costa et al. (2006) was not really surprising: they found that every one of 480 isolates of the diverse spore-forming genus *Streptomyces*, well known for its capacity to produce multiple antimicrobial agents, was also able to resist at least six to eight (and some up to 20) different antimicrobial agents that protect the bacteria against their own toxic products. This natural protective capacity could be harnessed for the synthesis of new antimicrobial compounds. Without such new departures we will steadily lose our ability to combat bacterial infections.

Biosphere's Integrity

Some forms of life have been on the Earth for well over 3 billion years. Complex organisms began diffusing about half a billion years ago, mammals became abundant 50 millions year ago, our hominid ancestors date to more than 5 million years ago, and our species has been evolving increasingly complex modes of existence (though not necessarily a greater sapience) during the past 100,000 years. On the civilizational time scale—10^3 years, with less than 10,000 years elapsing from the tentative beginnings of first settled cropping and less than 5,000 years since the establishment of the first complex social entities, precursors of states, in the Middle East—the biosphere is an astonishingly durable system, and worries about its unraveling seem overwrought. Such a conclusion is absolutely correct on the microbial level: a rich assemblage of viruses, archaea, bacteria, and microscopic fungi will survive any conceivable insult that human beings can inflict on the biosphere.

But a biosphere resembling an aquatic slime or consisting of a cyanobacterial layer on rocks, without millions of differentiated invertebrates (insects are the most biodiverse organisms) or any higher plants or aquatic and terrestrial animals, would not be a place fit for the evolution and complexification of hominids. This perspective illuminates the irreplaceable importance of ecosystemic services for human survival and hence, obviously, for any economic activity. The global economy is merely a subsystem of the biosphere, and it is easy to enumerate the natural services without which it would be impossible, but it is meaningless to rank their importance because they are interconnected with a multitude of feedbacks.

Categorizing natural environmental or ecosystemic services requires some arbitrary decisions because individual services overlap and interact and because it is difficult to separate form and function. In a pioneering account, Vernadsky (1931) listed these principal roles (conceptually corresponding to nature's services) played

Fig. 4.15
Roots of a pea plant with nodules containing the symbiotic bacterium *Rhizobium leguminosarum*, which converts nonreactive atmospheric N_2 to ammonia. Photo by Bert Luit, Department of Plant Science, University of Manitoba.

by the biosphere: gas function, the formation of atmospheric gases; oxygen function, the formation of free O_2; oxygenating function producing many inorganic compounds; binding of calcium by marine organisms; formation of sulfides by sulfate-reducing bacteria; bioaccumulation, the ubiquitous concentration of many elements from their dilute environmental presence; decomposition of organic compounds; and metabolism and O_2-consuming respiration resulting in CO_2 generation and biosynthesis (done by aerobic heterotrophs).

A differently formulated list must include at least the following nine service categories. Photosynthesis produces all our food, a large share of our fibers for clothes, all fiber (cellulose) for papermaking, and about one-tenth of the world's primary energy use (wood, crop residues). Nitrogen, the key macronutrient needed for plant growth, is constantly converted from its inert atmospheric form (N_2) into reactive NH_3 by diazotrophs (N-fixing microbes dominated by symbiotic *Rhizobium*) (fig. 4.15) and then into more water-soluble NO_3 (by nitrifying bacteria), which is readily absorbed by plant roots. Similarly, the cycling of sulfur is mediated by various bacteria.

Ecosystems regulate water runoff (and hence the severity of floods) and retain and purify water. Root systems of grasslands are particularly effective sponges, and forests soils can filter water as well as the best artificial filters. New York City relies on the Catskill forests for its high-quality drinking water; similarly, Winnipeg, Canada (where I live) relies on the forests surrounding the Lake of the Woods. Plant cover prevents or minimizes soil erosion and hence eliminates or reduces excessive silting of streams that contributes to floods and limits stream navigability. Forests help to regulate climate on regional scales. Even a small tree grove has a microclimatic impact, and all plants are highly effective controllers of many air pollutants. Coastal vegetation, particularly extensive wetlands, provides a highly effective means of reducing shore erosion and offering protection against storm surges. Healthy soils are assemblages of fine-grained minerals and a great deal of living matter; without soil bacteria we could not produce our crops. And we should not forget the soil invertebrates, above all, earthworms, Darwin's great favorites; his last published book was devoted to these "lowly organized creatures" (1881). The decomposition of organic matter (and hence of carbon sequestration) and the recycling of nutrients are almost totally dependent on microbes and micro- and macroscopic fungi. Without their services there would be neither agriculture nor animal grazing, neither coral reefs nor magnificent tropical rain forests. Natural biodiversity reduces the impact of disease and pest attacks. I have already noted the importance of pollination.

Even if expense were no object, none of these services could be performed at such scales and with such efficacy by any anthropogenic means. Our dependence on biospheric services is literally a matter of survival, and that is why the biosphere's integrity matters. Localized assaults exact a local price in degraded farmland or pasture, or in poor yields or skeletal animals (caused by inadequate recycling of organic matter or severe overgrazing), or in streams leaving their banks because of flooding aggravated by massive deforestation of a watershed. Regional impacts can influence the fortunes of a nation, but if the situation is desperate enough, people move. But major interference with ecosystemic services on a global scale is an entirely new challenge.

Never before have we been in a situation where we can concurrently affect so many biospheric functions on such a grand level and unleash a multitude of foreseeable as well as utterly unpredictable environmental consequences. The story of chlorofluorocarbons (CFCs) is a perfect illustration of such unintended perils. In the early 1920s, Thomas Midgley, Jr., a research chemist working for GM, introduced a lead-based additive for gasoline in order to eliminate engine knocking. At the time, the toxic effects of lead were well known. Midgley himself had lead poi-

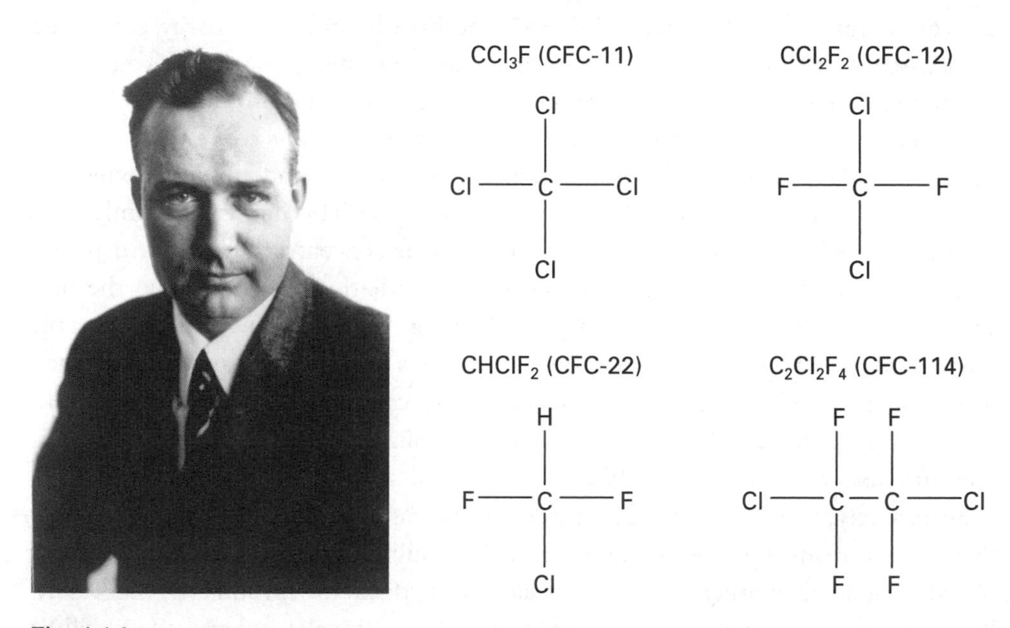

Fig. 4.16
Left, Thomas Midgley, Jr., who developed chlorofluorocarbons (CFCs). Photo courtesy Thomas Midgley IV. *Right*, the four most common CFCs.

soning in 1923, and the long-term problem with the use of this additive could have been anticipated. But a decade later Midgley selected CFCs as perfect refrigerants because they were inexpensive to synthesize, inert, noncorrosive, nonflammable, and "nontoxic" (fig. 4.16) (Midgley and Heene 1930).

Nobody anticipated any adverse effects when their use spread (after WW II) to hundreds of millions of refrigerators and air conditioners, and to propelling aerosol, blowing foams, cleaning electronic circuits, and extracting plant oils (Cagin and Dray 1993). Four decades after their introduction, experiments indicated that once these gases reach the stratosphere, they are dissociated by UVB waves of 290–320 nm, which are entirely screened from the troposphere by stratospheric ozone (Molina and Rowland 1974). (They are much heavier than air but turbulent mixing eventually transports them in small concentrations higher than 15 km above the ground.) This breakdown releases free chlorine atoms that then break down O_3 molecules and form chlorine oxide; in turn, ClO reacts with O to produce O_2 and frees the chlorine atom. Before it is eventually removed, a single chlorine atom can destroy about 10^5 O_3 molecules. In 1985, the British Antarctic Survey confirmed the existence of this process by discovering a severe thinning of ozone layer above Antarctica, the famous "hole" (Farman, Gardiner, and Shanklin 1985).

Human actions thus imperilled one of the key preconditions for complex life on Earth, the planet's ozone shield. Its continuing thinning and possible extension beyond Antarctica threatened all complex terrestrial life. This had evolved only after the oxygenated atmosphere gave rise to a sufficient concentration of stratospheric ozone to prevent all but a tiny fraction of UVB radiation from reaching the biosphere. Excessive UVB radiation drastically reduces the productivity of oceanic phytoplankton (the base of marine food webs), cuts crop yields, causes higher incidence of basal and squamous cell carcinomas, eye cataracts, conjunctivitis, photokeratitis of the cornea, and blepharospasm among animals and people, and also affects their immune systems (Tevini 1993).

Fortunately, CFCs were produced in quantity only by a handful of nations, and there was a ready solution: banning them (beginning with the Montreal Protocol of 1987) and substituting for them less harmful hydrofluorocarbons (UNEP 1995). Because of the long atmospheric lifetimes of CFCs, their stratospheric effect will be felt for decades to come, but atmospheric concentrations of these compounds have been falling since 1994, and the stratosphere may return to its pre-CFC composition before 2050. This experience offers no effective lessons for controlling greenhouse gases. All countries produce them, many economies depend on large-scale extraction and sales of fossil fuels, and all modernizing countries will be consuming more hydrocarbons and coal for decades to come. Moreover, there is no possibility of any simple and relatively rapid switch to noncarbon energy sources because the requisite conversions are not yet available at required scales and acceptable prices (see chapter 3).

At the same time, no country will be immune to global climate change, and no military capability, economic productivity, or orthodox religiosity can provide protection against its varied consequences. We might come to see such preoccupations as budget deficits and immigration or trade policies as trivial when compared to the climate's changing faster than it has at any time during the past million years. Unfortunately, our capacity to assess the impact of these changes is hampered not only by a imperfect understanding of the complex feedbacks involved in the process but also by uncertainty about the future pace of change.

Long-range forecasts of fossil fuel combustion and rates of economic growth, the primary drivers of greenhouse gas emissions, are notoriously unreliable even when

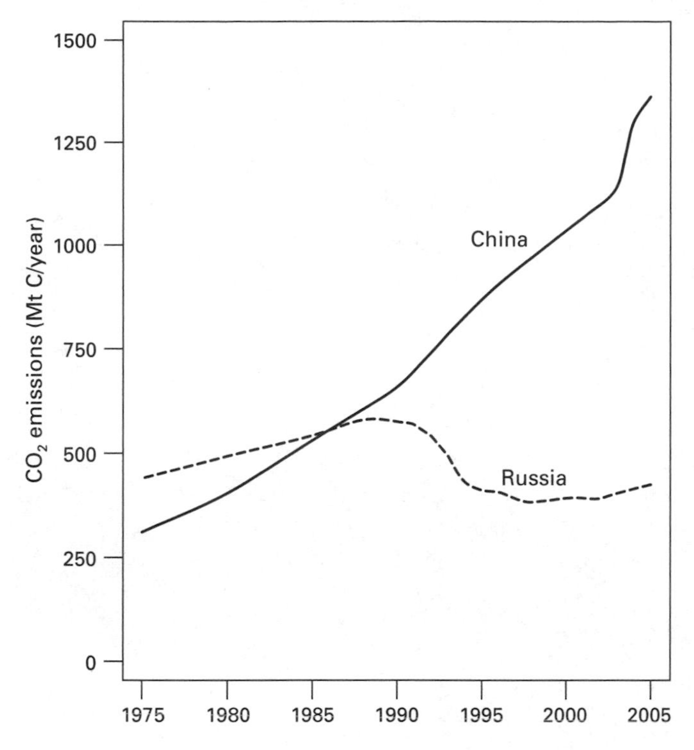

Fig. 4.17
CO_2 emissions in Russia and China, 1975–2005. Plotted from data in Marland, Boden, and Andres (2005).

countries are on fairly stable trajectories. The forecasts can be drastically pushed up or down by unpredictable discontinuities in the history of major consuming nations. As an example, China's carbon emissions rose sharply during Deng Xiaoping's post-1980 economic reforms (emissions more than tripled by 2005), and in contrast, carbon emissions declined (by 1998, 35% below their peak a decade earlier) after the 1991 demise of the USSR, which had been before its dissolution the world's second-largest emitter of greenhouse gases (fig. 4.17).

5

Dealing with Risk and Uncertainty

Quam multa fieri non posse prius quam sunt facta iudicantur?
(How many things are judged impossible before they actually occur?)
Gaius Plinius Secundus (Pliny the Elder), *Naturalis Historia*, VII.i, 6

A number of potential catastrophes that could transform the world in a matter of months (extraordinarily virulent pandemics, a sequence of volcanic mega-eruptions) or even minutes (collision with a massive extraterrestrial object, accidental nuclear war), and a much longer array of worrisome trends (whose outcome can be a new world order or a historically unprecedented global environmental change) add up, even when approached with a robust belief in the problem-solving capacity of our sapient species, to an enormous challenge. Any other attitude leads to responses that differ in form but agree in their dismal substance.

Fundamentally, there is little difference between an agnostically despondent wait for a (most unlikely) miraculous delivery from these seemingly inalterable perils and a religiously charged chiliastic expectation of the Judgment Day, or between historical reprise of a sudden demise of yet another civilization (Roman and Mayan precedents are the favorite examples, albeit in many ways irrelevant) and the scientifically justified (through "terror, error, and environmental disaster") inevitability of our final hour. All of these expectations are nothing but informal, qualitative ways of forecasting, and as I've said, I have a strong personal dislike of such efforts and plenty of historical evidence to demonstrate their ephemeral nature and their repeated failure to portray the complexity of future natural and human affairs.

A few of these attempts may capture important trends, but they cannot come anywhere close to the setting within which these trends will unfold. Moore's law of transistor packing on a microchip is an example of a rare accurate quantitative forecast, albeit in its revised form: the original period for doubling the number of transistors was 12 months, later extended to 18 months (Intel 2003). But when Moore formulated it in 1965, there were no microprocessors (devices made possible by this packing), and neither he nor anyone else could anticipate that four decades later the two leading customers for these remarkable devices would be the personal computers, devices nonexistent and entirely unanticipated in the mid-1960s, and car industries. (A typical U.S. car now has at least 50 embedded microprocessors, and their cost represents about 20% of the car's retail value.)

Ephemeral lifespans and spectacular failures are the normal lot of virtually all broader forward-looking appraisals, including those that are supported by the best scientific consensus. After all, we do not live in a society where everything is powered by nuclear fission (consensus about the early twenty-first century that prevailed into the 1970s), nor in one where global affairs are run from Moscow or Tokyo. And even widely accepted interpretations of past trends, made in order to derive lessons for the future, can suffer the same fate. I offer three illustrations, two prospective and one retrospective, all related to environmental catastrophes.

Scientific consensus during the early 1970s was that the planet faces a serious spell of cooling. In 1971, Stephen Schneider, later to become well known for his warnings about the perils of inevitable and intensive global warming, argued that even if the concentrations of CO_2 were to increase eightfold (an impossibility during the twenty-first century), the surface temperature would rise by less than 2°C, and a fourfold rise in aerosol (not to be ruled out during the twenty-first century) would lower the mean planetary temperature by 3.5°C. The consequence: "If sustained over a period of several years, such a temperature decrease over the whole globe is believed to be sufficient to trigger an ice age" (Schneider and Rasool 1971, 138).

By 1975, *Newsweek* reflected this consensus, writing about "The Cooling World" and "ominous signs . . . serious political implications . . . the drop in food output" based on "massively" accumulating scientific evidence (Gwynne 1975, 64). This planetary cooling trend was blamed, among other trends, for "the most devastating outbreak of tornadoes ever recorded" and for declines in agricultural production to be expected during the rest of the twentieth century. Now, of course, global warming gets blamed for the increase in violent weather, and after a busy 2005 hurricane season (with a record number of named storms), capped by Katrina and the destruction of New Orleans, we were told to expect even more catastrophic events in 2006, but the year did not see a single major hurricane hit the United States.

The second example concerns what is to be one of the most consequential outcomes of global warming, the fate of the Atlantic heat conveyor (fig. 5.1). As already noted, an abrupt disruption of this flow is quite unlikely, but a longer-term risk may exist. We are told (in a doomsday book by a leading British scientist) that the Gulf Stream, powered by thermohaline circulation, will be quenched by Greenland's melting ice sheet, and that the stream's truncation or reversal could plunge "Britain and neighboring countries . . . into near-Arctic winters" (Rees 2003, 111).

A couple years after Rees wrote those words came the first apparent confirmation of this undesirable trend. A comparison of five surveys of ocean temperature and salinity data in the Atlantic conducted between 1957 and 2004 showed that in 1998 and 2004 the conveyor weakened significantly as more of the northward-flowing Gulf Stream was recirculated in the subtropical gyre before reaching higher latitudes, and less cold water was returning southward at depth between 3–5 km (Bryden, Longworth, and Cunningham 2005). As the media reported the coming dramatic cooling of Europe, they ignored the lead author's caveat that the observed weakening was comparable to the estimated uncertainty of observations. Moreover, the long-term behavior of such a highly dynamic system cannot be judged by a few episodic measurements. Indeed, new data acquired between the Bahamas and the Canary Islands demonstrated that the flow reduction reported by Bryden et al. (2005) was well within the range of a very large natural seasonal variability; the conclusion: ocean circulation is noisy but it is *not* weakening (Cunningham et al. 2007).

And, most fundamentally, it is simply not true that the Gulf Stream is driven by thermohaline circulation, nor is it responsible for Britain's warm winters. The Gulf Stream, much like the Kuroshio Stream in the Pacific and the Agulhas Stream in the Indian Ocean, is a wind-driven flow produced by the torque exerted on the ocean and sustained by the conservation of angular momentum. We have known for decades that its prime movers are solar radiation (via wind) and the Earth's rotation (Stommell 1948). And Seager et al. (2002) demonstrated that the Gulf Stream is *not* responsible for Europe's mild winters. Loss of this heat flow would have a marginal effect on Atlantic Europe because its climate does not require a dynamic ocean. Most of the continent's winter heating is due to the seasonal release of heat previously absorbed by the ocean and advected by the prevailing winds, not by ocean heat-flux convergence (see also Seager 2006).

A fascinating retrospective example concerns the supposed choices societies make to engineer their own demise. Diamond (2004), in his book on collapse, uses the

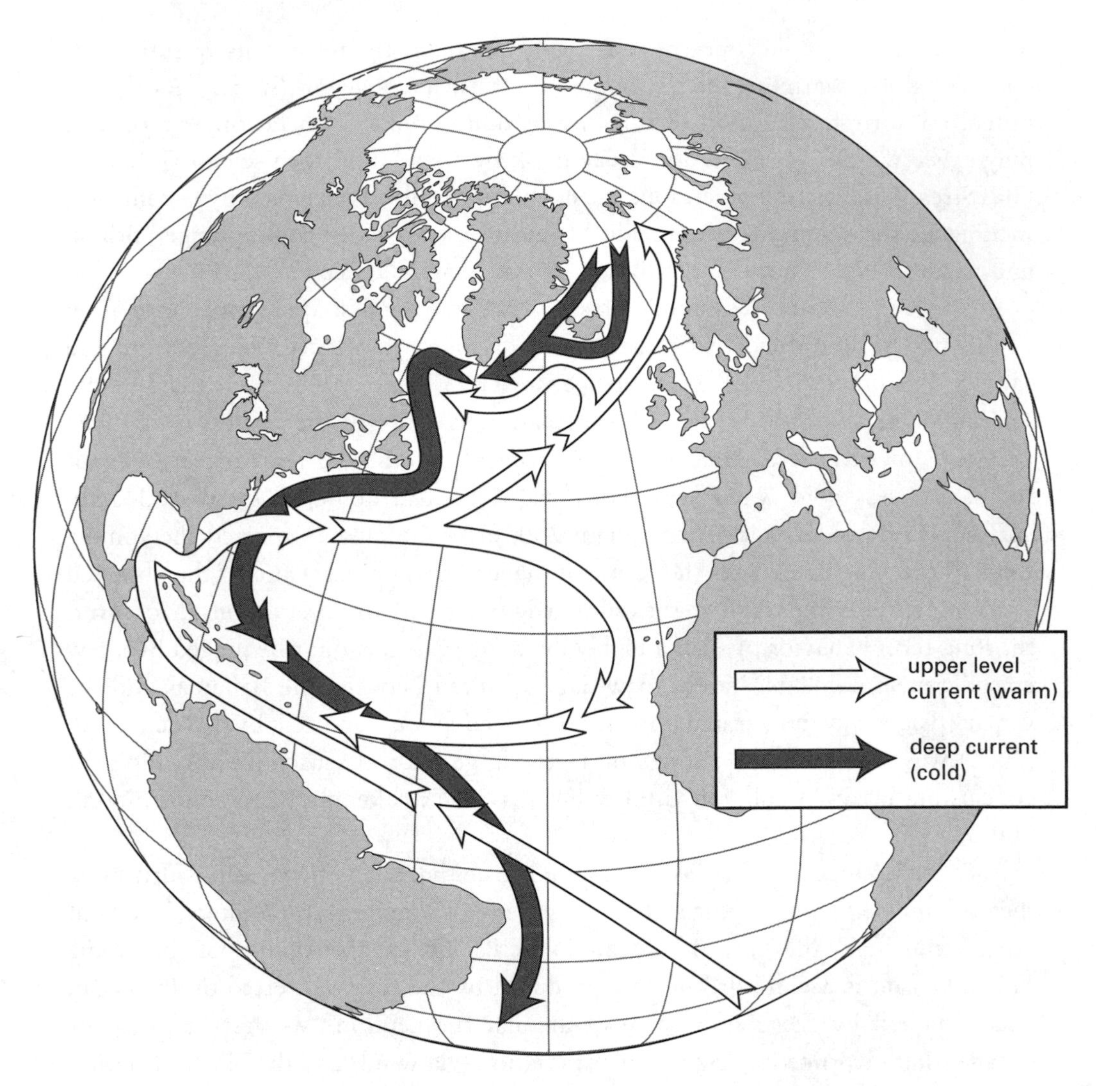

Fig. 5.1
Atlantic heat conveyor. Warm water (*light path*), detached from clockwise circulation in the subtropics, warms the North Atlantic, sinks in the northernmost region, and then continues as a deep southerly return flow (*dark path*). Based on Quadfasel (2005).

Fig. 5.2
Unfinished monumental *moai* at Rapa Nui (Easter Island). Photo courtesy of David Malone, Texas Tech University.

story of Rapa Nui (Easter Island) as one of the most prominent examples of anthropogenic environmental change that precipitated a societal collapse. The reckless deforestation of the island, he writes, caused a population crash and the disappearance of a mysterious *moai*-building civilization (fig. 5.2). This conveys a premonitory analogy to our reckless treatment of the environmental commons.

But Diamond's narrative has ignored everything else but the choices attributed to thoughtless humans. In contrast, Hunt's (2006) research identified Polynesian rats, which had been introduced to the island, as the real cause of destruction of most of the *Jubaea* palm forests, but this deforestation did not trigger the collapse of a small society (most likely just 3,000 people). The post-contact (after 1722) infectious diseases and enslavement did that. Thus a picture of man-induced environmental collapse rests on a simplistic explanation, but it will not be easily dislodged.

The point is made. There is no shortage of sudden and potentially catastrophic changes that could transform modern civilization. There is an even larger array of worrisome trends that could shape it. But none of these events and processes can be understood and appraised to a degree that would allow confident identification of future risks even when seen in a relative isolation. Our abilities to discern complex feedbacks (sometimes even just their net direction) and very long-term consequences are even weaker. Uncertainty rules, and there are no shortcuts to lead us from these dim perceptions to clearer understanding.

But we are not powerless either. Many risks can be quantified (though with significant margins of error), and many trends have relatively constrained outcomes. For example, the probability that the average fertility rate in affluent countries will double, or that the doubling of preindustrial greenhouse gase concentrations will produce a 15°C warming, is infinitesimally low. Consequently, by quantifying the odds of risky events, we can establish an approximate ranking of relative fears, and by looking at the most likely constraints of major trends, we can determine a range of possible outcomes. After that comes an even greater challenge: What can we do to lessen, if not eliminate, those risks? What steps can we take to change those worrisome trends or at least bend them in more benign directions?

Relative Fears

Appraisals of natural catastrophes that can have both a dramatic instantaneous effect and generations-long global consequences show low probabilities during the next half century, but, at the same time, such quantifications enter a realm that is alien even to those experts who routinely analyze risks. Leading hazards encountered in modern society have a fairly high frequency of fatalities, but they kill or injure people discretely and in small numbers, and many of the losses (mortality, injuries, or economic damage) are sustained through voluntary actions and exposures whose risks people almost uniformly underestimate. Annual mortality aggregates of such exposures may be relatively high, but they come to public attention only if a particular event of that kind is unusually large.

In confronting these risks we should not rely on imagination and fears. Instead, we should deploy, to the greatest extent possible, a revealing comparison of relative perils. While it is very difficult to find a uniform metric to compare risks of injuries or economic losses (both of these categories span a wide range of qualities), the finality of death makes it possible to compare the risk of catastrophic or accidental dying on a uniform basis. This evaluation is best done in terms of fatalities per person per hour of exposure, a risk assessment approach that was originally developed by Starr (1969; 1976). I use it to compare relative risks of all quantifiable events discussed in this book.

Starr concluded that acceptability of risk from an activity is roughly proportional to the third power of its benefits, and he posited a fundamental difference in risk appraisal. When people are engaged in voluntary activities—when they feel that they are in control of their actions and when repeated experiences have shown that the outcomes are predictable—they readily tolerate individual risks (driving, overeating, smoking) that are up to 3 OM higher than those arising from involuntary exposure to natural or anthropogenic catastrophes and providing a comparable level of benefits. Reexamination of Starr's work using psychometric studies of risk perception confirmed that people will tolerate higher risks if activities are seen as highly beneficial, but it suggested that familiarity and fear (dread) rather than the voluntary nature of exposure were the key mediators of acceptance (Fischoff et al. 1978; Slovic 1987; 2000).

That all of these three factors are at play is illustrated by car accidents, perhaps the best example of that peculiar attitude with which humans treat voluntary, well-known, and well-accepted risks that have a high frequency but a low fatality rate per event. As noted, car accidents cause worldwide nearly 1.2 million deaths per year (WHO 2004b), but more than 90% of individual events involve the demise of just one or two people and therefore do not attract media attention. Accidents are widely reported only when the per-event mortality rate suddenly rises (albeit in absolute terms it still remains fairly small). Fog- or ice-induced pileups of dozens or scores of cars causing a dozen or more casualties are the most common events that are invariably reported.

In sharp contrast is the attitude toward terrorist attacks. Their instant fatalities are sometimes large, their risks should not be minimized, but it is our utter ignorance regarding the time and mode of such attacks (psychometric "unknowability" factor) and the self-inflicted terrorizing perception of it (psychometric "dread" factor) that wildly exaggerate their likely frequency and impact.

Most of the risks arising from long-term trends remain beyond revealing quantification. What is the probability of China's spectacular economic expansion stalling or even going into reverse? What is the likelihood that Islamic terrorism will develop into a massive, determined quest to destroy the West? Probability estimates of these outcomes based on expert opinion provide at best some constraining guidelines but do not offer any reliable basis for relative comparisons of diverse events or their interrelations. What is the likelihood that a massive wave of global Islamic terrorism will accelerate the Western transition to non–fossil fuel energies? To what extent will the globalization trend be enhanced or impeded by a faster-than-expected sea level rise or by a precipitous demise of the United States? Setting such odds or multipliers is beyond any meaningful quantification.

Quantifying the Odds

The unavoidable yardstick for comparing the quantifiable odds of catastrophic events is general mortality, the crude death rate of a population measured annually per 1,000 people. Mortality figures have the advantage of being fairly accurate; only during war or famine is their accuracy questionable. During the first five years of the twenty-first century, the crude death rate in affluent countries ranged from 7.2 in Canada (thanks to its relatively young population) to 10.4 in Sweden, and the mean for all rich economies was 10.2. The year has 8,766 hours (corrected for leap years), and hence the average mortality of affluent nations (10/1,000) prorates to 0.000001, or 1×10^{-6}, per person per hour of exposure (which in this case means simply being alive).

Put another way, in affluent countries one person out of a million dies every hour. Cardiovascular diseases account for about one-third of this total, so the overall risk of succumbing to them is about 3×10^{-7} per person per hour of exposure. I must reiterate that this is the risk of mortality averaged across an entire population. The age-specific risks of dying for populations of premature babies or people over 90 years of age will be higher; those for populations of grade-school girls or diet-conscious adult Methodists will be lower. Few people are aware of the actual magnitude of this unavoidable risk, yet most people behave as if they were taking it into account in their everyday behavior. They have no second thoughts about routinely engaging in activities or living in environments that expose them to risks of death that are 1 OM or more lower than is the risk of general mortality. Figure 5.3 shows this for U.S. mortalities.

Scores of millions of people live in regions that are highly susceptible to such natural disasters as hurricanes or earthquakes, posing risks whose magnitude is only 10^{-10}–10^{-11} per person per hour of exposure. Even in the United States, with its poor rail transport (compared to Europe and Japan), people who travel every day by

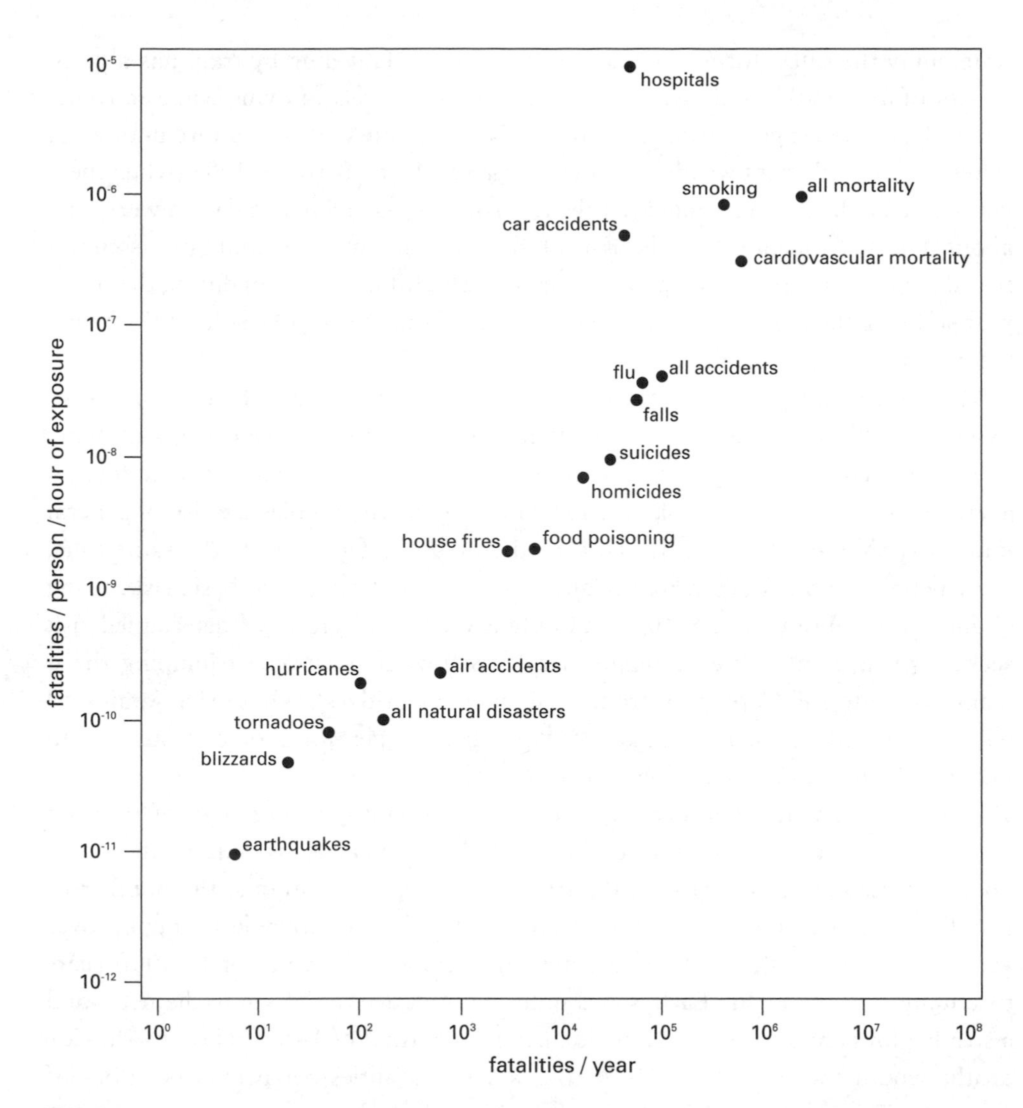

Fig. 5.3
U.S. fatalities per person per hour of exposure vs. average annual number of fatalities, 1991–2005. Both axes are logarithmic. Calculated from data published by Centers for Disease Control and Prevention, National Transportation Safety Board, National Weather Service, and U.S. Geological Survey.

train enjoy the safest form of public transportation. Traveling by train has a fatality risk of about 10^{-8}, adding a mere 1% to the overall risk of dying while en route. Similarly, the latest generation of jet planes is so reliable that only a rare pilot error (often during inclement weather) causes major accidents. Between 2002 (when there was not a single accident) and 2005 the risks of U.S. commercial aviation were only about 1×10^{-8} (identical to the risk of suicide) as some 600 million passengers boarded planes for trips averaging 2 hours (NTSB 2006). And even during the tragic year of 2001 the annual nationwide mean was about 3.3×10^{-7}, still 1 OM below the risk of general mortality.

Most people also tolerate activities that temporarily increase the overall risk of dying by 50% or that may even double it. Most notably, people drive, and many still smoke. Driving carries a risk of about 5×10^{-7} in the United States, adding on average 50% to the overall risk of dying. Smoking nearly doubles the risk of general mortality, to about 8×10^{-7} in the United States (see fig. 5.3). At the same time, most people shun actions whose relative risk is a multiple of the basic risk. Hang gliding rates more than 2×10^{-6}, and only a very small group of determined risk seekers engages in such exceedingly risky activities as fixed-object jumping (be it from Yosemite's El Capitan or from skyscrapers or bridges), whose U.S. fatalities I have calculated to be as high as 10^{-2} per person per hour of exposure (with individual exposure lasting usually less than 30 s).

The same statistical approach can be used for quantifying the risks of rare but highly destructive natural disasters. But all these calculations require three key uncertain assumptions regarding the frequency of specified events, the number of people exposed to these risks (this requires a highly questionable averaging over very long time spans), and total fatalities. Assuming a frequency of 100,000 years to 2 million years for the Earth's encounter with an asteroid 1 km in diameter and a steady global population of 10 billion people, the total of 1–1.5 billion worldwide deaths would translate to 2×10^{-10} to 5×10^{-12} fatalities per person per hour of exposure. The risk of residents of Tokyo or New York dying as a result of a smaller (200–400 m diameter) asteroid's striking their cities (assuming current populations and return frequency once in 100,000–300,000 years) would be about 3×10^{-12}.

A mega-eruption at Yellowstone, with an interval of 700,000 years and 5–50 million deaths among a future stationary population of 300 million people living in the United States downwind of the hotspot, would carry a risk of 3×10^{-12} to 3×10^{-11} per person per hour of exposure. A Toba-like event, with a frequency of 700,000 years and a worldwide (mostly Asian) death toll of at least 500 million people, would have a fatality risk rate of 1×10^{-11}. Various scenarios of mega-tsunamis generated by near-offshore asteroid impacts or volcanic mega-eruptions and destroying major coastal urban areas yield the same magnitudes (10^{-11}–10^{-12}) of risk to individuals. These risks are similar to those arising from such common natural disasters as Japanese or Californian earthquakes, but this comparison also shows the limits of this statistical approach. The common indicator of deaths per person per hour of exposure does not distinguish between relatively frequent low-mortality events and extremely rare events that could bring death to hundreds of millions or even more than a billion people.

Averaging makes perfect sense for such high-frequency risks as hurricanes and tornadoes (or airplane disasters) whose annual toll may differ by up to 1 OM but that take place every year. Averaging is also revealing when assessing the risks of phenomena with a recurrence rate of decades or centuries (major earthquakes, larger volcanic eruptions). But no society has any experience with preparing for a risk whose fatalities may prorate to a modest total of 1,000/year but which has an equal probability of taking place tomorrow or 100,000 years from now (or not materializing at all during the next million years) but when it does, it may kill hundreds of millions.

Different problems arise when this individual risk calculus is used to quantify the probabilities of violent deaths. In 2005 worldwide deaths in terrorist attacks totaled fewer than 15,000 (NCC 2006), which translates into 2.5×10^{-10} fatalities per person per hour of exposure, again on par with the risks of infrequent natural catastrophes. But that rate is not a meaningful average because this violence was concentrated. The rates in Finland or Japan were nil; for U.S. civilians (56 killed in Iraq), 2×10^{-11}; for Colombian citizens, it was 2×10^{-9}. And because more than half of all deaths due to terrorism (~8,300) took place in Iraq, the Iraqi rate was about 3×10^{-8} and the Baghdad rate (>5,000) was 1×10^{-7}.

The last number illustrates the problems of comparing different risks using individual exposure metrics. The Baghdad death risk was of the same order of magnitude as the risk faced by an average American while driving. But there are critical differences of exposure, understanding, and dread. The average time spent behind the wheel in the United States, voluntarily and with statistically well-known consequences, is less than 1 hour per day (and many people actually enjoy driving), whereas most Baghdadis could become involuntary victims of dreaded violence that can strike unexpectedly and in a variety of horrible ways (kidnapping of children, sectarian beheading, suicide bombing).

Similarly, the risk calculations reveal a very low probability of dying in a terrorist attack against U.S. citizens in the United States. The World Trade Center garage bombing in 1993 killed 6 people, the Oklahoma City bombing in 1995 caused 168 deaths, and the September 11, 2001, attacks killed 3,200 people in New York, Washington, and Pennsylvania. Consequently, during the 15 years between 1991 and 2005 the overall risk of death in a terrorist attack (using 275 million as mean population for the period) was about 1×10^{-10} per person per hour of exposure, or only slightly higher than dying in a blizzard in the snowy part of the United States. When the deaths of U.S. citizens killed in terrorist attacks abroad are also included—those in the 1996 bombing of Khobar Towers in Saudi Arabia (19 deaths), the bombing of embassies in Nairobi and Dar es Salaam (301 deaths, including 12 U.S. citizens), the Yemeni attack on the USS *Cole* (17 deaths)—the increase is insignificant.

Finally, because the invasions of Afghanistan and Iraq would probably not have taken place absent 9/11, it is plausible to argue that all U.S. military casualties (including noncombat deaths) in those countries (3,000 by the end of 2006) should be added to that total. The overall risk from terrorist attacks and military response to them then about doubles, to 2×10^{-10}, still 1 OM below the risk dying from homicide (7×10^{-9}) and 3 OM below the risk of fatal car accidents averaged over the same period of time (see fig. 5.3). During the first five years of the twenty-first century, the U.S. highway death toll exceeded the 9/11 fatalities every single month; at times it was higher in just three weeks.

Even one of the worst cases from leaked terrorist attack scenarios prepared by the Department of Homeland Security (Jakes 2005) does not imply an extreme risk. The spraying of anthrax from a truck driving through five cities over two weeks was estimated to kill 13,200 people. If these actions were to take place in metropolitan areas with populations of at least 2 million people each and be repeated every ten years, then even such an unlikely recurrence would prorate to only 1.5×10^{-8} fatalities per person per hour of exposure, a risk lower than the risk of dying from an accidental fall and less than one-thirtieth of the risk due to driving.

But, again, such perfectly valid comparisons are seen as odious, even utterly inappropriate. They pit voluntary activities undertaken by people who feel in control of actions they understand (presuming they can manage any risky situation) and who believe the outcomes are predictable (driving, walking downstairs) against involuntary exposures to unknown, unpredictable, and dreaded outbursts of violence. They also compare time-limited activities with a threat that is always there. These differences explain why the two sets of risks are viewed quite differently, and why this sort of violence is always seen as more dangerous than it really is.

Of course, this perception of a much dreaded event could become entirely justified, and the calculus of risk would profoundly change if fissionable materials were used in a terrorist attack. Fortunately, such an event would have to be preceded by many steps: a terrorist's gaining control of a state that possesses nuclear bombs, or stealing a bomb from a silo or submarine, transporting it undetected to a target, and being able to discharge it. Even so, some weapons experts argue that this could be done rather easily; others doubt it (Bunn and Weir 2005; Zimmerman and Lewis 2006).

Such irreconcilable expert appraisals are not uncommon in the assessment of risks. Some of them were noted in previous chapters, most notably, U.S. deficits as a harbinger of economic demise or a sign of economic strength, and global warming as a catastrophic or tolerable, perhaps even beneficial, change. In the absence of requisite detailed information it may be counterproductive to try to reconcile many of these disparities. What is needed instead is to have rational frameworks within which to respond to these risks and, most important, to insist on taking a number of well-known effective steps that would minimize these risks or at least moderate their consequences.

Rational Attitudes

A collective long-term probability approach must become a part of our reaction to sudden catastrophic events such as terrorist attacks or historically unprecedented natural catastrophes. Many of these events represent much lower dangers and have less profound and long-lasting consequences from the view point of national stability, economic damage, and standard of living than do many voluntary risk exposures (drinking, driving, smoking, overeating) and deliberate yet deleterious policy actions (enormous budget and trade deficits, wasteful subsidies, unchecked environmental destruction).

Indeed, Chapman and Harris (2002) argued that the disproportionate reaction to 9/11 was as damaging as the direct destruction of lives and property. In relative terms, the human toll and the economic damage were not unprecedented. As noted, discussing catastrophic premature deaths in statistical terms is a perilous enterprise, but this should not prevent us from realizing that the death of 3,200 people on 9/11 was equal to just four weeks of car fatalities averaged from the U.S. total of 38,615 road traffic deaths in 2001 (USDOT 2006) or less than two weeks' worth of U.S.

hospital deaths (44,000–98,000/year) caused by documented and preventable medical errors (Kohn, Corrigan, and Donaldson 2000).

Figure 2.24 made the same point in global terms, comparing the average annual fatalities from terrorist attacks with those from transportation accidents, major natural disasters, and errors during hospitalization for the years 1970–2005. As for the total (combat and noncombat) U.S. troop casualties in Iraq, Buzell and Preston (2007) calculated that between March 2003 and September 2006 they averaged 4.02/1,000, roughly three times the rate for a comparable age and sex cohort of young men and women in the United States (about 1.32/1,000) but significantly less than the deaths from homicide for young black males in Philadelphia.

Nor is the economic impact of the 9/11 attack entirely unprecedented. Evaluation of its costs depends on the spatial and temporal boundaries imposed on an analysis. New York City's Comptroller reported a year after the attack, on September 4, 2002, that the attack might cost the city up to $95 billion, including about $22 billion to replace the buildings and infrastructure and $17 billion in lost wages (Comptroller 2002). A broader national perspective, taking into account the longer-term decline in GDP, declines in stock values, losses incurred by airline and tourist industries, higher insurance and shipping rates, and increases in security and military spending, leads to totals in excess of $500 billion (Looney 2002).

In his November 2004 message to Americans, Usama bin-Lādin cited this total as he explained that the attacks were also chosen in order to continue the "policy of bleeding America to the point of bankruptcy," a goal aided by "the White House that demands the opening of war fronts" (bin-Lādin 2004, 3). Facetiously he added, it appears to "some analysts and diplomats that the White House and we are playing as one team" towards "the economic ruination of the country" ("even if the intentions differ"), and he cited the Royal Institute of International Affairs estimate that the attacks cost Al Qaeda just $500,000, "meaning that every dollar . . . defeated a million dollars."

This is an interesting but inaccurate interpretation. Most of the losses proved to be only temporary. Elevated monthly U.S. stock market volatility was as transitory in the wake of 9/11 as it was with other market disruptions (Bloom 2006). GDP growth, commercial flying, and tourism recovered to pre-9/11 levels with remarkable rapidity. And five years after the attack the U.S. budget deficit was a lower share of the country's GDP than in France or Germany. But even an overall loss of half a trillion dollars would not have been without precedent in the realm of nonviolent events. That much money was lost on a single day, October 19, 1987, when the New York stock market fell by a record 22.9%, a drop nearly twice as large as the infamous 1929 crash.

Risk studies offer revealing observations relevant to the perception and appraisal of catastrophes and trends (Morgan and Henrion 1990; Slovic 2000; Morgan et al. 2002; Sunstein 2003; Gigerenzer 2002; Renn 2006). First, in the immediate aftermath of catastrophic acts the stricken populations have strongly exaggerated perceptions of a repeat event. This has been true not only about constant post-9/11 expectations of another terrorist attack in the United States but also about almost psychotic fears of another massive hurricane in the aftermath of Katrina. Second, unfamiliar risks and those that appear to be impossible or very hard to control elicit a disproportionately high public fear. This can lead to patently excessive responses compared to reactions to old recurrent hazards or those that can produce further, indirect damage. The more dreadful and the more unexpected a risk, the more the public clamors for protection and a solution, and the more the public and governments are prepared to spend.

The post-9/11 haste to set up the Department of Homeland Security, an unwieldy and ineffective bureaucratic conglomerate of dubious utility, is a perfect example of such counterproductive public reaction, and reduced air travel and increased frequency of car trips illustrate irrational individual responses to the tragedy. Gigerenzer (2002) demonstrated that during the first 12 months after the 9/11 attack, nearly 1,600 U.S. drivers (six times the number of passengers killed on the four fatal hijacked flights) lost their lives on the road by trying to avoid the heightened risk of flying. Sunstein (2003, 121) makes another important point: "When strong emotions are involved, people tend to focus on the badness of the outcome rather than on the probability that the outcome will occur." This "probability neglect" helps to explain why societies have excessive reactions to low-probability risks of spectacular and horrific events. Terrorists operate with an implicit understanding of these realities.

This leads to some uncomfortable conclusions. Systematic post-9/11 appraisals have left us with a much better appreciation of the variety of threats we might be facing, but have done nothing to improve our capacity for even the roughest ranking of the most likely methods of globally significant future attacks and hence for a more rational deployment of preventive resources. Nor do we get the basic directional feeling regarding the future threats: Will they be more spectacular versions of successful attacks (e.g., striking New York's subway stations)? Or will they be copies of failed attempts, much as the prevented 2006 trans-Atlantic airliner attacks were copies of Ramzi Yusuf's plot prepared for 1995? Or will they be dreaded premieres of new ways of spreading death and mass fear (a dirty bomb, successful large-scale release of a pathogen), or even some as-yet-unidentified attacks?

All we know with certainty is that historical lessons are clear: ending terror everywhere is impossible. Terrorism is deeply rooted in modern culture, and even a virtual eradication of one of its forms or leading groups settles little in the long run because new forms and new groups may emerge unexpectedly, as did plane hijackings in the 1960s and al-Qaeda in the 1990s. Analogies between militant Islam and the Mafia are instructive. As Cottino (1999) explained, the Mafia's persistence is rooted in Sicily's culture of violence, which represents the historical memory of a particular world, a mind set resistant to change.

Analogously, militant Islam is rooted in deeply persistent historical memories. Westerners do not appreciate the extent to which the Crusades, the most explicit expression of ancient European aggression against the Muslim world, are alive in many disaffected Muslim minds (Hillenbrand 2000; Andrea 2003). Add to this the feelings of more recent humiliations brought about by the post–World War I European expansion and the carving up of the Ottoman Empire, the Western (particularly U.S.) support of Israel, and a deepening perception of unjust economic exploitation. As bin-Lādin (2002) said in his letter to Americans, the existing Middle Eastern governments "steal our '*umma's* wealth and sell it to you at a paltry price." What can the West do? Keep apologizing for events that took place 800–900 years ago, work for reinstating the caliphate, stop consuming Middle East oil, consent to the destruction of Israel? In the absence of such actions accumulated Muslim grievances will continue to spread (as Muslim populations grow) and deepen (as defeat of the West remains elusive).

And the Mafia analogy also suggests an important conclusion. Anti-terrorism strategy should be framed not as a war but as a repressive action against a cellular, secretive, networked, violent organization. A clear collateral of this reality is that there can be no meaningful end to this effort, and no victory (Schneider and Schneider 2002). Streusand and Tunnell (2006) made a thoughtful plea that in this effort we should use language that can actually help to fight Islamic terrorists. *Allah* should simply be translated as *God* and not transliterated because that treatment exaggerates the difference among the three monotheistic faiths. *Jihād* (whose primary meaning was explained in chapter 3) should not be used as a blanket label for brutal indiscriminate terrorist killings. *Hirabah*, sinful warfare waged contrary to Islamic law, is the proper term, and so is *mufsid* (evil, corrupt person), not *mujāhid*. Such terms, readily understood by Muslims, would remove any moral ambiguity from our dealings with *al-dar al-islām*.

But perhaps nothing is more important for the exercise of rational attitudes than always trying to consider events within longer historical perspectives and trying to avoid the chronic affliction of modern opinion makers who tend to favor extreme positions. The product of these ephemerally framed opinions is a range of attitudes and conclusions that resemble the cogitations of an unstable manic-depressive mind. Unrealistic optimism and vastly exaggerated expectations contrast with portrayals of irretrievable doom and indefensibly defeatist prospects. Examples of such extreme contrasts are easy to find in any area where future risks and trends have to be assessed rationally in order to come up with rational responses.

After 9/11, Fallaci (2002), who spent decades documenting political ambitions and wars and who deeply understood violence and personal danger, was unequivocal in her passionate appeal to the West that "the worst is still to come" and that Europe is committing collective suicide. But Fallows (2006), President Carter's former speechwriter, claiming that he voiced a broad consensus of Washington D.C. experts, concluded that the United States is succeeding in its struggle against terrorism, that "we have achieved a great victory," and that "the time has come to declare the war on terror over."

The fortunes of the dot-com economy exemplify how the same commentators can switch among two attitudes at the first indication of a shift. The new economy of the 1990s was to usher in an era of endless economic growth—heady valuations of companies that had nothing to sell, a bestselling book entitled *Dow 40,000*. But once the bubble burst, a new episode of U.S. economic difficulties (contrasted with the rise of China and the buoyant euro) swiftly became a harbinger of a collapsing dollar and an unstoppable demise. In reality, the new economy was not all that new, and it has not measured up to the truly great inventions of the past (Gordon 2000; Smil 2004). And an imminent U.S. economic collapse was yet again postponed.

While many commentators on the U.S. economy of the early 2000s indicate a system in considerable distress, there are also signs of relative resilience and even of continued unrivaled primacy. In 2005, U.S. government debt was equal to about 64% of GDP, a lower share than that of EU-12 (about 71%), slightly lower than the French or German share, and an almost 10% lower share than in 1996 (73%). A surprise: the U.S. government has actually been less reckless in living beyond its means than were the governments of the EU countries (fig. 5.4). And the United

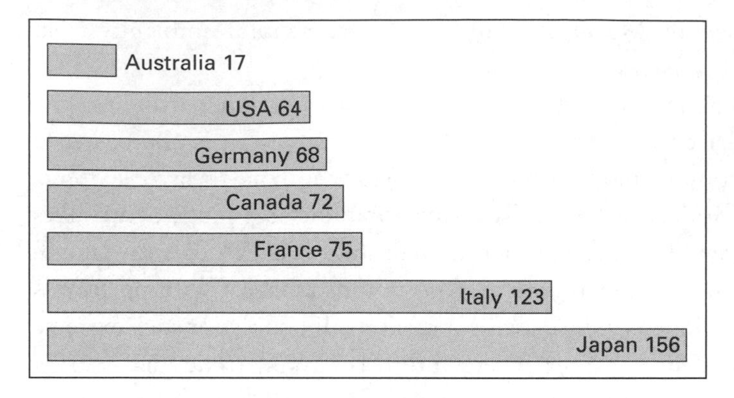

Fig. 5.4
Gross government debt as a percentage of GDP in seven countries, 1990–2005. From OECD (2006).

States still dominates the activities that have created the global electronics civilization. In 2005 the country had 116 of the world's top 250 companies (ranked by revenue) in this sector (electronics, telecommunications, and information equipment, software and allied services), Europe had 42, and China and India had none (OECD 2006).

Moreover, by 2004 30% of the world's scientific publications were still authored by U.S. researchers, compared to 6.5% in China (Zhou and Leydesdorff 2006). Adjusted for population size, this means that the U.S. output was 20 times higher than in China. Judged by these measures, China will not be taking over anytime soon. And by 2006 it also became quite clear that Boeing, whose fortunes fluctuated for years, was a success after all. Years of writing off Boeing and crowning Airbus as a new king of aviation are gone (for the time being). Boeing, with its new 787 and a new version of the venerable 747, is gathering record orders, while Airbus, with its superjumbo 840 schedule slipping, is sinking deeper into managerial ineptitude and technical and financial troubles.

To give a notable social example of reality that contravenes a preconceived conclusion, Huntington's (2004) argument about the dangers of Hispanic immigration—because of the common language, proximity of mother countries, regional congregation, residential segregation, and less interest in assimilation—is not supported by research done in Los Angeles and San Diego. Rumbaut et al. (2006) found that even in the country's largest Spanish-speaking region that is contiguous with

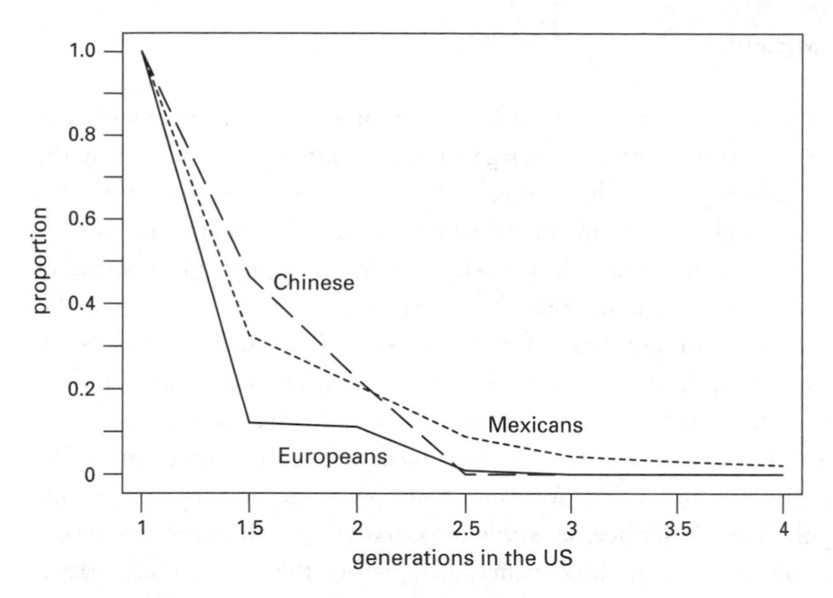

Fig. 5.5
Proportion of immigrants in Los Angeles and San Diego who speak their mother tongue at home, by generation. Based on Rumbaut et al. (2006).

(and historically a part of) Mexico, the preference for speaking Spanish at home declines only slightly slower than for other immigrant languages, getting close to a natural death by the third generation (fig. 5.5). So the prospect is not that good for the Mexican *reconquista* of the U.S. Southwest.

Rational attitudes should inform our deliberate decisions. What we get instead are increasingly splintered approaches to major challenges facing modern societies, perpetuated and accentuated by powerful special-interest lobbies and pressure groups, be they corporate relief seekers or Green warriors. The resulting adversarial and confrontational attitudes to modern policy making make it harder to formulate consensual steps. The key advantage of this prevailing approach is that it should help to minimize the probability of rash, overreaching decisions, but too many unfortunate decisions that have been made belie this hope. Consequently, I always find myself arguing that we should act as risk minimizers, as no-regrets decision makers who justify our actions by benefits that would accrue even if the original risk assessments were partial or even complete, failures.

Acting as Risk Minimizers

There is nothing we can do to avert such low-probability natural catastrophes as volcanic mega-eruptions or mega-tsunamis generated by asteroid impacts or by the most powerful earthquakes. In this sense, our civilization is no different from the cuneiform or hieroglyphic realms of the Middle East 5,000 years ago. We are getting better at anticipating some volcanic activity—clear warnings given ahead of Mount Pinatubo's 1991 eruption enabled the evacuation of more than 50,000 people and the limitation of fatalities to fewer than 400 (Newhall et al. 1997)—but discerning the likely magnitude of an eruption remains beyond our ability.

Earthquake prediction has been no less elusive. Minimal advance warnings have been sufficient to prevent any catastrophic derailment of Japan's rapid trains. The earthquake detection system devised for *shinkansen* picks up the very first seismic waves reaching the Earth's surface, uses this information to determine risk levels, and can halt trains or at least slow them down before the main shock arrives (Noguchi and Fujii 2000). USGS now makes routine 24-hour forecasts of the probability of aftershocks in California, and it continues its multidisciplinary research in the town of Parkfield, situated on the San Andreas fault, aimed at eventual better prediction of earthquakes.

But we can do much to be better prepared for a number of anticipated catastrophes, and we can take many steps to moderate negative impacts of some of the most worrisome trends. The precautionary principle should be invoked precisely when facing those high risks whose understanding is characterized by uncertainty or outright ignorance. Preventive, preparatory, or mitigating actions are called for in order to avoid extreme consequences of unmanaged outcomes, whether a viral pandemic, global warming, or the use of weapons of mass destruction by terrorists. In such situations imperfect scientific understanding or substantial uncertainties regarding the most likely mode and intensity of the next catastrophic event should not be used as an excuse for inaction or for postponing known effective measures.

We should act incrementally as prudent risk minimizers and pursue any effective no-regrets options. We do not have to wait for the formulation and acceptance of grand strategies, for the emergence of global consensual understanding, or for the universal adoption of more rational approaches. In order to illustrate the wide-ranging opportunities for effective or potentially promising no-regrets measures, I give examples for such disparate challenges as preventing an unlikely encounter with a sizable asteroid, preparing for a new influenza pandemic, eliminating alien invasive species from islands, and preserving the richest realms of biodiversity.

Surprisingly, we may soon be able to defend ourselves against an extraterrestrial object that is on a collision trajectory with the Earth. Once we have surveyed all NEOs larger than 1 km (the target date is the end of 2008) we will be able to calculate the risk of a major terrestrial impact with unprecedented completeness and accuracy. Moreover, at the current rate of discovery some 90% of all NEOs large enough to pose a global risk should be detected by the year 2020 (Rabinowitz et al. 2000). A low-probability surprise of encountering a new asteroid large enough to do globally significant damage will remain, and it is for this eventuality that serious proposals have been made to destroy the object or deflect it from its potentially destructive path (Milani 2003). Only a smaller body would be a candidate for the first option. Breaking up a large object would create a rain of still fairly large and randomly tumbling pieces and actually exacerbate the problem.

Deflection would also demand much less energy. Only 3 mJ/kg would be needed to change a body's velocity by about 10 cm/s, enough to convert a predicted hit into a miss a year later (Chapman, Durda, and Gold 2001). If we could dock with an object and push it, we could, with adequate warning, perform such tasks. For example, a few minutes' burn of the first stage of a Delta 2 rocket could deflect a 100-m-diameter object with six months' warning. Given the magnitude of damage that could be avoided by such a maneuver, it seems prudent for national space agencies to have well-funded long-term programs to develop and explore the most suitable techniques that might avert a potential global catastrophe (Schweickart and Chapman 2005). Compared to monies spent on risky, expensive, and repetitive manned missions, this relatively modest investment would be a most valuable form of planetary insurance.

As for a new viral pandemic, we are incomparably better prepared than in 1918. At that time there were no mechanical respirators, no readily available supplemental oxygen, and no antibiotics to treat secondary infections. And we are better off than we were even in 1968–1969, during the last flu pandemic. Our scientific understanding (virological, genetic, epidemiological) is vastly superior, new antiviral drugs afford some preventive capability, and there is a much better system for the near-instantaneous sharing of relevant information and for coordinating an effective response. Not surprisingly, major uncertainties remain.

None are more important than the pathogenicity and the morbidity profile of the next pandemic: a recurrence of the W-shaped pattern (see fig. 2.17) typical of the 1918–1919 event would have a severe impact even if overall pathogenicity remains fairly mild (MacKellar 2007). Complex mathematical models can help us to estimate the needed stockpiles of vaccines and their modification and targets as well as to optimize quarantine measures (D. J. Smith 2006). We cannot be certain about the effect of massively distributed antivirals, but stochastic simulations are encouraging. A large-scale epidemic simulation by Ferguson et al. (2006) found that international travel restrictions would only delay the spread of a pandemic (by 2–3 weeks) unless 99% effective, but if antivirals are administered to half the affected population within a day of symptom onset, they could, together with school closures, cut the clinical attack rates by 40%–50%, and a more widespread use of drugs could reduce the incidence by more than 75%.

Similarly, a simulation for Southeast Asia showed that as long as the basic reproductive number (the average number of secondary infections caused by a typical infected individual) remained below 1.6, a response with antiviral agents would contain the disease, and prevaccination could be effective even if the mean reproductive number were as high as 2.1 (Longini et al. 2005). Another simulation for the region confirmed that elimination of a nascent epidemic would be practical if the basic reproduction number remained below 1.8 and there were a stockpile of at least 3 million courses of antiviral drugs (Ferguson et al. 2005).

The limited supply of the most efficacious antiviral agent, Roche's Oseltamivir (Tamiflu), caused a great deal of temporary concern (even panic) in 2005, but new licensing agreements and new synthetic methods increased the drug's annual production rate by 2 OM between 2003 and 2007 (Enserink 2006). But the treatment (10 capsules) remains expensive, and it is only presumed to be as useful against an eventual pandemic strain as it is against seasonal viruses. There is also a high probability that any large-scale use for prophylaxis and treatment will lead to the evolution of drug-resistant strains (Regoes and Bonhoeffer 2006). The only effective prevention is vaccination, but because the form of the next pandemic cannot be predicted, the only foolproof strategy is the preparation of high-yield seed viruses of all 16 HA subtypes so they would be ready for potential mass production of vaccines. The ultimate, still elusive, goal—a vaccine that could protect against all human influenza strains—is being pursued by academic and big pharma researchers (Kaiser 2006).

Combating invasive species by simply removing them from infested islands is another no-regrets strategy that has already yielded some gratifying results. Eradication of alien terrestrial and aquatic species on continents is virtually impossible because the invaders readily migrate to adjacent habitats. Until the 1980s it was thought impossible to do the job even on relatively small islands. But by 2005, 234 islands had been cleared of invasive rats; 120 islands, of goats; 100 islands, of pigs; and 50 islands, of invasive cats and rabbits (Krajick 2005). A notable example of a recent success is ridding the Galápagos Islands of feral goats (already gone from Isabella, Santiago, and Pinta); now the aim includes cats and rats (Kaiser 2001; Guo 2006). Results are seen quickly as native vegetation returns and previously decimated song birds or lizards reclaim their habitats.

And it turns out that preserving the richest repositories of biodiversity does not have to be unrealistically expensive. James, Gaston, and Balmford (1999) put the annual cost of safeguarding the world's biodiversity at about $17 billion added to the inadequate $6 billion spent currently. Needed are adequate budgets for already protected areas and the purchase of additional land in order to extend coverage to a minimum standard of 10% of area in every major biodiversity region. Total cost would be equivalent to less than 0.1% of the combined GDP of the United States and the European Union, less than 5% of the monies now spent annually on generally environmentally harmful agricultural subsidies, and less than Westerners spend annually on yachts or perfume.

By far the best example of a rich opportunity to deploy a no-regrets strategy is minimizing the future magnitude of global warming. I hasten to emphasize that this strategy should not feature the currently fashionable carbon sequestration (IPCC 2005). To keep on generating ever larger amounts of CO_2 and to reduce its climate impacts by storing billions of tonnes of the compressed gas underground is decidedly a distant second-best approach. As long as we depend heavily on the combustion of fossil fuels, the reduction of atmospheric CO_2 levels would best be accomplished by striving for the lowest practicable energy flows through our societies, a strategy that would result in significant cuts of greenhouse gas emissions regardless of the imminence or the intensity of anticipated temperature change.

This risk-minimizing strategy would be insurance against prevailing uncertainties and dramatic surprises, but this benefit would not be the only, not even the most important, reason for its adoption. Most important, lower emissions of CO_2 would require reduced consumption of fossil fuels, a goal we should have been pursuing much more aggressively all along because of its multiple benefits. These gains include a large number of environmentally desirable changes and important health and socioeconomic benefits. Examples include less land destruction by surface coal mining, lower emissions of acid-forming gases, reduced chances for major oil spills, cleaner air in urban areas, improved visibility, and slower acidification of the ocean. By reducing the overall exposure to particulates, SO_2, NO_x, hydrocarbons, ozone, heavy metals, and ionizing radiation from coal burning, moderated fossil fuel combustion would lower the morbidity of exposed populations and improve their life expectancies. These changes would also improve the collective quality of life by lowering health care costs.

These conclusions are supported by large-scale epidemiological studies of excess mortality as well as by morbidity comparisons. For example, a multicity analysis of mortality found that a daily increment of 20 $\mu g/m^3$ of inhalable particulate matter (produced largely by power plants and vehicles) increases mortality by about 1% (Samet et al. 2000). And the benefits of reduced photochemical smog are illustrated by comparing Atlanta's acute asthma attacks and pediatric emergency admissions during the Olympic Games of 1996 (when measures that reduced traffic by some 30% were in effect) with the same periods during the previous and the following year (Friedman et al. 2001): asthma attacks fell by 40% and pediatric emergency admissions declined by 19%. Ask any asthmatic child or its parents if these are not good enough reasons to reduce the emissions. Moreover, lowering the energy intensity of economic output would increase a nation's competitiveness in foreign markets, a development that would benefit the balance of payments and create new employment opportunities.

But the historical lesson of the long-term impact of more efficient energy conversion is clear: it promotes rather than reduces aggregate energy consumption. Consequently, in all affluent countries where per capita energy use is already 1 OM higher than in populous modernizing nations (U.S., 350 GJ; China, 40 GJ; France, 170 GJ; India, 20 GJ), all future efforts to reduce specific carbon emissions (per vehicle, per kilometer driven, per kilowatt hour of electricity generated, per kilogram of steel smelted) must be combined with efforts to reduce overall per capita consumption of carbon-intensive commodities and services. Otherwise more efficient conversion will merely keep expanding the affluent world's already excessive use of energy.

I am tired of hearing that this cannot be done in a free market setting when it easily could be. The standard mantra is that one cannot regulate individual choice, that people yearning to drive a 4-tonne military assault machine to take them to a shopping center should be free to do so. This argument is risibly immature and utterly inconsistent because the purchase and use of such a vehicle is already subject to a multitude of restrictions and limits that are designed to increase safety, protect environmental quality, and promote social equity: seatbelts, airbags, unleaded low-sulfur gasoline, mandated minimum fuel consumption, scores of traffic rules, and taxes paid on the vehicle's purchase and with every tank fill-up.

If we accept as normal and civilized stopping at red lights to safeguard the lives of people crossing, and paying more for better and more heavily taxed gasoline to eliminate lead pollution and increase the revenue for social programs, should we not accept as normal and civilized putting an absolute limit on the size of vehicles in order to preserve the integrity of the only biosphere we have? Such steps would be merely rational extensions of restrictions and limits that have already become inevitable.

Missed opportunities for higher U.S. automotive efficiency illustrate the benefits of the approach. Between 1973 and 1987 all cars sold in the country had to comply with CAFE standards, which doubled the fleet's performance (halved the fuel consumption) to 27.5 mpg (8.6 L/100 km) (EIA 2005). The post-1985 slump in crude oil prices first stopped and then actually reversed this progress as vans, SUVs, and light trucks were exempted from the 27.5 mpg CAFE minimum. As these vehicles became more popular (by 2005 they accounted for nearly half of the passenger fleet), the nation's average fuel rate fell to only 22 mpg (EIA 2005). But if the 1973–1987 CAFE rate of improvement had been maintained (no great challenge from the technical point of view) and applied to *all* passenger vehicles, the average fleet performance in 2005 would have been close to 50 mpg. This means that automotive hydrocarbon, NO_x, and CO emissions as well as U.S. crude oil imports could have been cut by about two-thirds.

The Next 50 Years

I wish I could close with a crisp recapitulation that would neatly rank-order the probabilities of all plausible catastrophes during the next 50 years even as it tamed all the disparate trends surveyed in this book with a clever taxonomy offering measures of their likelihood, intensity, likely duration, and eventual impact. The former task is at least partly possible. Because I favor quantitative appraisals, I have offered as many as I could in assessing the probabilities of globally significant natural catastrophes, including a new viral pandemic. Here is a brief summation, necessarily punctuated, by many qualifying statements.

Fairly reliable judgments are possible regarding the major natural catastrophes. In order to leave a deep mark on global history they would have to be on scales not experienced during the historic era, and they would have to claim, almost instantly or within a few months, many millions of lives. Events of such a magnitude took place within the past million years, but none of them have probabilities higher than 0.1% during the next 50 years (fig. 5.6). But I must stress that the past record, while highly indicative, is basically one of singularities, events that are too few and mostly too far apart to allow for any meaningful statistical evaluation beyond the simplest calculations of highly uncertain return frequencies and approximate recurrence probabilities.

Still, we now know enough about NEOs to rank the danger from impacting asteroids as being the least likely discontinuity with the potential to change near-term history. Moreover, the overall risk may be revised substantially downward during the coming years and decades as our classification effort is completed and future trajectories are computed. The only reason we should not dismiss this worry entirely is that while we now know a great deal about the probability of a major impact, we cannot easily translate this knowledge into the number of immediate and delayed fatalities. There are simply too many factors to consider, and hence there is always a highly improbable possibility of a relatively minor impact's causing disproportionately consequential damage.

The Indian Ocean tsunami of December 2004 was a tragic reminder that the potential for large-scale natural catastrophes claiming 10^5–10^6 lives is always with us. But catastrophes able to kill 10^7 people—most likely a Toba-like mega-eruption that would affect directly (with lava flows and ejecta) not only a densely populated country but a large part of a hemisphere (with volcanic ash and possibly tsunamis) and that would cause multiyear global atmospheric cooling—appear to have a probability less than 0.01% during the next 50 years. By contrast, there is a high probability (1 OM higher than that of a new supereruption) of an influenza pandemic that would rival or surpass the greatest such event on the record. And a simple probabilistic assessment shows that the risk of a transformational mega-war is of the same order of magnitude (see fig. 5.6). But because the event of 9/11 has been the only terrorist attack that indisputably changed the course of global history, it is impossible to offer any meaningful probabilistic assessment of the frequency, intensity, and consequences of future attacks.

If we are to act as rational risk minimizers, the current preoccupation with terrorism should not blind us to what are historically two much more likely threats: another mega-war and another pandemic (possibly two) during the next 50 years. Early interventions to defuse emerging causes of potentially massive armed

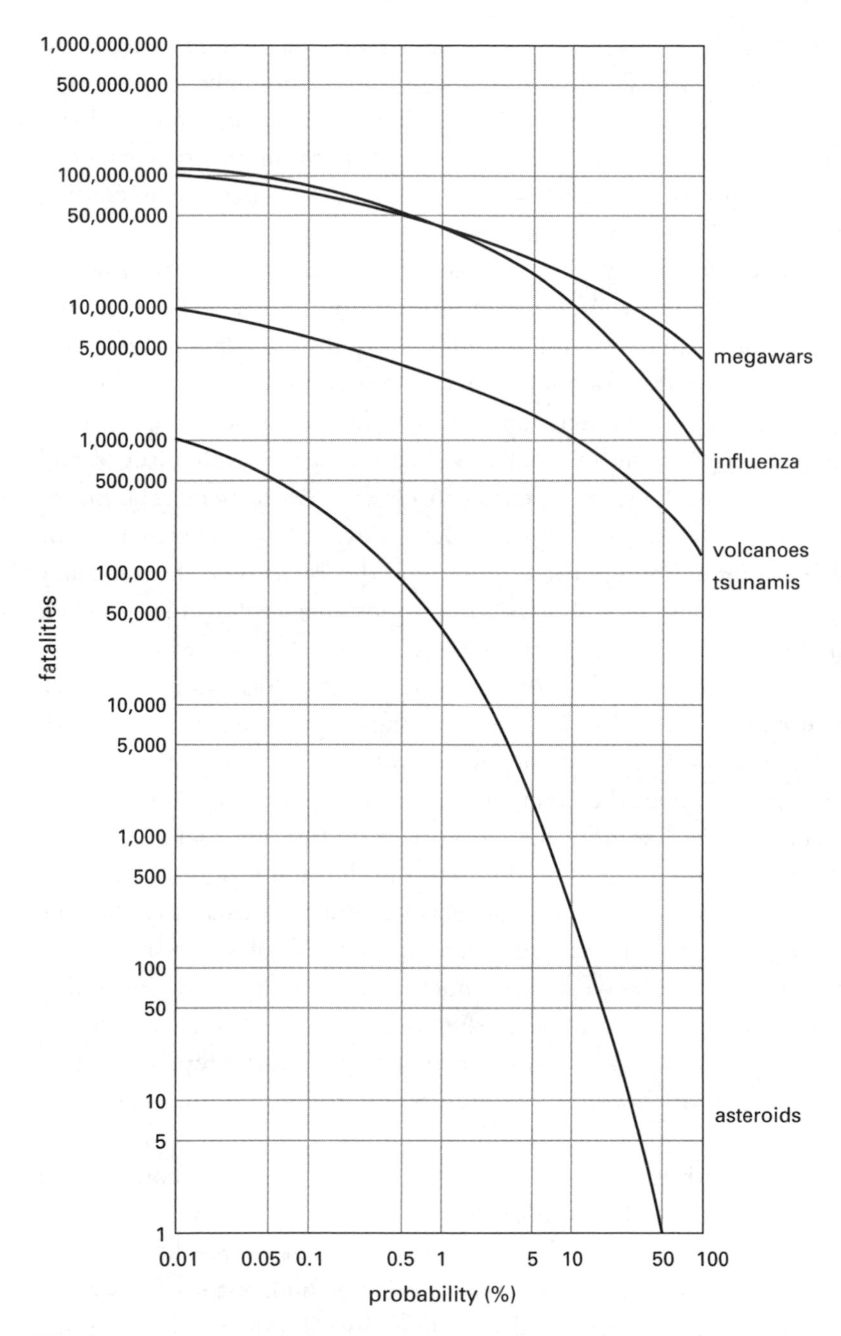

Fig. 5.6
Probabilities of fatal discontinuities during the first half of the twenty-first century, from an extremely low chance of catastrophic asteroid impacts to near-certainty of another virulent influenza epidemic. All curves are approximate but show correct orders of magnitude. Calculated from data presented in chapter 2.

confrontations, and better preparedness for a major pandemic, would be the most rewarding risk-reducing steps. We can forget (relatively speaking) about near-Earth asteroids, supervolcanoes, and monster tsunamis, but we must not underestimate the chances of another mega-war, and we must remember that unpredictably mutating viruses will be always with us. That is why we should be constantly upgrading preparedness to deal with a new pandemic.

In the past we have taken many steps to lessen the risk of a thermonuclear war (those that formed a part of the U.S.-Soviet détente, and the post-1991 mutual nuclear arms reductions), and we must continue with all possible efforts of this kind. We can be better prepared for another major terrorist attack. A combination of rational steps, such as better evaluation of available intelligence, more flexible armed response (keeping in mind the Mafia analogy), and the gradual social and political transformation of Muslim societies can clearly reduce the likelihood of terrorist attacks and moderate their impacts. But we will always confront major uncertainties. We cannot meaningfully quantify either the intensity or the frequency with which organized groups can sustain terrorist attacks whose global impacts put them in the category of 9/11 events (or worse).

At the same time, the historical realities are not entirely discouraging. While there is no chance of a terror-free future, the long-term record is not one of debilitating fear and despair or ubiquitous loss of life and destruction. Although we must remain agnostic about the eventual impact of Rapoport's fourth wave of terror, good arguments can be made for seeing it as a shocking (painful, costly) but manageable risk among other risks. We can note that the general tendency is to exaggerate the likelihood of new infrequent spectacular threats, and that the group participation, preparation, and organization, required for successful terrorist attacks could also work to prevent such attacks. At the same time, it is understandable why a responsible political leadership would tend to see these terrorist threats as an intolerable challenge to the perpetuation of modern open societies, and why it may overreact or choose (under pressure or from desperation) less than rational responses.

The necessity to live with profound uncertainties is a quintessential condition of our species. Bradbury got it right in *Fahrenheit 451* (1953): "Ask no guarantees, ask for no security, there never was such an animal." We, of course, keep asking. But we have no way of knowing if we are exaggerating or underestimating what is to come, be it from an inexplicable, accidental slide toward a mega-war, from the depth of the militant Islamic hatred, or from random mutations of viral genes. If we are grossly underestimating these risks, there is little we can do to make any fundamental difference. There is simply no way to prepare for a terrorist attack with hijacked nuclear-tipped missiles that could produce tens of millions of instant fatalities, or for a highly virulent pandemic that would produce more than 100 million deaths.

By contrast, there is a great deal of certainty regarding the duration and intensity of such fundamental trends as rapid aging of affluent populations, the need for a transition from fossil fuels, the rising economic importance of the world's most populous modernizing economy, the continuing impoverishment of the biosphere's diversity, or further atmospheric temperature increases. These trends continue to be probed by techniques that hope to forecast or model future events. But the greatest reward of the new quantitative models is the heuristic benefit their construction brings to the modelers, not their ability to capture anything approaching future complex realities. As for the standard applied forecasting models, they are nothing but series of ephemeral failures: thousands of forecasts are constantly issued by numerous chief economists and think tanks, only to be superseded by equally pointless exercises days or weeks later.

Exploratory scenarios are usually more sensible because they do not pretend to quantify the unquantifiable, but their limited utility springs from their inherently limited scope. A major project will issue a handful of scenarios that may make for an interesting reading but that do not add up to any helpful policymaking foundation. NIC's (2004) mapping of the global future to 2020 is a very good example. It offers an excellent conclusion, foreseeing "a more pervasive sense of insecurity" based as much on perceptions as physical threat, but it is extremely unlikely that any of its detailed scenarios will come to pass. We will not have either a Davos World (unlimited globalization) or a new caliphate with a world run by *sharīʿa* enforcers sitting in Baghdad or Kabul.

If trends cannot be easily quantified or captured by exploratory scenarios, they cannot be meaningfully ordered or ranked either. It would be largely a matter of guesswork, not an exercise based on the frequency of past events. Even rare discontinuities are more amenable to quantification than are the intensities and durations of gradually unfolding trends. Any meaningful taxonomy (or even just a simple ranking) is undercut by two incessant processes: the changing intensities of even the most embedded trends, and the shifting significance and concerns, which result not only from complex interactions among closely allied processes but also from often stunning impacts of previously underestimated or ignored trends.

An unexpected temporary upturn of U.S. economic fortunes during the 1990s is an example of the first category. It could not alter the fundamental trend of the country's declining weight in the global economy, but (coinciding with an equally unexpected retreat of Japan's economy and the socioeconomic unraveling of the post-Soviet states) it briefly interrupted and temporarily reversed that slide. Examples in the second category abound because the key drivers of major trends keep shifting. Radical Islam was not on anybody's list of factors threatening the United States during the 1980s. Indeed, during that decade, some protagonists of radical Islam were coopted by Washington's strategists, via the Saudi–Afghan connection, to fight the Soviet Empire, and even after the first World Trade Center attack in 1993, U.S. policymakers showed remarkable reluctance to tackle al-Qaeda.

But after 9/11 the threat of terrorism has asymmetrically infected all these policymakers' major actions, be it the setting of interest rates or lamenting the response to hurricane Katrina. As a result, a bearded, bespectacled, French-speaking Egyptian physician turned global terrorist is arguably as great a driver of the U.S. secular decline as is the excessive, deficit-raising consumption of the Wal-Mart-bound masses or the education system whose average graduates rank in problem-solving skills as poorly as they do in literacy and mathematics. How can we systematically compare or quantify these disparate, ever-shifting drivers?

Trends may thus seem obvious, but because they have a multitude of drivers whose importance shifts constantly, the resulting mix is beyond anyone's grasp. So, when contemplating the great U.S. retreat, it is useful to recall that historians have identified scores of causes for the decline of the Roman Empire (Rollins 1983; Tainter 1989), and only a naive mind would rank German tribes ahead of debased currency, or imperial overstretch ahead of ostentatious, consumption, in that typology of the causes of Rome's fall. This is also the reason that even when large corporations and large nations are fully aware of unfolding trends, they are rarely able to shape them to their long-term advantage.

Thus any verdicts must be circumspect, guided by an awareness of profound uncertainties. Perhaps the most assured conclusion is to say that the survival of modern civilization would be most severely tested by a nuclear war, be it through a failure of national controls, a madly deliberate launch by an irrational leadership, or a takeover of nuclear weapons by a terrorist group. The unknown risk of this catastrophe may be very low, but no anthropogenic act comes close to causing that many instant casualties. Threats from weapons of mass disruption (including any truly dangerous deployment of pathogens) would rank much lower. Risks of globally significant natural geocatastrophes should command sustained scientific attention but, relatively speaking, should not be a matter of great public concern. The threat of a new viral pandemic with its potentially massive mortality dictates the need for scientific progress and for greater public preparedness. By contrast, I would reserve the prospects of Earth's life being crippled during the next 50 years by entirely new pathogens or devouring nanobots for sci-fi accounts.

The relative diminishment of U.S. economic power has been under way since the end of World War II, and it is an entirely expected trend, not a cause for concern. What is more important is the rapidity of this decline and its eventual extent, and the concurrently unfolding fortunes of other principal contenders for the place on top. Russia and Japan will remain major economic powers for decades to come, but their chances of claiming the dominant spot are highly unlikely because both of them are rapidly aging and significantly depopulating. Neither Russia's vast resources nor Japan's productive proficiency will be enough to make them contenders for global primacy.

China is now a foremost contender for this position, and many foreigners (and Chinese) see it as an inevitable winner. These are imprudent projections in the face of a multitude of long-term complications and serious checks that delimit China's ascent: an extraordinarily aberrant gender ratio, serious environmental ills, the increasing inequality of economic rewards, and its weak soft-power appeal. India's quest, largely a race to catch up with China, also faces many natural, social, and economic limits.

This leaves us to consider the role of the resurgent Islam, but the very heterogeneity of that faith and the major social, cultural, and economic divides among its adherents preclude the emergence of a unified Muslim power. Violence emanating from politicized Islam is a major (and frightening) challenge to the West, to a large extent because its common practice often takes the form of a sacrificial fusion of murder and suicide, a delusion that does not leave any room for compromise, negotiation, or rational arguments. Even so, in the long run this violence will be only one (and perhaps not even the most important) ingredient of a complex challenge posed by a combination of the continuing relatively fast growth of Muslim populations, the modernization deficit of their economies, and the radicalized nature of their internal affairs.

The fortunes of all countries could be greatly affected by one of the most worrisome global trends, the growing inequality of incomes and opportunities. This gap is widening even as modernization has yet to bridge the enormous gap between the affluent Western world and rapidly advancing parts of Asia on one hand, and four score of insufficiently improving or backward-moving economies, above all those in Africa. This bridging will be made more difficult because of the world's fundamental energy problem, the stark poverty of choice for delivering large quanta of energy harnessed with high power densities to highly urbanized societies living increasingly in mega-cities.

We do not currently have any resource or conversion technique that can supply the nearly 10 Gt of fossil carbon that we now extract from the Earth's crust every year. Neither wind nor biomass energy will do as fundamental fossil fuel substitutes. Their low power densities (and wind's stochasticity) are utterly mismatched with the needs of today's industries and settlements, and are unfit for ever-higher base load needs of pervasively electrified societies. Nuclear energy is the best high-power-density choice, but its unpopularity and the persistent absence of permanent waste storage facilities militate against its grand resurrection in the West. Our continuing reliance on fossil fuels (and there is no shortage of coal and unconventional hydrocarbons) may be leading us to an unprecedented increase in tropospheric temperatures.

Compared to this challenge, all other unwelcome trends may come to be seen as relatively unimportant. But it also may be that by 2050 we will find that global warming is a minor nuisance compared to something that we are as yet unable to identify even as a remote threat. Stepping back again, in 1955 few were concerned about the CFC-driven destruction of stratospheric ozone, which by 1985 panicked the planet. But even if global warming turns out to be a manageable challenge, its uneven regional and national consequences may weaken, change, or terminate a number of seemingly embedded economic and strategic trends.

This combination of catastrophic risks and unfolding worrisome trends appears insurmountable to many who think of civilization's fortunes. There are, of course, irrepressible high-tech enthusiasts who keep on envisioning ubiquitous computing, maglev trains swishing past cities, or minifusion reactors distributed in basements (if not the complete takeover of humanity by sapient machines). They would do well to ponder the ubiquity of catastrophes throughout history, to think about viruses, bacteria, irrationality, hatred, the drive for power and dominance, and the different consequences of the global who's-on-top game or what an unprecedented global environmental change could do to our species.

Doing that might lead one to conclude that despite many localized problems, the second half of the twentieth century was an exceptionally stable and an unusually benign period in global terms, and that the probabilities of more painful events will greatly increase during the next 50 years. During this time of uncertainty, trials, and doubts, when feelings of loss, despondency, and despair affect not only intellectuals but diffuse through a society, it is natural to turn to historical precedents and to parade the examples of collapsed civilizations and lost empires. The unsubtle message is, we are inevitably next in line.

The beginning of a new millennium has offered many of these prophecies (Lovelock 2006; Diamond 2004; Kunstler 2005; Rees 2003). One is Ferguson's (2005; 2006) extended comparison of the West, particularly the United States, with the Roman Empire, a piece in the venerable tradition of Spengler (1919). Ferguson's account lists a number of items that I introduced in my description of the U.S. retreat, but I do not share his conclusion of an early transformation of Washington's Capitol into a picturesque ruin. I see things differently, and always less assuredly. The success of our species makes it clear that humans beings, unlike all other organisms, have evolved not to adapt to specific conditions and tasks but to cope with change (Potts 2001).

This ability makes us uniquely fit to cope with assorted crises and to transform many events from potentially crippling milestones to resolved challenges. This strength is inevitably also our weakness because it often leads us to overestimate our capacities. In the West, our wealth, the extent of our scientific knowledge, and the major areas of our lives where we have successfully asserted control over the environment mislead us into believing that we are more in charge of history than we can ever be. But this does not mean that our future is catastrophically preordained. That is why I have been deliberately agnostic about the civilization's fortunes in this survey.

My intent was to identify, illuminate, and probe what I believe to be the key risks that global civilization faces in the coming decades, and to explain and assess some key trends that may contribute to its fate. My strong preference was to do so without engaging in counterproductive rankings, classifications, verdicts, and predictions. And my determination was to keep making clear the complexities, contradictions, and uncertainties of our understanding. The depth and extent of our ignorance make it imperative to chant a mantra that sounds discouraging but that is true and honest. There is so much we do not know, and pretending otherwise is not going to make our choices clearer or easier.

None of us knows which threats and concerns will soon be forgotten and which will become tragic realities. That is why we repeatedly spend enormous resources

Fig. 5.7
Basilica of Santa Sabina, Aventine Hill, Rome. A luminous parable of an end as a new beginning: a dozen years after Alaric's sack of Rome (410 C.E.), Petrus of Illyria initiated construction of this basilica (422 C.E.). *Top*, Santa Sabina's main nave with a round apse; *bottom*, the basilica's exterior. Photos by V. Smil.

in the pursuit of uncertain (even dubious) causes and are repeatedly unprepared for real threats or unexpected events. The best example of this reality is the trillions spent on thousands of nuclear warheads, with not a single one of them being of any use as a deterrent or defense against the sacrificial terrorism. But the admirable human capacity to adapt and change offers a great deal of encouragement and justifies a great deal of skepticism about an imminent end to modern civilization.

As for endings, being an admirer of Kafka and Borges, I believe in the power of parables, and here is an apposite one. My favorite place in Rome is the Aventine Hill, high above the Tiber's left bank. At its top, next to a rather unkempt orange garden, stands the basilica of Santa Sabina (fig. 5.7). Its unadorned brick exterior looks mundane, but once you approach its magnificent carved cypress doors, the perception changes. And then you enter one of the most perfect spaces ever put under a roof. The plan is simple, a wide nave and two aisles separated by rows of large fluted columns; the walls are bare, but the building is filled with light, bright but not dazzling, penetrating yet ethereal.

Construction of the basilica, built at the request of Petrus of Illyria, began in 422, just a dozen years after the Goths under Alaric sacked Rome, destroying much of its imperial and Christian splendor. Not far from the Aventine, on the Celian Hill, is Santo Stefano Rotondo, another splendid building with a central rotunda surrounded by two concentric circles of plain columns. It was completed in 483, seven years after the date that is generally considered the final end of the Western Roman Empire. These two magnificent structures remind us, gracefully and forcefully, of the continuity of history, of the fact that such terms as *demise* or *collapse* or *end* are often merely categories of our making, and that catastrophes and endings are also opportunities and beginnings.

Appendix A

Units and Abbreviations, Prefixes

Units and Abbreviations

bbl	barrels
°C	degree Celsius
g	gram
Ga	giga annum; billion years
h	hour
ha	hectare (10,000 m^2)
J	joule
L	liter
m	meter
Ma	mega annum; million years
mpg	miles per gallon
OM	order of magnitude
ppm	parts per million
s	second
t	tonne (metric ton)
W	watt
¥	yen

Prefixes

μ	micro	10^{-6}
m	milli	10^{-3}

h	hecto	10^2
k	kilo	10^3
M	mega	10^6
G	giga	10^9
T	tera	10^{12}
P	peta	10^{15}
E	exa	10^{18}

Appendix B
Acronyms

CAFE	corporate automobile fuel efficiency
CDIAC	Carbon Dioxide Information Analysis Center (U.S.)
EU	European Union
EU-25	European Union, after the 2004 enlargement to 25 countries.
G-5, G-6	group of five (six) nations that are economic leaders
GDP	gross domestic product
GHG	greenhouse gas
GWP	gross world product
HDI	Human Development Index
MRSA	methicillin-resistant *Staphylococcus aureus*
NEA	near-Earth asteroid
NEO	near-Earth object
NPP	net primary productivity
OECD	Organisation for Economic Co-operation and Development
OPEC	Organization of Petroleum Exporting Countries
PPP	purchasing power parity
PV	photovoltaics
PWR	pressurized water reactor
R/P	reserve/production
SUV	sports utility vehicle
TPES	total primary energy supply
UN	United Nations

VEI	volcanic explosivity index
WHO	World Health Organization
WW I	World War I
WW II	World War II

References

Abraham, T. 2005. *Twenty-First Century Plague: The Story of SARS*. Baltimore: Johns Hopkins University Press.

Acharya, S. K., and P. K. Basu. 1992. Toba ash on the Indian subcontinent and its implications for correlation of late Pleistocene alluvium. *Quaternary Research* 40: 10–19.

Adelman, M. 1992. Oil resource wealth of the Middle East. *Energy Studies Review* 4 (1): 7–21.

Ager, D. 1993. *The New Catastrophism: The Importance of the Rare Event in Geological History*. Cambridge: Cambridge University Press.

Airbus. 2004. <http://www.airbus.com/en/>.

Albritton, C. C. 1989. *Catastrophic Episodes in Earth History*. London: Chapman and Hall.

Aldhous, P. 2005. Melioidosis? Never heard of it. . . . *Nature* 434: 692–693.

Alexander, Y., and K. A. Myers, eds. 1982. *Terrorism in Europe*. London: Croom Helm.

Al-Hassani, S., ed. 2006. *1001 Inventions: Muslim Heritage in Our World*. Manchester: Foundation for Science, Technology and Civilisation.

Alley, R. B. 2000. The Younger Dryas cold interval as viewed from central Greenland. *Quaternary Science Reviews* 19: 213–226.

Alley, R. B., P. U. Clark, P. Huybrechts, and I. Joughin. 2005. Ice-sheet and sea-level changes. *Science* 310: 456–460.

Allison, G., and P. Zelikow. 1999. *Essence of Decision: Explaining the Cuban Missile Crisis*. New York: Longman.

Allon, Y. 1970. *Shield of David: The Story of Israel's Armed Forces*. New York: Random House.

Alvarez, L. W., W. Alvarez, F. Asaro, and H. V. Michel. 1980. Extraterrestrial cause for the Cretaceous-Tertiary extinction. *Science* 360: 1095–1108.

Amábile-Cuevas, C. F. 2003. New antibiotics and new resistance. *American Scientist* 91: 138–149.

Ambrose, S. H. 1998. Late Pleistocene human population bottlenecks: Volcanic winter and the differentiation of modern humans. *Journal of Human Evolution* 34: 623–651.

———. 2003. Did the super-eruption of Toba cause a human population bottleneck? Reply to Gathorne-Hardy and Harcourt-Smith. *Journal of Human Evolution* 45: 231–237.

Ammann, C. M., and P. Naveau. 2003. Statistical analysis of tropical explosive volcanism occurrences over the past six centuries. *Geophysical Research Letters* 30: 14/1–14/4.

Amyes, S. G. B. 2001. *Magic Bullets, Lost Horizons: The Rise and Fall of Antibiotics*. London: Taylor and Francis.

Andrea, A. J. 2003. The Crusades in perspective: The Crusades in modern Islamic perspective. *History Compass* 1 (1): doi:10.1111/1478-0542.019.

Andriolo, K. 2002. Murder by suicide: Episodes from Muslim history. *American Anthropologist* 104: 736–742.

Angell, J. K., and J. Korshover. 1985. Surface temperature changes following the six major volcanic episodes between 1780 and 1980. *Journal of Climate and Applied Meteorology* 24: 937–951.

Annan, J. D., and J. C. Hargreaves. 2006. Using multiple observationally based constraints to estimate climate sensitivity. *Geophysical Research Letters* 33: L06704.

Applebaum, A. 2005. In search of Pro-Americanism. *Foreign Policy*, July/August, 32–40.

Arita, I., M. Nakane, and F. Fenner 2006. Is polio eradication realistic? *Science* 312: 852–855.

Arrhenius, S. 1896. On the influence of carbonic acid in the air upon the temperature of the ground. *Philosophical Magazine* 41: 237–276.

Ashton, B., K. Hill, A. Piazza, and R. Zeitz. 1984. Famine in China, 1958–61. *Population and Development Review* 10: 613–645.

Astrogor, A. 2001. *Mitika Putina [Putin's Myth]*. Moscow: Rostkniga.

Atkinson, H., C. Tickell, and D. Williams, eds. 2000. *Report of the Task Force on Potentially Hazardous Near Earth Objects*. London: British National Space Centre.

Atran, S. 2003. Genesis of suicide terrorism. *Science* 299: 1534–1539.

———. 2004. Individual factors in suicide terrorism. *Science* 304: 47–49.

Aunan, K., T. K. Berntsen, and H. M. Seip. 2000. Surface ozone in China and its possible impact on agricultural crop yields. *Ambio* 29: 294–301.

Ausubel, J. H. 1996. Can technology spare the Earth? *American Scientist* 84: 166–178.

Ayres, R. U., and L. A. Febre. 1999. *Nitrogen Consumption in the United States*. Fontainebleau: INSEAD.

Baillie, J. E. M., C. Hilton-Taylor, and S. N. Stuart. 2004. IUCN red list of threatened species: A global species assessment. Gland, Switzerland: International Union for Conservation of Nature and Natural Resources.

Baker, D. F. 2007, Reassessing carbon sinks. *Science* 316: 1708–1709.

Barnett, T. P., et al. 2005. Penetration of human-induced warming into the world's oceans. Science 309: 284–287.

Barry, J. M. 2004. *The Great Influenza: The Epic Story of the Deadliest Plague in History*. New York: Viking.

Baskin, Y. 2002. *A Plague of Rats and Rubbervines: The Growing Threat of Species Invasions*. Washington, D.C.: Island Press.

Battle, M., et al. 2000. Global carbon sinks and their variability inferred from atmospheric O_2 and $\delta^{13}C$. *Science* 287: 2467–2470.

Baumgartner, U. 1995. *Saving Behavior and the Asset Price "Bubble" in Japan: Analytical Studies*. Washington, D.C.: International Monetary Fund.

BEA (Bureau of Economic Analysis). 2006. U.S. international investment position. <http://bea.gov/bea/newsrel/intinvnewsrelease.htm>.

Beer, C., W. Lucht, C. Schmullius, and A. Shvidenko. 2006. Small net carbon dioxide uptake by Russian forests during 1981–1999. *Geophysical Research Letters* 33 (15): L15403.

Bellamy, P. H., et al. 2005. Carbon losses from all soils across England and Wales 1978–2003. *Nature* 437: 245–248.

Bellouin, N., O. Boucher, J. Haywood, and M. S. Reddy. 2005. Global estimate of aerosol direct radiative forcing from satellite measurements. *Nature* 438: 1138–1141.

Bennett, W. J. 2003. *Why We Fight: Moral Clarity and the War on Terrorism*. Washington, D.C.: Regnery.

Bentele, G., and P. Rosner, eds. 1997. *Deutsche Einheit: Irritationen, Probleme, Perspektiven*. Bamberg: Zentrum für wirtschaftliche Weiterbildung.

Bernanke, Ben S. 2005. *The Global Saving Glut and the U.S. Current Account Deficit*. Sandridge Lecture at the Virginia Association of Economists, Richmond, Va., April 14, 2005. <http://www.federalreserve.gov/boarddocs/speeches/2005/200503102/default.htm>.

Berner, R. A. 1998. The carbon cycle and CO_2 over Phanerozoic time: the role of land plants. *Philosophical Transaction of the Royal Society London B* 353: 75–82.

Bhagat, C. 2005. *One Night @ the Call Center*. New Delhi: Rupa.

bin-Lādin, U. 2002. Letter to the American people. <http://www.globalsecurity.org/security/library/report/2002/021120-ubl.htm>.

———. 2004. Message to the American people. <http://english.aljazeera.net/English/archive/archive?ArchiveId=7403>.

Billari, F. C., and H.-P. Kohler. 2002. *Patterns of Lowest-Low Fertility in Europe*. Rostock: Max-Planck-Institut für demografische Forschung.

Binzel, R. P. 2000. The Torino impact hazard scale. *Planetary and Space Science* 48: 297–303.

Blanc, H. 2004. *KGB Connexion: Le Système Poutine*. Paris: Hors Commerce.

Bland, P. A., and N. A. Artemieva. 2003. Efficient disruption of small asteroids by Earth's atmosphere. *Nature* 424: 288–291.

Blasing, T. J., and K. Smith. 2006. Recent greenhouse gas concentrations. <http://cdiac.esd.ornl.gov/pns/current_ghg.html>.

Blight, J. G., and D. A. Welch. 1989. *On the Brink: Americans and Soviets Reexamine the Cuban Missile Crisis*. New York: Hill and Wang.

Bloom, N. 2006. The impact of uncertainty shocks: Firm level estimation and a 9/11 simulation. <http://econpapers.repec.org/paper/cepcepdps/dp0718.htm>.

Boeing. 2006. <http://www.boeing.com>.

Bogen, K. T., and E. D. Jones. 2006. Risks of mortality and morbidity from worldwide terrorism: 1968–2004. *Risk Analysis* 26: 45–59.

Bollmann, S., ed. 2002. *Patient Deutschland: Eine Therapie*. Stuttgart: Deutsche Verlags-Anstalt.

Bongaarts, J. 2004. Population aging and the rising cost of public pensions. *Population and Development Review* 30: 1–23.

Bostrom, N. 2002. Existential risks: Analyzing human extinction scenarios and related risks. *Journal of Evolution and Technology* 9: 1–36.

Bosworth, B., and G. Burtless, eds. 1998. *Aging Societies*. Washington, D.C.: Brookings Institution.

Bousquet, P., et al. 2000. Regional changes in carbon dioxide fluxes of land and oceans since 1980. *Science* 290: 1342–1346.

Bowen, J. R. 2004. Does French Islam have borders? Dilemmas of domestication in a global religious field. *American Anthropologist* 106: 43–55.

Box, J. E., D. H. Bromwich, and L. Bai. 2004. Greenland ice sheet surface mass balance 1991–2000: Application of Polar MM5 mesoscale model and in situ data. *Journal of Geophysical Research* 109: D16105.

BP (British Petroleum). 1979. *Oil Crisis Again*. London: BP.

———. 2006. *Statistical Review of World Energy 2006*. London: British Petroleum. <http: //www.bp.com/>.

Bradbury, R. 1953. *Fahrenheit 451*. New York: Simon and Schuster.

Brandt, J. C., and R. D. Chapman. 2004. *Introduction to Comets*. Cambridge: Cambridge University Press.

Brecke, P. 1999. Violent conflicts A.D. 1400 to the present in different regions of the world. In *Proceedings of the Annual Meeting of the Peace Science Society*, Ann Arbor.

Breeze, P. 2004. *The Future of Biomass Power Generation*. London: Business Insights.

Briffa, K. R., P. D. Jones, F. H. Schweingruber, and T. Osborn. 1998. Influence of volcanic eruptions on Northern Hemisphere summer temperature over the past 600 years. *Nature* 393: 450–454.

Britten, S. 1983. *The Invisible Event*. London: Menard Press.

Brockerhoff, M. P. 2000. An urbanizing world. *Population Bulletin* 55 (3): 44.

Broecker, W. S. 1997. Thermohaline circulation, the Achilles heel of our climate system: Will man-made CO_2 upset the current balance? *Science* 278: 1582–1588.

———. 2003. Does the trigger for abrupt climate change reside in the ocean or in the atmosphere? *Science* 300: 1519–1522.

———. 2006. Was the Younger Dryas triggered by a flood? *Science* 312: 1146–1148.

Brown, P., et al. 2002. The flux of small near-Earth objects colliding with the Earth. *Nature* 420: 294–296.

Bryden, H. L., H. R. Longworth, and S. A. Cunningham. 2005. Slowing of the Atlantic meridional overturning circulation at 25°N. *Nature* 438: 655–657.

Bühring, C., and M. Sarnthein. 2000. Toba ash layers in the South China Sea: Evidence of contrasting wind directions during eruption ca. 74 ka. *Geology* 28: 275–278.

Bukharaev, R. 2000. *Islam in Russia: The Four Seasons*. New York: St. Martin's Press.

Bundy, M. 1988. *Danger and Survival: Choices about the Bomb in the First Fifty Years*. New York: Random House.

Bunn, M., and A. Wier. 2005. *Securing the Bomb 2005: The New Global Imperatives*. Managing the Atom Project. Cambridge, Mass.: Nuclear Threat Initiative, Harvard University. <http://www.nti.org/e_research/report_cnwmupdate2005.pdf>.

Bush, R. C. 2005. *Untying the Knot: Making Peace in the Taiwan Strait*. Washington, D.C.: Brookings Institution.

Butler, D. 2006. The data gap. *Nature* 442: 26–27.

Butler, D., and J. Ruttimann. 2006. Avian flu and the New World. *Nature* 441: 137–139.

Buzell, E., and S. H. Preston. 2007. Mortality of American troops in the Iraq War. *Population and Development Review* 33: 555–566.

Cagin, S., and P. Dray. 1993. *Between Earth and Sky: How CFCs Changed Our World and Endangered the Ozone Layer*. New York: Pantheon.

Cain, P. J., and A. G. Hopkins. 2002. *British Imperialism 1688–2000*. Harlow: Longman.

Caldeira, K., and M. E. Wickett. 2003. Anthropogenic carbon and ocean pH. *Nature* 425: 365–365.

Calder, N. 1980. *Nuclear Nightmares*. New York: Viking.

Calhoun, C., P. Price, and A. Timmer, eds. 2002. *Understanding September 11*. New York: Norton.

Callen, T. 2003. *Japan's Lost Decade: Policies for Economic Revival*. Washington, D.C.: International Monetary Fund.

Callendar, G. S. 1938. The artificial production of carbon dioxide and its influence on temperature. *Quarterly Journal of the Royal Meteorological Society* 64: 223–237.

Cameron, D. 2005. Robot, kindly bring me a beer from the fridge. *Fairfax Digital*, December 3, 2005. <http://www.fairfax.com.au/index.ac/>.

Campbell, C. J. 1997. *The Coming Oil Crisis*. Brentwood, UK: Petroconsultants and Multi-Science Publishing.

Campbell, C. J., and J. Laherrère. 1998. The end of cheap oil. *Scientific American* 278(3): 78–83.

Čapek, K. 1921. *R.U.R. Rossum's Universal Robots*. Praha: Aventinum.

Carey, E. V., A. Sala, R. Keane, and R. M. Callaway. 2001. Are old forests underestimated as global carbon sinks? *Global Change Biology* 7: 339–344.

Carlile, L. E., and M. C. Tilton, eds. 1998. *Is Japan Really Changing Its Ways? Regulatory Reform and the Japanese Economy*. Washington, D.C.: Brookings Institution.

Carr, C. 2002. *The Lessons of Terror: A History of Warfare against Civilians*. New York: Random House.

Cassman, K. G., A. Dobermann, and D. T. Walters. 2002. Agroecosystems, nitrogen-use efficiency, and nitrogen management. *Ambio* 31: 132–140.

Casti, J. L. 2004. Synthetic thought. *Nature* 427: 680.

CBRF (Central Bank of the Russian Federation). 2006. *Social and Economic Situation in 2005*. Moscow.

CCM (Committee for the Compilation of Materials on Damage Caused by the Atomic Bombs in Hiroshima and Nagasaki). 1981. *Hiroshima and Nagasaki: The Physical, Medical, and Social Effects of the Atomic Bombings*. Trans. E. Ishikawa and D. L. Swain. New York: Basic Books.

Centers for Disease Control and Prevention. 2005. *Mortality Data from the National Vital Statistics System*. Atlanta, GA: CDC. <http://www.cdc.gov/nchs/about/major/dvs/mortdata.htm>.

Chang, J., and J. Halliday. 2005. *Mao: The Unknown Story*. London: Jonathan Cape.

Chapman, C. R. 2004. The hazard of near-Earth asteroid impacts on Earth. *Earth and Planetary Science Letters* 222: 1–15.

Chapman, C. R., D. D. Durda, and R. E. Gold. 2001. *The Comet/Asteroid Impact Hazard: A Systems Approach*. Boulder, Colo.: Southwest Research Institute.

Chapman, C. R., and A. W. Harris. 2002. A skeptical look at September 11th: How we can defeat terrorism by reacting to it more rationally. *Skeptical Inquirer* 26 (5): 29–34.

Chapman, C. R., and D. Morrison. 1994. Impacts on the Earth by asteroids and comets: Assessing the hazard. *Nature* 367: 33–40.

Chen, G., and C. Wu. 2004. *Zhongguo nongmin diaocha* [*A Survey of Chinese Peasants*]. Beijing: People's Literature Publishing House.

Chen, H., et al. 2004. The evolution of H5N1 influenza viruses in ducks in southern China. *Proceedings of the National Academy of Sciences* 101: 10452–10457.

Chen, Y., et al. 1988. *The Great Tangshan Earthquake of 1976: An Anatomy of Disaster*. New York: Pergamon Press.

Chesley, S., and S. Ward. 2006. A quantitative assessment of the human and economic hazard from impact-generated tsunami. *Natural Hazards* 38: 355–374.

China Daily. 2005. Luxury goods chase Chinese travelers. May 24, 2005, 1.

Chotpitayasunondh, T., et al. 2005. Human disease from influenza A (H5N1), Thailand. *Emerging Infectious Diseases* 11: 201–209.

CIA (Central Intelligence Agency). 1979. *The World Oil Market in the Years Ahead*. Washington, D.C.: CIA.

Cline, W. R. 2005. *The United States as a Debtor Nation: Risks and Policy Reform*. Washington, D.C.: Institute of International Economics.

Coale, A. 1985. On the demographic impact of nuclear war. *Population and Development Review* 9: 562–568.

Coffin, M. F., and O. Eldholm. 1994. Large igneous provinces: Crustal structure, dimensions, and external consequences. *Reviews of Geophysics* 32: 1–36.

Cohen, M. L. 2000. Changing patterns of infectious disease. *Nature* 406: 762–767.

Cohen, W. I. 2005. *America's Failing Empire: U.S. Foreign Relations since the Cold War*. Oxford: Blackwell.

Comptroller, New York City. 2002. Thompson releases report on fiscal impact of 9/11 on New York City. Press release. <http://comptroller.nyc.gov/press/2002_releases/02-09-054.shtm>.

Conquest, R. 1990. *The Great Terror: A Reassessment*. London: Hutchinson.

Cottino, A. 1998. *Vita da Clan*. Torino: EGA.

———. 1999. Sicilian cultures of violence: The interconnections between organized crime and local society. *Crime, Law and Social Change* 32: 103–113.

Covey, C., S. L. Thompson, P. R. Weissman, and M. C. MacCracken. 1994. Global climatic effects of atmospheric dust from an asteroid or comet impact on Earth. *Global and Planetary Change* 9: 263–273.

Cowan, R. 1990. Nuclear power reactors: a study in technological lock-in. *Journal of Economic History* 50: 541–567.

Cox, G. W. 2004. *Alien Species and Evolution: The Evolutionary Ecology of Exotic Plants, Animals, Microbes, and Interacting Native Species*. Washington, D.C.: Island Press.

Crews, R. D. 2006. *For Prophet and Tsar: Islam and Empire in Russia and Central Asia*. Cambridge, Mass.: Harvard University Press.

Crosby, A. 1989. *America's Forgotten Pandemic*. Cambridge: Cambridge University Press.

Cunningham, S. A., et al. 2007. Temporal variability of the Atlantic meridional overturning circulation at 26.5°N. *Science* 317: 935–937.

Cushman, T., and S. G. Meštrovic, eds. 1996. *This Time We Knew: Western Response to Genocide in Bosnia*. New York: New York University Press.

D'Costa, V. M., K. M. McGrann, D. W. Hughes, and G. D. Wright. 2006. Sampling the antibiotic resistome. *Science* 311: 374–377.

Dansgaard, W., J. W. C. White, and S. J. Johnsen. 1989. The abrupt termination of the Younger Dryas climate event. *Nature* 339: 532–534.

Darwin, C. 1881. *The Formation of Vegetable Mould: Through the Action of Worms, with Observation on Their Habitats*. London: John Murray.

DaVanzo, J. 2001. *Dire Demographics: Population Trends in the Russian Federation*. Santa Monica: Rand Corporation.

Davidson, E. A., and I. A. Janssens. 2006. Temperature sensitivity of soil carbon decomposition and feedbacks to climate change. *Nature* 440: 165–173.

Davidson, L., and R. M. Miller. 1998, *Islamic Fundamentalism*. Westport, Conn.: Greenwood Press.

Davies, P. 1999. *The Devil's Flu: The World's Deadliest Influenza Epidemic and the Scientific Hunt for the Virus That Caused It*. New York: Henry Holt.

Davis, C. H., et al. 2005. Snowfall-driven growth in East Antarctic ice sheet mitigates recent sea-level rise. *Science* 308: 1898–1901.

Davis, D. 2005. Urban consumer culture. *China Quarterly* 183: 692–709.

Deffeyes, K. S. 2001. *Hubbert's Peak: The Impending World Oil Shortage*. Princeton, N.J.: Princeton University Press.

Demeny, P. 2003. Population policy dilemmas in Europe at the dawn of the twenty-first century. *Population and Development Review* 29: 1–28.

Demirdöven, N., and J. Deutsch. 2004. Hybrid cars now, fuel cell cars later. *Science* 305: 974–976.

Démurger, S. 2000. *Economic opening and growth in China*. Paris: Development Centre of the Organisation for Economic Co-operation and Development.

DeParle, J. 2007. The American prison nightmare. *The New York Review of Books* 54(6). <http: //www.nybooks.com/articles/20056>.

Derickson, A. 1998. *Black Lung: Anatomy of a Public Health Disaster*. Ithaca, NY: Cornell University Press.

De Vries, W., G. J. Reinds, P. Gundersen, and H. Sterba. 2006. The impact of nitrogen deposition on carbon sequestration in European forests and forest soils. *Global Change Biology* 12: 1151–1173.

Dew-Becker, I., and R. J. Gordon. 2005. Where did the productivity growth go? Inflation dynamics and the distribution of Income. *Brookings Papers on Economic Activity* 36(2): 67–127.

Diamond, J. 2004. *Collapse: How Societies Choose to Fail or Succeed*. New York: Viking.

Ding, Q., and T. Hesketh. 2006. Family size, fertility preferences, and sex ratio in China in the era of the one-child family policy: Results from national family planning and reproductive health survey. *British Medical Journal* 333: 371–373.

Dobson, W. J. 2006. The day nothing much changed. *Foreign Policy*, September/October, 22–32.

Dolgov, Y. A., ed. 1984. *Meteoritnye issledovaniia v Sibiri: 75 let tungusskomu fenomenu*. Novosibirsk: Izdatel'stvo Nauka.

Dong, Z., C. W. Hoven, and A. Rosenfield. 2005. Lessons from the past. *Nature* 433: 573–574.

Dooley, M. P., D. Folkers-Landau, and P. M. Garber. 2004. *The US Current Account Deficit and Economic Development: Collateral for a Total Return Swap*. Washington, D.C.: National Bureau for Economic Research.

Doran, P. T., et al. 2002. Antarctic climate cooling and terrestrial ecosystem response. *Nature* 415: 517–520.

Dorget, F. 1984. *Le choix nucléaire français*. Paris: Economica.

Douglass, C. B., ed. 2005. *Barren States: The Population "Implosion" in Europe*. New York: Oxford University Press.

Durand, M., and J. Grattan. 1999. Extensive respiratory health effects of volcanogenic fog in 1783 inferred from European documentary sources. *Environmental Geochemistry and Health* 21: 371–376.

DWIA (Danish Wind Industry Association). 2006. *Future Energy Supply*. <http://www.windpower.org/en/futuresupply.htm>.

Edenhofer, O., et al. 2006. Induced technological change: Exploring its implications for the economics of atmospheric stabilization. *Energy Journal* 1: 57–107.

Ehrlich, P., and A. Ehrlich. 2004. *One with Nineveh: Politics, Consumption, and the Human Future*. Washington, D.C.: Island Press.

EIA (Energy Information Administration). 2005. *Annual Energy Review*. <http://www.eia.doe.gov/emeu/aer/contents.html>.

Eldredge, N., and S. J. Gould. 1972. Punctuated equilibria: An alternative to phyletic gradualism. In *Models in Paleobiology*, ed. T. J. M. Schopf, 82–115. San Francisco: Freeman.

Emmott, W. 2005. The sun also rises: A survey of Japan. *Economist*, October 8. <http://www.economist.com/displaystory.cfm?story_id=4454244>.

Engelberger, J. F. 1989. *Robotics in Service*. Cambridge, Mass.: MIT Press.

England, R. S. 2005. *Aging China: The Demographic Challenge to China's Economic Prospects*. Westport, Conn.: Praeger.

———, ed. 2002. *The Macroeconomic Impact of Global Aging: A New Era of Economic Frailty?* Washington, D.C.: Center for Strategic and International Studies.

Enserink, M. 1999. Biological invaders sweep in. *Science* 285: 1834–1836.

———. 2006. Oseltamivir becomes plentiful, but still not cheap. *Science* 312: 382–383.

Enserink, M., and J. Kaiser. 2005. Has biodefense gone overboard? *Science* 307: 1396–1398.

Epstein, P. R., and E. Mills, eds. 2005. *Climate Change Futures: Health, Ecological and Economic Consequences*. Cambridge, Mass.: Harvard Medical School.

Erdogan, R. T. 2005. Turkey's historic journey. In *The World in 2006*. Economist. <http://www.economist.com/theWorldIn/europe/?d=2006>.

Esposito, J., ed. 1999. *The Oxford History of Islam*. New York: Oxford University Press.

Eurostat. 2005a. European Commission. Structural Indicators: Employment.

———. 2005b. EU25 population up by 0.5% in 2004. Press release, October 26.

Evans, M. N. 2006. The woods fill up with snow. *Nature* 440: 1120–1121.

Fagin, J. A. 2006. *When Terrorism Strikes Home: Defending the United States*. Boston: Pearson/Allyn & Bacon.

Fagone, J. 2006. Horsemen of the esophagus. *Atlantic Monthly* 297 (5): 86–93.

Failed States Index. 2005. *Foreign Policy*/Fund for Peace. July/August. <http://www.foreignpolicy.com/story/cms.php?story_id=3098>.

Falkowski, P., et al. 2000. The global carbon cycle: a test of our knowledge of Earth as a system. *Science* 290: 291–296.

Fallaci, O. 2002. *The Rage and Pride*. New York: Rizzoli.

Fallows, J. 2005. Countdown to a meltdown. *Atlantic Monthly* 296: 51–68.

———. 2006. Declaring victory. *Atlantic Monthly* 298 (2): 60–73.

Fan, S., et al. 1998. A large terrestrial carbon sink in North America implied by atmospheric and oceanic carbon dioxide data and models. *Science* 282: 442–446.

Fang, J. Y., et al. 2005. Biomass accumulation by Japan's forests from 1947 to 1995. *Global Biogeochemical Cycles* 19: doi:10.1029/2004GB002253.

FAO. 2006. *FAOSTAT*. Rome: FAO. <http://app.fao.org>.

Farman J. C., B. G. Gardiner, and J. D. Shanklin. 1985. Large losses of total ozone in Antarctica reveal seasonal ClO_x/NO_x interaction. *Nature* 315: 207–210.

Federal Statistical Office Germany. 2003. In 2050 every 3rd person will be 60 or older in Germany. Press release, June 6.

Fergus, R., et al. 2006. Migratory birds and avian flu. *Science* 312: 845.

Ferguson, N. 2002. *Colossus: The Price of America's Empire*. New York: Penguin.

———. 2004. A world without power. *Foreign Policy* (July/August): 32–39.

———. 2006. Empires with expiration dates. *Foreign Policy*, September/October, 46–42.

Ferguson, N. M., D. A. Cummings, Cauchernez, et al. 2005. Strategies for containing an emerging influenza pandemic in Southeast Asia. *Nature* 437: 209–214.

Ferguson, N. M., D. A. Cummings, C. Fraser, et al. 2006. Strategies for mitigating an influenza pandemic. *Nature* 442: 448–452.

Ferguson, R. W. 2005. Current Account Deficit: Causes and Consequences. Lecture to the Economics Club of the University of North Carolina, April 20, 2005. <http://www.federalreserve.gov/boarddocs/speeches/2005/20050420/default.htm>.

Fiqh Council of North America. 2005. U.S. Muslim Religious Council Issues Fatwa against Terrorism. <www.cair-net.org/includes/Anti-TerrorList.pdf>.

Fischhoff, B., et al. 1978. How safe is safe enough? A psychometric study of attitudes towards technological risks and benefits. *Policy Sciences* 9: 127–152.

Fisher, R. V., G. Heiken, and J. B. Hulen. 1997. *Volcanoes: Crucibles of Change*. Princeton, N.J.: Princeton University Press.

Fishman, T. C. 2005. *China Inc.: How the Rise of the Next Superpower Challenges America and the World*. New York: Simon and Schuster.

Fitter, A. H., and R. S. R. Fitter. 2002. Rapid changes in flowering time in British plants. *Science* 296: 1689–1691.

Flynn, S. E. 2004. *America the Vulnerable: How Our Government Is Failing to Protect Us from Terrorism*. New York: HarperCollins.

Forrow, L., et al. 1998. Accidental nuclear war: A post–Cold War assessment. *New England Journal of Medicine* 338: 506–511.

Fourier, J. B. J. 1822. *Théorie Analytique de la Chaleur*. Paris: Firmin Didot.

Frank, A. G. 1998. *ReOrient: Global Economy in the Asian Age*. Berkeley: University of California Press.

Frazier, M. W. 2006. Pensions, public opinion, and the graying of China. *Asia Policy* 1: 43–68.

Freedman, L. 2001. *The Cold War: A Military History*. London: Cassell & Company.

Freedom House. 2006. *Freedom in the World*. Freedom House: New York. <http://www.freedomhouse.org/template.cfm?page=363&year=2006>.

Friedman, E., P. G. Pickowicz, and M. Selden. 2005. *Revolution, Resistance, and Reform in Village China*. New Haven: Yale University Press.

Friedman, M. S., et al. 2001. Impact of changes in transportation and commuting behaviors during the 1996 Summer Olympic Games in Atlanta on air quality and childhood asthma. *Journal of American Medical Association* 285: 897–905.

Friedman, T. L. 2006. The first law of petropolitics. *Foreign Policy*, May/June, 28–36.

Friedrich, W. L. 2000. *Fire in the Sea. The Santorini Volcano: Natural History and the Legend of Atlantis*. Trans. A. R. McBirney. Cambridge: Cambridge University Press.

Fromkin, D. 2001. *A Peace to End All Peace: The Fall of the Ottoman Empire and the Creation of the Modern Middle East*. New York: Henry Holt.

Fukuyama, F. 1992. *The End of History and the Last Man*. New York: Free Press.

Funabashi, Y.1988. *Managing the Dollar: From the Plaza to the Louvre*. Washington, D.C.: Institute for International Economics.

Futures of artificial life. 2004. *Nature* 431. <http://www.nature.com/nature/journal/v431/n7009/full/431613b.html>.

Gagneux, S., et al. 2006. The competitive cost of antibiotic resistance in *Mycobacterium tuberculosis*. *Science* 312: 1944–1946.

Galloway, J. N., and E. B. Cowling. 2002. Reactive nitrogen and the world: 200 years of change. *Ambio* 3: 64–71.

Gambetta, D. 1992. *La Mafia siciliana, un'industria della protezione privata*. Torino: Einaudi.

Gattuso, J.-P. and R. W. Buddemeier. 2000. Calcification and CO_2. *Nature* 407: 311–313.

Gedney, N., et al. 2006. Detection of a direct carbon dioxide effect on continental river runoff records. *Nature* 439: 835–838.

Gehrels, T., ed. 1994. *Hazards Due to Comets and Asteroids*. Tucson: University of Arizona Press.

Geifman, A. 1993. *Thou Shalt Kill: Revolutionary Terrorism in Russia, 1894–1917*. Princeton, N.J.: Princeton University Press.

Gelbspan, R. 1998. A good climate for investment. *Atlantic Monthly* 283 (5): 22–28.

Gewin, V. 2003. Agriculture shock. *Nature* 421: 106–108.

Gibbon, E. 1776–1788. *The Decline and Fall of the Roman Empire*. London: Strachan and Cadell.

Gigerenzer, G. 2002. *Reckoning with Risk: Learning to Live with Uncertainty*. London: Penguin.

Gillis, J. R., L. A. Tilly, and D. Levine, eds. 1992. *The European Experience of Declining Fertility, 1850–1970: The Quiet Revolution*. Oxford: Blackwell.

Gilliver, M. A., et al. 1999. Antibiotic resistance found in wild rodents. *Nature* 401: 233.

Giorgini, J. D., et al. 2002. Asteroid 1950 DA's encounter with Earth in 2880: Physical limits of collision probability prediction. *Science* 296: 132–136.

Glass, L. M., and D. Phillips. 2006. The Kalkarindji continental flood basalt province: A new Cambrian large igneous province in Australia with possible links to fauna extinctions. *Geology* 34: 461–464.

Glibert, P. M., et al. 2006. Escalating worldwide use of urea: A global change contributing to coastal eutrophication. *Biogeochemistry* 77: 441–463.

Glickman, S. W., et al. 2006. A portfolio model of drug development for tuberculosis. *Science* 311: 1246–1247.

Goes, J. I., P. G. Choppily, H. Gnomes, and J. T. Pullovers. 2005. Warming of the Eurasian landmass is making the Arabian Sea more productive. *Science* 308: 545–547.

Goldman, M. 2006. Russia's middle class muddle. *Current History* 105: 321–326.

Gordon, Robert J. 2000. Does the "New Economy" measure up to the great inventions of the past? *Journal of Economic Perspectives* 14: 49–74.

Gorenburg, D. 2006. Russia confronts radical Islam. *Current History* 105: 334–340.

Gowan, J. E. 2001. Economic impact of antimicrobial resistance. *Emerging Infectious Diseases* 7: 1–12.

Grais, R. F., J. H. Ellis, A. Tress, and G. E. Glass. 2004. Modelling the spread of annual influenza epidemics in the U.S.: The potential role of air travel. *Health Care Management Science* 7: 127–134.

Gregor, C. B., et al., eds. 1988. *Chemical Cycles in the Evolution of the Earth*. New York: John Wiley.

Grieve, R. A. F. 1987. Terrestrial impact structures. *Annual Reviews of Earth and Planetary Science* 15: 245–270.

Grimond, J. 2002. What ails Japan? *Economist*, October 6. <http://www.economist.com/surveys/displaystory.cfm?story_id=1076723>.

Guo, J. 2006. The Galápagos Islands kiss their goat problem goodbye. *Science* 313: 1567.

Gust, I. D., A. W. Hampson, and D. Lavanchy. 2001. Planning for the next pandemic of influenza. *Review in Medical Virology* 11: 59–70.

Gwynne, P. 1975. The cooling world. *Newsweek*, April 29, 64.

Halmer, M. M., and H.-U. Schmincke. 2003. The impact of moderate-scale explosive eruptions on stratospheric gas injections. *Bulletin of Volcanology* 65: 433–440.

Hammond, A. 1998. *Which World? Scenarios for the 21st Century*. Washington, D.C.: Island Press.

Hansen, J., L. Nazarenko, et al. 2005. Earth's energy imbalance: Confirmation and implications. *Science* 308: 1431–1435.

Hansen, J., M. Sato, et al. 2000. Global warming in the twenty-first century: An alternative scenario. *Proceedings of the National Academy of Sciences* 97: 9875–9880.

———. 2006. Global temperature change. *Proceedings of the National Academy of Sciences* 103: 14288–14293.

Hanson, G. H. 2005. *Why Does Immigration Divide America? Public Finance and Political Opposition to Open Borders*. Washington, D.C.: Institute for International Economics.

Hardy, D. 1987. *Land and Freedom: The Origins of Russian Terrorism, 1876–1879*. New York: Greenwood Press.

Harpending, H. C., S. T. Sherry, A. R. Rogers, and M. Stoneking. 1993. The genetic structure of ancient human populations. *Current Anthropology* 34: 483–496.

Harper, S. A., et al. 2000. Prevention and Control of Influenza: Recommendations of the Advisory Committee on Immunization Practices. <http://www.cdc.gov/mmwr/preview/mmwrhtml/rr54e713a1.htm>.

Harvell, C. D., et al. 2002. Climate warming and disease risks for terrestrial and marine biota. *Science* 296: 2158–2162.

Hatfield, H. 1928. *Automaton: Or, the Future of the Mechanical Man*. London: Trench, Trubner.

Hausmann, R., and R. Sturzenegger. 2006. U.S. current account deficit is sustainable. *International Finance* 9 (2): 1–18.

Hawken, P., A. Lovins, and L. H. Lovins. 1999. *Natural Capitalism*. Boston: Little, Brown.

Hawkes, N. 1987. *Chernobyl: The End of the Nuclear Dream*. New York: Vintage.

Hayes, B. 2002. Statistics of deadly quarrels. *American Scientist* 90: 10–15.

Hegerl, G. C., T. J. Crowley, W. T. Hyde, and D. J. Frame. 2006. Climate sensitivity constrained by temperature reconstructions over the past seven centuries. *Nature* 440: 1029–1032.

Heilprin, A. 1903. *Mont Pelée and the Tragedy of Martinique: A Study of the Great Catastrophes of 1902, with Observations and Experiences in the Field*. Philadelphia: J. B. Lippincott.

Henao, J., and C. Baanante. 2006. *Agricultural Production and Soil Nutrient Mining in Africa*. Muscle Shoals, Ala.: International Fertilizer Development Center.

Henisz, W., et al. 2005. Global risks: A perspective through hurricane Katrina. Philadelphia: Wharton School, University of Pennsylvania.

Herlihy, P. 2002. *The Alcoholic Empire: Vodka and Politics in Late Imperial Russia*. New York: Oxford University Press.

Herspring, D. R., ed. 2006. *Putin's Russia: Past Imperfect, Future Uncertain*. Lanham, Md.: Rowman and Littlefield.

Hildenbrand, A., P.-Y. Gillot, and A. Bonneville. 2006. Offshore evidence for a huge landslide of the northern flank of Tahiti-Nui (French Polynesia). *Geochemistry Geophysics Geosystems* 7 (Q03006): doi:10.1029/2005GC001003.

Hillenbrand, C. 2000. *The Crusades: Islamic Perspectives*. London: Routledge.

Hills, J. H., and M. P. Goda. 1993. Fragmentation of small asteroids in the atmosphere. *Astronomical Journal* 105: 1114–1144.

Homer-Dixon, T. 2002. The rise of complex terrorism. *Foreign Policy* 128 (1): 52–62.

Hou, Y., and Y. Shu. 2006. *Strategic Preparedness for Potential Influenza Pandemic Originating in Developing Countries*. Geneva: International Risk Governance Council.

Houghton, R. A. 2003. Why are estimates of terrestrial carbon balance so different? *Global Change Biology* 9: 500–509.

Huang, Y., and T. Khanna. 2003. Can India overtake China? *Foreign Policy*, July/August, 74–78.

Hubbert, M. K. 1969. Energy resources. In *Resources and Man: A Study and Recommendations by the Committee on Resources and Man of the Division of Earth Sciences, National Academy of Sciences–National Research Council*, 157–242. San Francisco: Freeman.

Huber, P. W., and M. P. Mills. 2004. Can terrorists turn out Gotham's lights? *City Journal* Autumn 2004, 1–7.

Hudson, V. M., and A. M. den Boer. 2003. *Bare Branches: The Security Implications of Asia's Surplus Male Population*. Cambridge, Mass.: MIT Press.

Huixian, L., G. W. Housner, X. Lili, and H. Duxin. 2002. *The Great Tangshan Earthquake of 1976*. Pasadena: California Institute of Technology. <http://caltecheerl.library.caltech.edu/353/>.

Human Security Centre. 2006. *Human Security Report 2005: War and Peace in the 21st Century*. New York: Oxford University Press.

Hunt, T. L. 2006. Rethinking the fall of Easter Island. *American Scientist* 94: 412–419.

Huntington, S. P. 2004. *Who Are We: The Challenge to America's National Identity*. New York: Simon & Schuster.

Hutchison, M. M., and F. Westermann, eds. 2006. *Japan's Great Stagnation*. Cambridge, Mass.: MIT Press.

Hutton, W. 2006. *The Writing on the Wall: China and the West in the 21st Century*. Boston: Little, Brown.

ICFTU (International Confederation of Free Trade Unions). 2005. *Whose Miracle? How China's Workers Are Paying the Price for Its Economic Boom*. Brussels. <http://www.eldis.org/go/display/?id=12743&type=Document>.

IFR (International Federation of Robotics). 2005. *The World Market of Industrial Robots*. <http://www.ifr.org/statistics/keyData2005.htm>.

IISS (International Institute of Strategic Studies). 2003. *Armed Conflict Database*. <http://acd.iiss.org/>.

Imhoff, M. L., et al. 2004. Global patterns in human consumption of net primary production. *Nature* 429: 870–873.

Intel. 2003. Moore's law. <http://www.intel.com/technology/mooreslaw/index.htm>.

International Energy Agency. 2002. *Russia Energy Survey*. Paris: OECD/IEA.

International Iron and Steel Institute (IISI). 2006. *World Steel in Figures*. Brussels: IISI.

IPCC (Intergovernmental Panel on Climate Change). 1990. *Scientific Assessment of Climate Change*, ed. J. T. Houghton, G. J. Jenkins, and J. J. Ephraums. Cambridge: Cambridge University Press.

———. 1995. *Climate Change 1995: The Science of Climate Change*, ed. J. T. Houghton, L. G. Meira Filho, B. A. Callendar, et al. Cambridge: Cambridge University Press.

———. 2001. *Climate Change 2001: The Scientific Basis*, ed. J. T. Houghton, Y. Ding, D. J. Griggs et al. Cambridge: Cambridge University Press.

———. 2005. *Special Report on Carbon Dioxide Capture and Storage*, ed. B. Metz, O. Davidson, H. de Coninck, et al. Cambridge: Cambridge University Press.

———. 2007. *Climate Change 2007: The Physical Science Basis*, ed. R. B. Alley, T. Berntsen, N. L. Bindoff, et al.

IUCN (World Conservation Union). 2001. *IUCN Red List Categories and Criteria*. Geneva: IUCN.

Ivanhoe, L. F. 1995. Future world oil supplies: There is a finite limit. *World Oil* 216 (10): 77–79.

Jackson, R., and N. Howe. 2004. *The Graying of the Middle Kingdom: The Demographics and Economics of Retirement Policy in China*. Washington, D.C.: Center for Strategic and International Studies.

Jacoby, G. A. 1996. Antimicrobial-resistant pathogens in the 1990s. *Annual Review of Medicine* 47: 169–179.

Jakes, L. 2005. Security report outlines terror scenarios. ABC News, March 16.

James, A., K. Gaston, and A. Balmford. 1999. Balancing the Earth's accounts. *Nature* 401: 323–324.

Janssens, I. A., et al. 2005. The carbon budget of terrestrial ecosystems at country-scale: A European case study. *Biogeosciences* 2: 15–26.

Jerardo, A. 2004. The U.S. Ag trade balance, more than just a number. *Amber Waves* 2 (1): 36–41.

Jevons, W. S. 1865. *The Coal Question: An Inquiry Concerning the Progress of the Nation, and the Probable Exhaustion of Our Coal Mines*. London: Macmillan.

Jewitt, D. 2000. Eyes wide shut. *Nature* 403: 145–148.

Johannessen, O. M., K. Khvorostovsky, M. W. Miles, and L. P. Bobylev. 2005. Recent ice-sheet growth in the interior of Greenland. *Science* 310: 1013–1016.

Johnson, C. 2004. *The Sorrows of Empire: Militarism, Secrecy, and the End of the Republic*. New York: Metropolitan Books.

Johnson, N. P., and J. Mueller. 2002. Updating the accounts: Global mortality of the 1918–1920 "Spanish" influenza pandemic. *Bulletin of the History of Medicine* 76: 105–115.

Johnson, R. G., and B. T. McClure. 1976. A model for Northern Hemisphere continental ice sheet variation. *Quaternary Research* 6: 825–858.

Johnston, W. R. 2005. *Nuclear Weapon Milestones.* <http://www.johnstonsarchive.net/nuclear/wrjp205.html>.

Jones, E. L. 2001. China: Are comparisons odious? <http://www.chass.utoronto.ca/~brandt/Jones.pdf> (no longer accessible).

Jones, H. M. 1971. *The Age of Energy: Varieties of American Experience, 1865–1915*. New York: Viking.

Joy, B. 2000. Why the future doesn't need us. *Wired* 8.04. <http://www.wired.com/wired/archive/8.04/joy.html>.

JREI (Japan Real Estate Institute). 2006. *Monthly JREI Report*. Tokyo.

Kagitçibasi, C. 2003. Science on the rise in Turkey. *TWAS Newsletter* 15 (2): 24–27.

Kaiser, J. 2001. Galápagos takes aim at alien invaders. *Science* 293: 590–592.

———. 2006. A one-size-fits-all flu vaccine? *Science* 312: 380–382.

Kaneda, T. 2006. China's concern over population aging and health. Washington, D.C.: Population Reference Bureau. <http://www.prb.org/Articles/2006/ChinasConcernOverPopulationAgingandHealth.aspx>.

Kaplan, R. D. 2005. How we would fight China. *Atlantic Monthly* 295 (5): 49–64.

Kaye, G. D., D. A. Grant, and E. J. Emond. 1985. *Major Armed Conflict: A Compendium of Interstate and Intrastate Conflict, 1720 to 1985*. Ottawa: Department of National Defense.

Khaled, A. E. 2005. *The Great Theft: Wrestling Islam from the Extremists*. San Francisco: Harper.

Khan, A., and C. Riskin. 2001. *Inequality and Poverty in China in the Age of Globalization.* New York: Oxford University Press.

———. 2005. China's household income and its distribution, 1995 and 2002. *China Quarterly* 182: 356–384.

Khomeini, A. R. 1991. Interpretation of Surah al-Hamd. In *Light within Me*, ed. S. M. Mutahhari, A. M. H. Tabatabai, and A. R. Khomeini. Trans. M. A. Ansari. Karachi: Islamic Seminary Publications. <http://al-islam.org/LWM/>.

Kilbourne, E. D. 2006. Influenza pandemics of the 20th century. *Emerging Infectious Diseases* 12: 9–14.

Kim, N. 2005. The end of Britain? Challenges from devolution, European integration, and multiculturalism. *Journal of International and Area Studies* 12: 61–80.

Kinzer, S. 2006. *Overthrow: America's Century of Regime Change, from Hawaii to Iraq.* New York: Times Books.

Klironomos, J. N., et al. 2005. Abrupt rise in atmospheric CO_2 overestimates community responses in a model plant-soil system. *Nature* 433: 621–624.

Kohn, L. T., J. M. Corrigan, and M. S. Donaldson, eds. 2000. *To Err Is Human: Building a Safer Health System*. Washington, D.C.: National Academies Press.

Kolata, G. 1999. *Flu: The Story of the Great Influenza Pandemic of 1918 and the Search for the Virus That Caused It*. New York: Farrar, Straus, Giroux.

Komatsu, S., and K. Tani. 2006. *Nihon chinbotsu dainibu [Japan Sinks, Part II]*. Tokyo: Shogakukan.

Kotlikoff, L. J., and S. Burns. 2004. *The Coming Generational Storm: What You Need to Know About America's Economic Future*. Cambridge, Mass.: MIT Press.

Krajick, K. 2005. Winning the war against island invaders. *Science* 310: 1410–1413.

Kremen, C., N. M. Williams, and R. W. Thorp. 2002. Crop pollination from native bees at risk from agricultural intensification. *Proceedings of the National Academy of Sciences* 99(26): 16812–16816.

Kunstler, J. H. 2005. *The Long Emergency: Surviving the End of the Oil Age, Climate Change, and Other Converging Catastrophes of the Twenty-first Century*. Boston: Atlantic Monthly Press.

Kurlantzick, J. 2005. The decline of American soft power. *Current History* 104: 419–424.

Kurzweil, R. 1999. *The Age of Spiritual Machines*. New York: Viking Penguin.

———. 2005. *The Singularity Is Near: When Humans Transcend Biology*. New York: Viking Penguin.

Kuter, L. S. 1973. *The Great Gamble: The Boeing 747*. Tuscaloosa: University of Alabama Press.

Kynge, J. 2006. *China Shakes the World: A Titan's Breakneck Rise and Troubled Future, and the Challenge for America*. Boston: Houghton Mifflin.

Lacquer, W. 2001. *A History of Terrorism*. New Brunswick, N.J.: Transaction.

Laherrère, J. H. 1997. Future sources of crude oil supply and quality considerations. In *Proceedings of the DRI/McGraw-Hill/French Petroleum Institute Conference: Oil Markets over the Next Two Decades: Surplus or Shortage?* <http://www.hubbertpeak.com/laherrere/supply.htm>.

Langford, C. 2005. Did the 1918–19 influenza pandemic originate in China? *Population and Development Review* 31: 473–505.

Lardner, J., and D. A. Smith, eds. 2005. *Inequality Matters: The Growing Economic Divide in America and Its Poisonous Consequences*. New York: W.W. Norton.

Larres, K., ed. 2001. *Germany Since Unification: The Development of the Berlin Republic*. New York: Palgrave.

Lay, T., et al. 2005. The Great Sumatra-Andaman earthquake of 26 December 2004. *Science* 308: 1127–1133.

Lee, J. 2004. *Waking the Bear: Prospects for Russian Oil Production*. London: Centre for Global Energy Studies.

Leeb, M. 2004. A shot in the arm. *Nature* 431: 892–893.

Leijonhielm, J., and R. L. Larsson. 2004. *Russia's Strategic Commodities*. Stockholm: Swedish Defence Research Agency.

Leonard, M. 2004. *Why Europe Will Run the 21st Century*. New York: HarperCollins.

Levy, S. B. 1998. The challenge of antibiotic resistance. *Scientific American* 275: 46–53.

Lewis, B. 1968. *The Assassins: A Radical Sect in Islam*. New York: Basic Books.

———. 2006. August 22: Does Iran have something in store? *Wall Street Journal,* August 8.

———, ed. 1992. *The World of Islam: Faith, People, Cultures*. New York: Thames and Hudson.

Lewis, J. S. 1995. *Rain of Iron and Ice: The Very Real Threat of Comet and Asteroid Bombardment*. Reading, Mass.: Addison-Wesley.

———. 2000. *Comet and Asteroid Impact Hazards on Populated Earth: Computer Modeling*. San Diego: Academic Press.

Li, K. S., et al. 2004. Genesis of a highly pathogenic and potentially pandemic H5N1 influenza virus in eastern Asia. *Nature* 430: 209–213.

Lincoln, E. J. 2001. *Arthritic Japan: The Slow Pace of Economic Reform*. Washington, D.C.: Brookings Institution.

Linder, F. E., and R. D. Grove. 1943. *Vital Statistics Rates in the United States: 1900–1940*. Washington, D.C.: Government Printing Office.

Lipman, P. W., and D. R. Mullineaux, eds. 1981. *The 1980 Eruptions of Mount St. Helens*. Washington, D.C.: U.S. Geological Survey.

Liu, C. 1998. Environmental issues and the South-North water transfer scheme. *China Quarterly* 156: 899–910.

Lombard, A., et al. 2005. Thermosteric sea level rise for the past 50 years: Comparison with tide gauges and inference on water mass contribution. *Global and Planetary Change* 48: 303–312.

Long, S. 2006. A survey of business in India. *Economist*, June 3.

Longini, I. M., et al. 2005. Containing pandemic influenza at the source. *Science* 309: 1083–1087.

Looney, R. 2002. *Economic Costs to the United States Stemming from the 9/11 Attacks*. Monterey, Calif.: Center for Contemporary Conflict.

Lovejoy, T. E., and L. Hannah, eds. 2005. *Climate Change and Biodiversity*. New Haven: Yale University Press.

Lovelock, J. 2006. *The Revenge of Gaia: Earth's Climate Crisis and the Fate of Humanity*. New York: Basic Books.

Lowry, H. W. 2003. *The Nature of the Early Ottoman State*. Albany: State University of New York Press.

Luthcke, S. B., et al. 2006. Recent Greenland ice mass loss by drainage system from satellite gravity observations. *Science* 314: 1286–1289.

Lutz, W., V. Skirbekk, and M. R. Testa. 2005. *The Low Fertility Trap Hypothesis: Forces That May Lead to Further Postponement and Fewer Births in Europe*. Vienna: Institute of Demography of the Austrian Academy of Sciences.

MacKellar, L. 2007. Pandemic influenza: a review. *Population and Development Review* 33: 429–451.

MacKellar, L., T. Ermolieva, D. Horlacher, and L. Mayhew. 2004. *The Economic Impacts of Population Ageing in Japan*. Northampton, Mass.: Edward Elgar.

Maddison, A. 2001. *The World Economy: A Millennial Perspective*. Paris: OECD.

Mader, C. L. 1998. Modeling the Eltanin asteroid impact. *Science of Tsunami Hazards* 16 (1): 17–20.

Malik, J. 1985. *The Yields of the Hiroshima and Nagasaki Nuclear Explosions*. Los Alamos, N.M.: Los Alamos National Laboratory.

Mandelbaum, M. 2005. *The Case for Goliath: How America Acts as the World's Government in the 21st Century*. New York: Public Affairs.

———. 2006. David's friend Goliath. *Foreign Policy*, January/February, 50–56.

Manion, M. 2004. *Corruption by Design: Building Clean Government in Mainland China and Hong Kong*. Cambridge, Mass.: Harvard University Press.

Mann, C. L. 2002. Perspectives on the U.S. current account deficit and sustainability. *Journal of Economic Perspectives* 16: 131–152.

Mann, M. E., R. S. Bradley, and M. K. Hughes. 1999. Northern Hemisphere temperatures during the past millennium: Inferences, uncertainties, and limitations. *Geophysical Research Letters*, 26: 759–762.

Marchetti, C., and N. Nakićenović. 1979. *The Dynamics of Energy Systems and the Logistic Substitution Model*. Laxenburg: IIASA.

Margulies, P., ed. 2006. *The Rise of Islamic Fundamentalism*. San Diego: Thomson Gale.

Margulis, L. 1998. *Symbiotic Planet: A New Look at Evolution*. New York: Basic Books.

Marland, G., T. Boden, and R. J. Andres. 2005. *Global CO2 Emissions from Fossil-Fuel Burning, Cement Manufacture, and Gas Flaring, 1751–2002*. Oak Ridge, Tenn.: Oak Ridge National Laboratory.

Marris, E. 2006. Extreme TB strain threatens HIV victims worldwide. *Nature* 443: 131.

Marshall, P., ed. 2005. *Radical Islam's Rules: The Worldwide Spread of Extreme Shari'a Law*. Lanham, Md.: Rowman and Littlefield.

Mason, B. G., D. M. Pyle, and C. Oppenheimer. 2004. The size and frequency of the largest explosive eruptions on Earth. *Bulletin of Volcanology* 66: 735–748.

Maxwell, B. 2003. *Terrorism: A Documentary History*. Washington, D.C.: CQ Press.

Mazza, P., and R. Hammerschlag. 2004. *Carrying Energy Future: Comparing Hydrogen and Electricity for Transmission, Storage and Transportation*. Seattle, Wash.: Institute for Lifecycle Environmental Assessment.

McCaffray, S. P., and M. Melancon, eds. 2005. *Russia in the European Context, 1789–1914: A Member of the Family*. New York: Palgrave Macmillan.

McCloskey, J. M., and H. Spalding. 1989. A reconnaissance-level inventory of the amount of wilderness remaining in the world. *Ambio* 18: 221–227.

McCormick, M. P., L. W. Thomason, and C. R. Trepte. 1995. Atmospheric effects of Mt. Pinatubo eruption. *Nature* 373: 399–404.

McKee, M. 1999. Alcohol in Russia. *Alcohol and Alcoholism* 34: 824–829.

McMenamin, M. A. S., and D. L. McMenamin. 1990. *The Emergence of Animals: The Cambrian Breakthrough*. New York: Columbia University Press.

McMorrow, K. 2004. *The Economic and Financial Market Consequences of Global Ageing*. Berlin: Springer Verlag.

McNeely, J. A. 1996. The great reshuffling: How alien species help feed the global economy. In: *Proceedings of the Norway/UN Conference on Alien Species, July 1–5*, ed. O. T. Sandlund, P. J. Schei, and Å. Viken. Tondheim, Norway: Directorate for Nature Management/Norwegian Institute for Nature Research.

McSween, H. Y. 1999. *Meteorites and Their Parent Planets*. Cambridge: Cambridge University Press.

Mearsheimer, J. J. 2006. China's unpeaceful rise. *Current History* 105: 160–162.

Meehl, G. A., and C. Tebaldi. 2004. More intense, more frequent, and longer lasting heat waves in the 21st century. *Science* 305: 994–997.

MEMRI (Middle Eastern Media Research Institute). 2005a. Reformist Saudi author: Religious cassettes advocate Jihad by emphasizing martyr's sexual reward. Special Dispatch Series. No. 1032. November 23.

———. 2005b. Saudi columnist: Jihadist Salafist ideology is like Nazism. Special Dispatch Series. No. 1007. October 17.

———. 2006a. Official Saudi fatwa of July 2000 forbids construction of churches in Muslim countries; Kuwaiti MP concurs. Special Dispatch Series. No. 1123. March 24.

———. 2006b. Renowned Syrian poet "Adonis": "We, in Arab society, do not understand the meaning of freedom." Special Dispatch Series. No. 1393. December 14.

Merenkov, A. P. 1999. *Toplivno-energeticheskii kompleks Rossii: sovremennoe sostoianie i vzgliad v budushchee* (*Fuel and Energy Complex of Russia: Current Situation and Outlook for the Future*). Moskva: Nauka.

Merry, R. W. 2005. *Sands of Empire: Missionary Zeal, American Foreign Policy, and the Hazards of Global Ambition*. New York: Simon and Schuster.

Mesquida, C. G., and N. I. Wiener. 1996. Human collective aggression: A behavioral ecology perspective. *Ethology and Sociobiology* 17: 247–262.

———. 1999. Male age composition and severity of conflicts. *Politics and the Life Sciences* 18: 181–189.

Midgley, T., and A. L. Heene. 1930. Organic fluorides as refrigerants. *Industrial and Engineering Chemistry* 22: 542–545.

Milani, A. 2003. Extraterrestrial material: Virtual or real hazards? *Science* 300: 1882–1883.

Milanovic, B. 2002. *Worlds Apart: Inter-National and World Inequality, 1950–2000*. Washington, D.C.: World Bank.

Millennium Ecosystem Assessment. 2005. *Ecosystems and Human Well-being: Current State and Trends*. Washington, D.C.: Island Press.

Moaddel, M., and K. Talattof, eds. 2002. *Modernist and Fundamentalist Debates in Islam*. New York: Palgrave Macmillan.

Mokdad, A., et al. 2003. Obesity in the United States: A fresh look at its high toll. *JAMA* 289: 229–230.

Molina, M. J., and F. S. Rowland. 1974. Stratospheric sink for chlorofluoromethanes: chlorine atom catalyzed destruction of ozone. *Nature* 249: 810–812.

Monaghan, A. J., et al. 2006. Insignificant change in Antarctic snowfall since the International Geophysical Year. *Nature* 313: 827–830.

Mooney, H. A., and R. J. Hobbs. 2000. *Invasive Species in a Changing World*. Washington, D.C.: Island Press.

Moore, J. G., W. R. Normark, and R. T. Holcomb. 1994. Giant Hawaiian landslides. *Annual Review of Earth and Planetary Sciences* 22: 119–144.

Moore, P. 2006. Going nuclear. *Washington Post*, April 16.

Moravec, H. P. 1999. *Robot: Mere Machine to Transcendent Mind*. New York: Oxford University Press.

Morens, D. M., G. K. Folkers, and A. S. Fauci. 2004. The challenge of emerging and re-emerging infectious diseases. *Nature* 430: 242–249.

Morgan, M. G., B. Fischhoff, A. Bostrom, and C. J. Atman. 2002. *Risk Communication: A Mental Models Approach*. New York: Cambridge University Press.

Morgan, M. G., and M. Henrion. 1990. *Uncertainty: A Guide to Dealing with Uncertainty in Quantitative Risk and Policy Analysis*. Cambridge: Cambridge University Press.

Mörner, N., M. Tooley, and G. Possnert. 2004. New perspectives for the future of the Maldives. *Global and Planetary Change* 40 (1–2): 177–182.

Morrison, D., ed. 1992. *The Spaceguard Survey: Report of the NASA International Near-Earth-Object Detection Workshop*. Pasadena: Jet Propulsion Laboratory.

Mukherji, C. 1983. *From Graven Images: Patterns of Modern Materialism*. New York: Columbia University Press.

Murakami, H. 2001. *Underground*. New York: Vintage International.

Murray, S. 2006. Challenges of tuberculosis. *Canadian Medical Association Journal* 174 (1): 33–34.

Nabuurs, G.-J., M. J. Schelhaas, G. M. Mohren, and C. B. Field. 2003. Temporal evolution of the European forest carbon sink from 1950 to 1999. *Global Change Biology* 9: 152–160.

Nader, R. 2003. Signs of societal decay. *The Nader Page*, October 24. <http://www.nader.org/interest/102403.html>.

NASA (National Aeronautics and Space Administration). 2003. *Study to Determine the Feasibility of Extending the Search for Near-Earth Objects to Smaller Limiting Diameters*. <http://neo.jpl.nasa.gov/neo/neoreport030825.pdf>.

———. 2006. *Historic Comet Close Approaches prior to 2006.* <http://neo.jpl.nasa.gov/ca/historic_comets.html>.

———. 2007. Near-Earth Object Program. <http://neo.jpl.nasa.gov/>.

NBS (National Bureau of Statistics of China). 2006. Statistical communiqué of the People's Republic of China on the 2005 national economic and social development. Beijing.

NCC (National Counterterrorism Center). 2006. *Report on Incidents of Terrorism 2005.* <http://wits.nctc.gov/reports/crot2005nctcannexfinal.pdf>.

Nee, S. 2004. More than meets the eye. *Nature* 429: 804–805.

Nemani, R., et al. 2002. Recent trends in hydrologic balance have enhanced the terrestrial carbon sink in the United States. *Geophysical Research Letters* 29: 1468+.

Newhall, C. G., J. W. Hendley, and P. J. Stauffer. 1997. Benefits of volcano monitoring far outweigh costs: The case of Mount Pinatubo. U.S. Geological Survey Fact Sheet 115–97. <http://pubs.usgs.gov/fs/1997/fs115–97/>.

Newhall, C. G., and R. S. Punongbayan. 1996. *Fire and Mud: Eruptions and Lahars of Mount Pinatubo, Philippines.* <http://pubs.usgs.gov/pinatubo/>.

Newhall, C. G., and S. Self. 1982. The volcanic explosivity index (VEI): An estimate of explosive magnitude for historical volcanism. *Journal of Geophysical Research* 87: 1231–1238.

NIC (National Intelligence Council). 2004. *Mapping the Global Future.* <http://www.dni.gov/nic/NIC_globaltrend2020.html>.

Nippon Keidanren (Japan Business Federation). 2005. *Looking to Japan's Future: Keidanren's Perspective on Constitutional Policy Issues.* Tokyo.

NIPSSR (National Institute of Population and Social Security Research). 2002. *Population Projections for Japan: 2001–2050.* <http://www.ipss.go.jp/index-e.html>.

Noguchi, T., and T. Fujii. 2000. Minimizing the effect of natural disasters. *Japan Railway & Transport Review* 23: 52–59.

Nordhaus, W. D. 2006. Geography and macroeconomics: New data and new findings. *Proceedings of the National Academy of Sciences* 103: 3510–3517.

Nordhaus, W. D., and J. Boyer 2001. *Warming the World: Economic Models of Global Warming.* Cambridge, Mass.: MIT Press.

Norris, R. S., and H. M. Kristensen. 2006. Global nuclear stockpiles, 1945–2006. *Bulletin of the Atomic Scientists* 62 (4): 64–66.

Notzon, F. C., et al. 1998. Causes of declining life expectancy in Russia. *Journal of American Medical Association* 279: 793–800.

NRC (National Research Council). 2006a. *Status of Pollinators in North America.* Washington, D.C.: National Academies Press.

———. 2006b. *Surface Temperature Reconstructions for the Last 2,000 Years.* Washington, D.C.: National Academies Press.

NRCanada (Natural Resources Canada). 2007. Impact cratering on Earth. Earth Impact Database. Planetary and Space Science Centre, University of New Brunswick, Canada. <http://www.unb.ca/passc/ImpactDatabase/>.

NRDC (Natural Resources Defense Council). 2006. Archive of Nuclear Data. <http://www.nrdc.org/nuclear/nudb/datainx.asp>.

NTSB (National Transportation Safety Board). 2006. Aviation Accident Statistics. <http://www.ntsb.gov/aviation/aviation.htm>.

Nuclearfiles.org: Project of the Nuclear Age Peace Foundation. <http://www.nuclearfiles.org/>.

Nye, J. S. 2002. *The Paradox of American Power: Why the World's Only Superpower Can't Go It Alone*. New York: Oxford University Press.

O'Brien, K. J., and L. Li. 2006. *Rightful Resistance in Rural China*. Cambridge: Cambridge University Press.

Obstfeld, M., and K. Rogoff. 2004. *The Unsustainable US Current Account Position Revisited*. Washington, D.C.: National Bureau of Economic Research.

OECD. 2003. *PISA (Program for Student Assessment) 2003 Database*. Paris: OECD.

———. 2006. *OECD Information Technology Outlook 2006*. Paris.

O'Keefe, M. 2001. *Francophone Minorities: Assimilation and Community Vitality*. Ottawa: Department of Canadian Heritage.

Oki, T., and S. Kanae. 2006. Global hydrological cycles and world water resources. *Science* 313: 1068–1072.

Olah, G. A., A. Goeppert, and G. K. S. Prakash. 2006. *Beyond Oil and Gas: The Methanol Economy*. Weinheim: Wiley-VCH.

OMB (Office of Management and Budget). 2006. *Budget of the United States Government: FY 2007*. Washington, D.C.

Ometto, J.-P., et al. 2005. Amazonia and the modern carbon cycle: Lessons learned. *Oecologia* 143: 483–500.

Orr, J. C., et al. 2006. Anthropogenic ocean acidification over the twenty-first century and its impact on calcifying organisms. *Nature* 437: 681–686.

Ota, K. 2005. Rise in earnings inequality in Japan: A sign of bipolarization? <http://www.esri.go.jp/jp/workshop/050914/050914Ota.pdf>.

Pamuk, O. 2004. *Snow*. New York: Knopf.

Pan, Y. D., et al. 2004. New estimates of carbon storage and sequestration in China's forests: Effects of age-class and method on inventory-based carbon estimation. *Climatic Change* 6: 211–236.

Paraje, G., R. Sadana, and G. Karam. 2005. Increasing international gaps in health-related publications. *Science* 308: 959–960.

Parkins, W. E. 2006. Fusion power: Will it ever come? *Science* 311: 1380.

Parmesan, C., and G. Yohe. 2003. A globally coherent fingerprint of climate change impacts across natural systems. *Nature* 421: 37–42.

Parry, A. 2006. *Terrorism: From Robespierre to the Weather Underground*. Mineola, N.Y.: Dover Publications.

Pattan, J. N., et al. 2001. An occurrence of approximately 74 ka Youngest Toba tephra from the western continental margin of India. *Current Science* 80: 1322–1326.

Patterson, K. D., and G. F. Pyle 1991. The geography and mortality of the 1918 influenza pandemic. *Bulletin of the History of Medicine* 65: 4–21.

Pei, M. 2006a. *China's Trapped Transition: The Limits of Developmental Autocracy*. Cambridge, Mass.: Harvard University Press.

———. 2006b. The dark side of China's rise. *Foreign Policy*, March/April, 32–40.

Petit, J. R., D. Raynaud, J. Jouzel, and S. Duparcq. 1999. Climate and atmospheric history of the past 420,000 years from the Vostok ice core, Antarctica. *Nature* 399: 429–426.

Pew Research Center. 2006. *The Great Divide: How Westerners and Muslims View Each Other*. <http://pewglobal.org/reports/display.php?ReportID=253>.

Phillips, A. F. 1998. Twenty mishaps that might have started accidental nuclear war. <http://www.wagingpeace.org/articles/1998/01/00_phillips_20-mishaps.htm>.

Phillips, H., and D. Killingray, eds. 2003. *The Spanish Influenza Pandemic of 1918–19: New Perspectives*. London: Routledge.

Pickett, S. E. 2002. Japan's nuclear energy policy: From firm commitment to difficult dilemma addressing growing stocks of plutonium, program delays, domestic opposition and international pressure. *Energy Policy* 30: 1337–1355.

Pilkington, M., and R. A. F. Grieve. 1992. The geophysical signature of terrestrial impact craters. *Reviews of Geophysics* 30: 161–181.

Pirard, P., et al. 2005. Summary of the mortality impact assessment of the 2003 heat wave in France. *Eurosurveillance* 10: 153–156.

Plass, G. N. 1956. The carbon dioxide theory of climatic change. *Tellus* 8: 140–154.

Podmazo, A. A. 2003. *Bol'shaia evropeiskaia voina 1812–1815: khronika sobytii* (*The Great European War 1812–1815: The Chronicle of Events*). Moskva: ROSSPEN.

Pomeranz, K. 2001. *The Great Divergence: China, Europe, and the Making of the Modern World Economy*. Princeton, N.J.: Princeton University Press.

Pope, K. O. 2002. Impact dust not the cause of the Cretaceous-Tertiary mass extinction. *Geology* 30: 99–102.

Posner, R. A. 2004. *Catastrophe: Risk and Response*. New York: Oxford University Press.

Potter, C., et al. 2005. Variability in terrestrial carbon sinks over two decades. Part 2. Eurasia. *Global and Planetary Change* 49: 177–186.

Potts, R. 2001. Complexity and adaptability in human evolution. Paper presented at the AAAS conference Development of the Human Species and Its Adaptation to the Environment, July, Cambridge, MA.

Powell, D. E., 2002. Death as way of life: Russia's demographic decline. *Current History* 101: 344–348.

Prestowitz, C. 2005. *Three Billion New Capitalists: The Great Shift of Wealth and Power to the East*. New York: Basic Books.

PRIO (Peace Research Institute, Oslo). 2004. *UCDP/PRIO Armed Conflicts Dataset.* <http://new.prio.no/CSCW-Datasets/>.

Proctor, M., P. Yeo, and A. Lack. 1996. *The Natural History of Pollination.* Portland, Ore.: Timber Press.

Putin, V. V. 2000. Annual State of the Nation Address to the Federal Assembly, July 8.

———. 2006. Annual State of the Nation Address to the Federal Assembly, May 10.

Pyle, K. B. 2007. *Japan Rising: The Resurgence of Japanese Power and Purpose.* New York: Public Affairs.

Quadfasel, D. 2005. The Atlantic heat conveyor slows. *Nature* 438: 565–566.

Qianlong. 1793. Letter to George III. In *Annals and Memoirs of the Court of Peking.* Trans. E. Backhouse and J.P.P. Bland. London, 1914, 322–334. <http://academic.brooklyn.cuny.edu/core9/phalsall/texts/qianlong.html>.

Rabalais, N. N. 2002. Nitrogen in aquatic ecosystems. *Ambio* 31: 102–112.

Rabinowitz, D., et al. 2000. A reduced estimate of the number of kilometer-sized near-Earth asteroids. *Nature* 403: 165–166.

Raghu, S., et al. 2006. Adding biofuels to the invasive species fire? *Science* 313: 1742.

Rahman, R. D., and J. M. Andreu. 2006. *China and India: Towards Global Economic Supremacy?* New Delhi: Academic Foundation.

Rampino, M. R., and S. Self. 1992. Volcanic winter and accelerated glaciation following the Toba super-eruption. *Nature* 359: 50–52.

Rao, B. N. 1974. An overview of the fertility trends in Ontario and Quebec. *Canadian Studies in Population* 1: 37–42.

Raper, S. C. B., and R. J. Braithwaite. 2006. Low sea level rise projections from mountain glaciers and icecaps under global warming. *Nature* 439: 311–313.

Rapoport, D. C. 2001. The fourth wave: September 11 in the history of terrorism. *Current History* 100: 419–424.

Ravallion, M., and S. Chen. 2004. *China's (Uneven) Progress Against Poverty.* Washington, D.C.: World Bank.

Raynaud, D., et al. 1993. The ice record of greenhouse gases. *Science* 259: 926–934.

Rees, M. 2003. *Our Final Hour: A Scientist's Warning: How Terror, Error, and Environmental Disaster Threaten Humankind's Future in This Century—On Earth and Beyond.* New York: Basic Books.

Regoes, R. R., and S. Bonhoeffer. 2006. Emergence of drug-resistant influenza virus: Population dynamical considerations. *Science* 312: 389–391.

Reid, T. R. 2004. *The United States of Europe: The New Superpower and the End of American Supremacy.* New York: Penguin.

Renn, O. 2006. *Risk Governance: Towards an Integrative Approach.* Geneva: International Risk Governance Council.

Renne, P. R., and A. R. Basu. 1991. Rapid eruption of the Siberian Traps flood basalts at the Permo-Trisassic boundary. *Science* 253: 176–179.

Revelle, R., and H. E. Suess. 1957. Carbon dioxide exchange between atmosphere and ocean and the question of an increase of atmospheric CO_2 during the past decades. *Tellus* 9: 18–27.

Revenga, C., et al. 2000. *Pilot Analysis of Global Ecosystems: Freshwater Systems.* Washington, D.C.: World Resources Institute.

Rhodes, R. 1988. Man-made death: A neglected mortality. *Journal of American Medical Association* 260: 686–687.

Richardson, L. F. 1960. *Statistics of Deadly Quarrels.* Pacific Grove, Calif.: Boxwood Press.

Riebesell, U., et al. 2000. Reduced calcification of marine plankton in response to increased atmospheric CO_2. *Nature* 407: 364–367.

Rifkin, J. 2004. *The European Dream: How Europe's Vision of the Future Is Quietly Eclipsing the American Dream.* New York: Tarcher Penguin.

Richie, D. 2004. *Japan Journals, 1947–2004.* Berkeley, Calif.: Stone Bridge Press.

RMI (Rocky Mountain Institute). 1997. Climate protection for fun and profit. *Rocky Mountain Institute Newsletter*, Fall/Winter, 1–3.

Roberts, L. 2006. Polio eradication: Is it time to give up? *Science* 312: 832–835.

Robock, A. 1999. Volcanoes and climate. *Reviews of Geophysics* 38 (2): 191–219.

———. 2002. The climatic aftermath. *Science* 295: 1242–1244.

Robock, A., and C. Oppenheimer, eds. 2003. *Volcanism and the Earth's Atmosphere.* Washington, D.C.: American Geophysical Union.

Rodrigues, A. S. L., et al. 2004. Effectiveness of the global protected area network in representing species diversity. *Nature* 428: 640–643.

Rojstaczer, S., S. M. Sterling, and N. J. Moore. 2001. Human appropriation of photosynthesis products. *Science* 294: 2549–2551.

Rollins, A. 1983. *The Fall of Rome: A Reference Guide.* Jefferson, N.C.: McFarland & Company.

Root, T. L., et al. 2003. Fingerprints of global warming on wild animals and plants. *Nature* 421: 57–63.

Rose, W. I., and C. A. Chesner. 1990. Worldwide dispersal of ash and gases from Earth's largest known eruption, Toba, Sumatra, 75 ka. *Global and Planetary Change* 89: 269–275.

Rosegrant, M., et al. 2002. *Global Water Outlook to 2025: Averting an Impending Crisis.* Washington, D.C.: International Food Policy Research Institute.

Royal Society. 2005. *Ocean Acidification Due to Increasing Atmospheric Carbon Dioxide.* London.

Ruiz, G. M., et al. 2000. Global spread of microorganisms by ships. *Nature* 408: 49.

Rumbaut, R. G., et al. 2006. Linguistic life expectancies: Immigrant language retention in Southern California. *Population and Development Review* 32: 447–460.

Rydell, I. 2003. *Demographic Patterns from the 1960s in France, Italy, Spain and Portugal.* Stockholm: Institutet för Framtidsstudier.

Sagan, S. D. 1993. *The Limits of Safety.* Princeton, N.J.: Princeton University Press.

Sakharov, A. 1983. The danger of thermonuclear war. *Foreign Affairs* 61: 1001–1016.

Sala-i-Martin, X. 2002. *The Disturbing "Rise" of Global Income Inequality.* Cambridge, Mass.: NBER.

Samet, J. M., et al. 2000. Fine particulate air pollution and mortality in 20 U.S. cities, 1987–1994. *New England Journal of Medicine* 343 (24): 1742–1749.

Sapp, J. 1994. *Evolution by Association: A History of Symbiosis.* New York: Oxford University Press.

Saudi Committee. 2004. Saudi religious body issues fatwa against terrorism and urging public to report suspects. Press release, June 7. Royal Embassy of Saudi Arabia.

Sax, D. F., et al., eds. 2005. *Species Invasions: Insights into Ecology, Evolution and Biogeography.* Sunderland, Mass.: Sinauer.

Scavia, D., and S. B. Bricker. 2006. Coastal eutrophication in the United States. *Biogeochemistry* 79: 187–208.

Schneider, J., and P. Schneider. 2002. The Mafia and al-Qaeda: Violent and secretive organizations in comparative and historical perspective. *American Anthropologist* 104: 776–782.

Schneider, P., and J. Schneider. 2003. *Reversible Destiny: Mafia, Antimafia, and the Struggle for Palermo.* Berkeley: University of California Press.

Schneider, S., and S. Rasool. 1971. Atmospheric carbon dioxide and aerosols: Effects of large increases on global climate. *Science* 173: 138–141.

Schröter, D., et al. 2005. Ecosystem service supply and vulnerability to global change in Europe. *Science* 310: 1333–1337.

Schwartz, S. 2002. *The Two Faces of Islam: The House of Sa'ud from Tradition to Terror.* New York: Doubleday.

Schweickart, R. L., and C. R. Chapman. 2005. Better collision insurance. *American Scientist* 93: 392–394.

Seager, R. 2006. The source of Europe's mild climate. *American Scientist* 94: 334–341.

Seager, R., et al. 2002. Is the Gulf Stream responsible for Europe's mild winters? *Quarterly Journal of the Royal Meteorological Society* 128: 2563–2586.

Service, R. F. 2004. The hydrogen backlash. *Science* 305: 958–961.

Severgnini, B. 2005. Italy's squirrel syndrome. In *The World in 2006.* Economist. <http://www.economist.com/theWorldIn/europe/?d=2006>.

Shaffer, M. J., L. Ma, and S. Hansen, eds. 2001. *Modeling Carbon and Nitrogen Dynamics for Soil Management.* Boca Raton, Fla.: CRC Press.

Shahrūr, M. 1990. *al-kitāb wa al-qurān: qirā'a muāsira* [*The Book and the Qurān: A Contemporary Reading*]. Damascus: al-Ahali.

———. 1994. *dirāsat islāmīya muāsira fi al-dawla wa al-mujtama* [*Contemporary Islamic Studies on State and Society*]. Damascus: al-Ahali.

———. 2000. *Proposal for an Islamic Covenant*. Trans. D. F. Eickelman and I. S. Abu Shehadeh. Damascus: al-Ahali.

———. 2004. We Urgently Need Religious Reform. Interview by Ahmad Hissou. *Deutsche Welle/DW-World.de Newsletter*. <http://www.qantara.de/webcom/show_article.php/_c-476/_nr-226/i.html?PHPSESSID=/>.

———. 2005. A call for reformation. *Current History* 104: 39–43.

Sharpton, V. L., et al. 1993. Chicxulub multiring impact structure: Size and other characteristics derived from gravity analysis. *Science* 261: 1564–1567.

Shell Group. 2006. *Global Scenarios to 2025*. <http://www.shell.com/scenarios>.

Shenkar, O. 2005. *The Chinese Century: The Rising Chinese Economy and Its Impact on the Global Economy, the Balance of Power, and Your Job*. Philadelphia, Pa.: Wharton School Publishing.

Shiklomanov, I. A., and J. L. Rodda. 2003. *World Water Resources at the Beginning of the Twenty-first Century*. New York: Cambridge University Press.

Sidahmed, A. S., and A. Ehteshami, eds. 1996. *Islamic Fundamentalism*. Boulder, Colo.: Westview Press.

Siegel, J. J. 2005. *The Future for Investors*. New York: Crown Business.

Siegenthaler, U., et al. 2005. Stable carbon cycle–climate relationship during the late Pleistocene. *Science* 310: 1313–1317.

Sigurdsson, H., and S. Carey. 1989. Plinian and co-ignimbrite tephra fall from the 1815 eruption of Tambora volcano. *Bulletin of Volcanology* 51: 243–270.

Sigurdsson, H., S. Carey, W. Cornell, and T. Pescatore. 1985. The eruption of Vesuvius in A.D. 79. *National Geographic Research* 1: 332–387.

Silvers, R. B., and B. Epstein, eds. 2002. *Striking Terror: America's New War*. New York: New York Review of Books.

Simkin, T. 1993. Terrestrial volcanism in space and time. *Annual Review of Earth and Planetary Sciences* 21: 427–452.

Simkin, T., L. Siebert, and R. Blong. 2001. Volcano fatalities: Lessons from the historical record. *Science* 291: 255.

Simpson, G. G. 1983. *Fossils and the History of Life*. New York: Scientific American Library.

Sims, L. D., et al. 2002. Avian influenza in Hong Kong 1997–2002. *Avian Diseases* 47: 832–838.

Sinclair, A. 2003. *An Anatomy of Terror: A History of Terrorism*. London: Macmillan.

Singer, J. D., and M. Small. 1972. *The Wages of War, 1816–1965: A Statistical Handbook*. New York: Wiley.

Skerry, P. 2006. How not to build a fence. *Foreign Policy*, September/October, 64–67.

Slovic, P. 1987. Perception of risk. *Science* 236: 280–285.

———. 2000. *The Perception of Risk*. London: Earthscan.

Smil, V. 1984. *The Bad Earth: Environmental Degradation in China*. Armonk, N.Y.: M. E. Sharpe.

———. 1993. *Global Ecology*. London: Routledge.

———. 1994. *Energy in World History*. Boulder, Colo.: Westview Press.

———. 1996. *Environmental Problems in China: Estimates of Economic Costs*. Honolulu, Hawaii: East-West Center.

———. 1999. China's great famine: 40 years later. *British Medical Journal* 319: 1619–1621.

———. 2000. *Feeding the World*. Cambridge, Mass.: MIT Press.

———. 2001. *Enriching the Earth*. Cambridge, Mass.: MIT Press.

———. 2002. *The Earth's Biosphere: Evolution, Dynamics, and Change*. Cambridge, Mass.: MIT Press.

———. 2003. *Energy at the Crossroads: Global Perspectives and Uncertainties*. Cambridge, Mass.: MIT Press.

———. 2004. *China's Past, China's Future*. New York: RoutledgeCurzon.

———. 2005a. *Creating the Twentieth Century: Technical Innovations of 1867–1914 and Their Lasting Impact*. New York: Oxford University Press.

———. 2005b. The next 50 years: Fatal discontinuities. *Population and Development Review* 31: 201–236.

———. 2005c. The next 50 years: Unfolding trends. *Population and Development Review* 31: 605–643.

———. 2006. *Transforming the Twentieth Century: Technical Innovations and Their Consequences*. New York: Oxford University Press.

———. 2008. *Energy in Nature and Society: General Energetics of Complex Systems*. Cambridge, Mass.: MIT Press.

Smith, D. J. 2006. Predictability and preparedness in influenza control. *Science* 312: 392–394.

Smith, P., et al. 2005. Carbon sequestration potential in European croplands has been overestimated. *Global Change Biology* 11: 2153–2163.

Smith, R. B., and L. W. Braile. 1994. The Yellowstone hotspot. *Journal of Volcanology and Geothermal Research* 61: 121–188.

Snacken, R., et al. 1999. The next influenza pandemic: Lessons from Hong Kong, 1997. *Emerging Infectious Diseases* 5: 195–203.

Soden, B. J., R. T. Wetherald, G. L. Stenchikov, and A. Robock. 2002. Global cooling after the eruption of Mount Pinatubo: A test of climate feedback by water vapor. *Science* 296: 727–733.

Solinger, D. 2004. *The Creation of a New Urban Underclass in China and Its Implications*. Irvine, Calif.: Department of Political Science, University of California.

Sparks, S., et al. 2005. *Super-eruptions: Global Effects and Future Threats*. 2d ed. London: Geological Society.

Spence, J. D. 1996. *God's Chinese Son: The Taiping Heavenly Kingdom of Hong Xiuquan*. New York: W.W. Norton.

Spengler, O. 1919. *Der Untergang des Abendlandes: Umrisse einer Morphologie der Weltgeschichte*. Munich: Beck.

Srinivasan, T. N., and D. Suresh. 2002. *Reintegrating India into the World Economy*. Washington, D.C.: Institute for International Economics.

SSB (State Statistical Bureau). 2002. *China Statistical Yearbook*. Beijing: SSB.

Statistics Bureau. 2006. *Japan Statistical Yearbook 2006*. Tokyo: Statistics Bureau.

Stainforth, D. A., et al. 2005. Uncertainty in predictions of the climate response to rising levels of greenhouse gases. *Nature* 433: 403–406.

Starobin, P. 2006. Misfit America. *Atlantic Monthly* 297 (1): 144–149.

Starr, C. 1969. Social benefit versus technological risk. *Science* 165: 1232–1238.

Starr, C., R. Rudman, and C. Whipple. 1976. Philosophical basis for risk analysis. *Annual Review of Energy* 1: 629–662.

Statistics Bureau (Japan). 2005.

———. 2006.

Stenseth, N. C., et al. 2002. Ecological effects of climate fluctuations. *Science* 297: 1292–1296.

Stephens, B. B., et al. 2007. Weak northern and strong tropical land carbon uptake from vertical profiles of atmospheric CO_2. *Science* 316: 1732–1735.

Stern, D. I. 2005. Global sulfur emissions from 1850 to 2000. *Chemosphere* 58: 163–175.

Stern, N. 2006. *Stern Review of the Economics of Climate Change*. London: HM Treasury.

Stöhr, K., and M. Esveld. 2004. Will vaccines be available for the next influenza pandemic? *Science* 306: 2195–2196.

Stokstad, E. 2007. Puzzling decline of U.S. bees linked to virus from Australia. *Science* 317: 1304–1305.

Stommel, H. M. 1948. The westward intensification of wind-driven ocean currents. *Transactions of American Geophysical Union* 29: 202–206.

Stone, R., and H. Jia. 2006. Hydroengineering: Going against the flow. *Science* 313: 1034–1037.

Stothers, R. B. 1984. The great Tambora eruption in 1815 and its aftermath. *Science* 224: 1191–1198.

———. 1996. The great dry fog of 1783. *Climatic Change* 32: 79–89.

Streusand, D. E., and H. D. Tunnell. 2006. Choosing words carefully: Language to help fight Islamic terrorism. Washington, D.C.: National Defense University Center for Strategic Communications.

Stuart, J. S. 2001. A near-Earth asteroid population estimate from the LINEAR survey. *Science* 294: 1691–1693.

Stuiver, M., and P. M. Grootes. 2000. GISP2 oxygen isotope ratios. *Quaternary Research* 53: 277–284.

Sull, D. N., and Y. Wang. 2005. *Made in China: What Western Managers Can Learn from Trailblazing Chinese Entrepreneurs*. Cambridge, Mass.: Harvard Business School.

Summers, L. H. 2004. *The U.S. Current Account Deficit and the Global Economy*. The Per Jacobsson Lecture. Washington, D.C., October 3, 2004. <http://www.perjacobsson.org/2004/100304.pdf>.

Sunstein, C. R. 2003. Terrorism and probability neglect. *Journal of Risk and Uncertainty* 26: 121–136.

Swiss Re (Swiss Reinsurance Company). 2003. *Natural Catastrophes and Reinsurance*. Zurich.

———. 2004. Natural catastrophes and man-made disasters 2003: Many fatalities, comparatively moderate insured losses. *Sigma* 1: 1–44.

———. 2005. Natural catastrophes and man-made disasters 2004: More than 300,000 fatalities, record insured losses. *Sigma* 1: 1–40.

———. 2006a. Natural catastrophes and man-made disasters 2005: High earthquake casualties, new dimension in windstorm losses. *Sigma* 2: 1–40.

———. 2006b. *The 40 Worst Catastrophes in Terms of Victims, 1970–2005*. Zurich.

Taheri, A. 1986. *Holy Terror: Inside the World of Islamic Terrorism*. Bethesda, Md.: Adler and Adler.

Taitner, J. A. 1989. *The Collapse of Complex Societies*. Cambridge: Cambridge University Press.

Tandon, R. 2005. *The Japanese Economy and the Way Forward*. New York: Palgrave Macmillan.

Tartakovsky, J. 2006. Vodka, elixir of the masses. *St. Petersburg Times*, April 18.

Taubenberger, J. K., and D. M. Morens. 2006. Influenza revisited. *Emerging Infectious Diseases* 12: 1–2.

Taubenberger, J. K., A. H. Reid, A. E. Krafft, et al. 1997. Initial genetic characterization of the 1918 "Spanish" influenza virus. *Science* 275: 1793–1796.

Taubenberger, J. K., A. H. Reid, R. M. Lourens, et al. 2005. Characterization of the 1918 influenza virus polymerase genes. *Nature* 437: 889–893.

Teller, E., et al. 1996. *Completely Automated Nuclear Reactors for Long-Term Operation. II: Toward a Concept-Level Point-Design of a High-Temperature, Gas-Cooled Central Power Station System*. Livermore, Calif.: Lawrence Livermore National Laboratory.

Teller, J. T., and L. Clayton, eds. 1983. *Glacial Lake Agassiz*. St. John's, Canada: Geological Association of Canada.

Terry, K. D., and W. H. Tucker. 1968. Biologic effects of supernovae. *Science* 159: 421–423.

Tevini M, ed. 1993. *UV-B Radiation and Ozone Depletion: Effects on Humans, Animals, Plants, Microorganisms, and Materials*. Boca Raton, Fla.: Lewis Publishers.

Thomas, C. D., et al. 2004. Extinction risk from climate change. *Nature* 427: 145–148.

Thordarson, T., and S. Self. 2003. Atmospheric and environmental effects of the 1783–1784 Laki eruption: A review and reassessment. *Journal of Geophysical Research* 108 (D1): doi:10.1029/2001JD002042.

Thordarson, T., et al. 1996. Sulfur, chlorine and fluorine degassing and atmospheric loading by the 1783–1784 AD Laki (Skaftár Fires) eruption in Iceland. *Bulletin of Volcanology* 58: 205–225.

Threlfall, E. J. 2002. Antimicrobial drug resistance in *Salmonella*: Problems and perspectives in food- and water-borne infections. *FEMS Microbiology Reviews* 26: 141–148.

Titov, V., et al. 2005. The global reach of the 26 December 2004 Sumatra tsunami. *Science* 309: 2045–2048.

Toon, O. B., et al. 1997. Environmental perturbations caused by the impacts of asteroids and comets. *Reviews of Geophysics* 35: 41–78.

Tozzi, A. E., et al. 2005. Diagnosis and management of pertussis. *Canadian Medical Association Journal* 172: 509–515.

Transparency International. 2006. Corruption Perceptions Index. <http://www.transparency.org/>.

Treydte, K. S., et al. 2006. The twentieth century was the wettest period in northern Pakistan over the past millennium. *Nature* 440: 1179–1182.

Trombley, R. B., and J.-P. Toutain. 2001. *Long-Range Volcano Eruption Forecasting Programme, ERUPTION Pro 10.2*. Geneva: Société Volcanologique Européenne.

Tucker, N. B., ed. 2005. *Dangerous Strait: The U.S.-Taiwan-China Crisis*. New York: Columbia University Press.

Tumpey, T. M., et al. 2005. Characterization of the reconstructed 1918 Spanish influenza pandemic virus. *Science* 310: 77–80.

Turco, R. P., et al. 1983. Nuclear winter: Global consequences of multiple nuclear explosions. *Science* 222: 1283–1292.

———. 1991. Nuclear winter: Physics and physical mechanisms. *Annual Review of Earth and Planetary Sciences* 19: 383–422.

UNDP (United Nations Development Programme). 2006. *Human Development Report 2006*. New York: UNDP. <http: //www.undp.org/hdr2006/>.

UNEP (United Nations Environment Programme). 1995. *Montreal Protocol on Substances That Deplete the Ozone Layer*. Nairobi.

———. 2001. Impact of climate change to cost the world $US 300 billion a year. Press release. February 3.

UNICEF. 2005. Bad water kills 4,000 children a day. <http://www.unicef.org/wes/index_23606.html>.

United Nations. 1991. *World Population Prospects 1990*. New York.

———. 2000. *Replacement Migration: Is It a Solution to Declining and Ageing Populations?* New York.

———. 2003. *Water for People, Water for Life: First United Nations World Water Development Report*. Paris: UNESCO.

———. 2005. *World Population Prospects: The 2004 Revision*. New York.

U.S. Bureau of the Census (USBC). 1975. *Historical Statistics of the United States, Colonial Times to 1970*. Washington, D.C.: U.S. Department of Commerce.

———. 2006. *U.S. International Trade in Goods and Services: Annual Revision for 2005*. <http://www.census.gov/foreign-trade/Press-Release/2005pr/final_revisions/>.

USDA (U.S. Department of Agriculture). 2006. <http: www.usda.gov/>.

USDL (U.S. Department of Labor). 2006. *Comparative Civilian Labor Force Statistics, 10 Countries, 1960–2005*. Washington, D.C.

USDOD (Department of Defense). 2004. *Military Power of the People's Republic of China*. Washington, D.C.

USDOE (U.S. Department of Energy). 1996. *Closing the Circle on the Splitting of the Atom*. Washington, D.C.

———. 2005. *Major Russian Oil and Natural Gas Pipeline Projects*. Washington, D.C.

USDOT (Department of Transport). 2006. *The Intermodal Transportation Database*. <http: //www.transtats.bts.gov/>.

USDT (U.S. Department of the Treasury). 2006. *Gross Federal Debt History*. <http://www.treas.gov/education/fact-sheets/taxes/fed-debt.shtml>.

———. 2007. *Major Foreign Holders of Treasury Securities*. <http://www.treas.gov/tic/mfh.txt>.

USGS (U.S. Geological Survey). 2005. *Steam Explosions, Earthquakes and Volcanic Eruptions: What's in Yellowstone's Future?* <http://pubs.usgs.gov/fs/2005/3024/>.

———. 2006. Zebra Mussel Page. <http://nas.er.usgs.gov/taxgroup/mollusks/zebramussel/>.

Vaughan, D. G., and R. Arthern. 2007. Why it is hard to predict the future of ice sheets? *Science* 315: 1503–1504.

Vernadsky, V. I. 1931. O biogeokhimicheskom izuchenii iavleniia zhizni. *Izvestiia AN SSSR* 6: 633–653.

Vecchi, G. A., et al. 2006. Weakening of tropical Pacific atmospheric circulation due to anthropogenic forcing. *Nature* 441: 73–76.

Velicogna, I., and J. Wahr. 2006a. Acceleration of Greenland ice mass loss in spring 2004. *Nature* 443: 329–331.

———. 2006b. Measurements of time-variable gravity show mass loss in Antarctica. *Science* 311: 1754–1758.

Vitousek, P., et al. 1986. Human appropriation of the products of photosynthesis. *BioScience* 36: 368–373.

Vodka Museum. 2006. *History of Vodka*. <http://www.vodkamuseum.ru/english/>.

Vogel, E. 1979. *Japan as Number One: Lessons for America*. New York: Harper and Row.

von Weizsäcker, E., A. B. Lovins, and L. H. Lovins. 1998. *Factor Four: Doubling Wealth, Halving Resource Use*. New ed. London: Earthscan.

Vörösmarty, C. J., et al. 2005. Geospatial indicators of emerging water stress: An application to Africa. *Ambio* 34: 230–236.

WAES (Workshop on Alternative Energy Strategies). 1977. *Energy Supply-Demand Integrations to the Year 2000*, ed. P. S. Basile. Cambridge, Mass.: MIT Press.

Walker, M. 2006. The geopolitics of sexual frustration. *Foreign Policy,* March/April, 60–61.

Walker, R., and F. W. Peters. 2002. *The Commission on the Future of the U.S. Aerospace Industry*. Washington, D.C.: www.aerospacecommission.gov

Wallerstein, I. 2002. The eagle has crash landed. *Foreign Policy*, July/August, 60–68.

Walsh, T. R., and R. A. Howe. 2002. The prevalence and mechanisms of vancomycin resistance in *Staphylococcus aureus. Annual Review of Microbiology* 56: 657–675.

Walter, K. M., et al. 2006. Methane bubbling from Siberian thaw lakes as a positive feedback to climate warming. *Nature* 443: 71–75.

Walther, G.-R., et al. 2002. Ecological responses to recent climate change. *Nature* 416: 389–395.

Ward, J. 2002. Aztlan: A warped vision of history. <http://www.propertyrightsresearch.org/aztlan.htm>.

Ward, S. N., and E. Asphaug. 2000. Asteroid impact tsunami: A probabilistic hazard assessment. *Icarus* 145: 64–78.

Ward, S. N., and S. Day. 2001. Cumbre Vieja volcano: Potential collapse and tsunami at La Palma, Canary Islands. *Geophysical Research Letters* 28: 3397–3400.

Watanabe, M. E. 1994. Pollination worries rise as honey-bees decline. *Science* 265: 1170.

Wayne, B. 1997. *The Reluctant Superpower: United States Policy in Bosnia, 1991–1995*. New York: St. Martin's Press.

WBCSD (World Business Council for Sustainable Development). 1997. Global Scenarios 2000–2050: Exploring Sustainable Development. <http://www.wbcsd.org/>.

WCD (World Commission on Dams). 2001. *Dams and Development: A New Framework for Decision-Making*. London: Earthscan.

WCI (World Coal Institute). 2006. *How Coal Is Used*. <http://www.worldcoal.org/>.

Webster, R. G. 1997. Predictions for future human influenza pandemics. *Journal of Infectious Diseases* 176: S14–S19.

Wedeman, A. 2004. Great disorder under heaven: Endemic corruption and rapid growth in contemporary China. *China Review* 4: 1–32.

Westerling, A. L., et al. 2006. Warming and earlier spring increase western U.S. forest wildfire activity. *Science* 313: 940–943.

Wheeler, J. C. 2000. *Cosmic Catastrophes*. Cambridge: Cambridge University Press.

Wheelis, M., R. Casagrande, and L. V. Madden. 2002. Biological attack on agriculture: Low-tech, high-impact bioterrorism. *BioScience* 52: 569–576.

White, M. 2003. *Historical Atlas of the Twentieth Century*. <http://users.erols.com/mwhite28/20centry.htm>.

White, N. J., and J. R. Hunter. 2006. Sea-level rise at tropical Pacific and Indian Ocean islands. *Global and Planetary Change* 53: 155–168.

White House. 2006. *National Drug Control Strategy*. Washington, D.C.

Whittaker, C. H., ed. 2003. *Russia Engages the World, 1453–1825*. Cambridge, Mass.: Harvard University Press.

WHO (World Health Organization). 2000. *World Health Report 2000*. Geneva.

———. 2002. *Antimicrobial Resistance*. Geneva.

———. 2004a. *Global Status Report on Alcohol 2004*. Geneva.

———. 2004b. *World Report on Road Traffic Injury Prevention*. Geneva.

———. 2005. *Avian Influenza: Assessing the Pandemic Threat*. Geneva.

———. 2006. *World Health Report 2006*. Geneva.

———. 2007. *Cumulative Number of Confirmed Cases of Avian Influenza (A/(H5N1) Reported to WHO*. Geneva.

Wigley, T. M. L., and D. S. Schimel, eds. 2000. *The Carbon Cycle*. Cambridge: Cambridge University Press.

Wilkinson, D. 1980. *Deadly Quarrels*. Los Angeles: University of California Press.

Wilson, D., and R. Purushothaman. 2003. Dreaming with BRICs: The path to 2050. Goldman Sachs Global Economic Paper 99. <http://www.gs.com/insight/research/reports/report6.html>.

Wilson, E. O. 2006. *The Creation: An Appeal to Save Life on Earth*. New York: W.W. Norton.

Winters, L. A., and S. Yusuf. 2007. *Dancing with Giants: China, India, and the Global Economy*. Washington, D.C.: World Bank.

Witham, C. S., and C. Oppenheimer. 2005. Mortality in England during the 1783–84 Laki Craters eruption. *Bulletin of Volcanology* 67: 15–26.

Wolfe, T. 1968. What if he is right? In *The Pump House Gang*, 119–154. New York: Farrar, Straus, Giroux.

Wolferen, K. van. 1990. *The Enigma of Japanese Power: People and Politics in a Stateless Nation*. New York: Knopf.

Wood, C. 1992. *The Bubble Economy: Japan's Extraordinary Speculative Boom of the '80s and the Dramatic Bust of the '90s*. New York: Atlantic Monthly Press.

World Audit. 2006. Political Rights Checklist, Civil Liberties Checklist. <http://www.worldaudit.org/>.

World Bank. 1997. *Clear Water, Blue Skies: China's Environment in the New Century*. Washington, D.C.

———. 2005. *Dying Too Young: Addressing Premature Mortality and Ill Health Due to Non-communicable Diseases and Injuries in the Russian Federation.* Washington, D.C.

———. 2006. *World Development Indicators 2006.* Washington, D.C.: World Bank.

———. 2007. *The 2005 International Comparison Program: Preliminary Results.* Washington, D.C.: World Bank. <http://siteresources.worldbank.org/ICPINT/Resources/summary-tables.pdf>.

WPP. 2006. *Annual Report 2005.* London: WPP.

WRI (World Resources Institute). 2000. *World Resources 2000–2001: People and Ecosystems: The Fraying Web of Life.* Washington, D.C.

———. 2006. *World Resources 2006.* Washington, D.C.

Wright, D. G. 1990. *Revolution and Terror in France, 1789–1795.* London: Longman.

Wu, Z., K. Viisainen, and E. Hemminki. 2006. Determinants of high sex ratio among newborns: A cohort study from rural Anhui province. *Reproductive Health Matters* 14: 172–180.

Wunsch, C. 2006. Abrupt climate change: An alternative view. *Quaternary Research* 65: 191–203.

Wynn, R. B., and D. G. Masson. 2004. Canary Islands landslides and tsunami generation: Can we use turbidite deposits to interpret landslide processes? In *Submarine Mass Movements and Their Consequences*, ed. J. Locat and J. Mienert, 325–332. London: Kluwer.

Yamamoto, T., et al. 1995. Emergence of tetracycline resistance due to a multiple drug resistance plasmid in *Vibrio cholerae* O139. *FEMS Immunology and Medical Microbiology* 11: 131–136.

Yamani, S. A. Z. 2000. Interview. September 5. Planet Ark. <http://www.planetark.org/dailynewsstory.cfm?newsid=8054>.

Yang, B., et al. 2002. General characteristics of temperature variation in China during the last two millennia. *Geophysical Research Letters* 29: 1–4.

Yang, D. L. 2005. China's looming labor shortage. *Far Eastern Economic Review* January/February. <http://www.feer.com/articles1/2005/0501/free/p019.html>.

Yates, T. L., et al. 2002. The ecology and evolutionary history of an emergent disease: Hantavirus pulmonary syndrome. *BioScience* 52: 989–998.

Yemelianova, G. M. 2002. *Russia and Islam: A Historical Survey.* New York: Palgrave.

Yeomans, D., S. Chesley, and P. Chodas. 2004. Possibility of an Earth impact in 2029 ruled out for asteroid 2004 MN4. Near Earth Object Program. NASA. <http://neo.jpl.nasa.gov/news/news148.html>.

Ying, S. 2004. Regime and curbing corruption. *China Review* 4: 99–128.

Yoda, T. 2001. *Millennial Japan: Rethinking the Nation in the Age of Recession.* Durham, N.C.: Duke University Press.

Zamoyski, A. 2004. *Moscow 1812: Napoleon's Fatal March.* New York: HarperCollins.

Zdorovov, I. A. 2004. *Nashe svetlo budushchee, ili, Putin na vsegda* [*Our Future Light, or, Putin Forever*]. Moscow: Integral-Inform.

Zhang, Y., and F. W. Goza. 2006. Who will care for the elderly in China? A review of the problems caused by China's one-child policy and their potential solutions. *Journal of Aging Studies* 20: 151–164.

Zhao, Z. 2006. Income inequality, unequal health care access, and mortality in China. *Population and Development Review* 32: 461–483.

Zheng, B. 2005. China's "peaceful rise" to great-power status. *Foreign Affairs* 84 (5): 18–24.

Zhou, P., and L. Leydesdorff. 2006. The emergence of China as a leading nation in science. *Research Policy* 35(1): 83–104.

Zhu, Z. 2005. Power transition and U.S.-China relations: Is war inevitable? *Journal of International and Area Studies* 12: 1–24.

Zielonka, J. 2006. *Europe as Empire: The Nature of the Enlarged European Union*. Oxford: Oxford University Press.

Zimmer, M., ed. 1997. *Germany: Phoenix in Trouble?* Edmonton, Canada: University of Alberta Press.

Zimmerman, P. D., and J. G. Lewis. 2006. The bomb in the backyard. *Foreign Policy*, November/December, 33–39.

Zwally, H. J., et al. 2005. Mass changes of the Greenland and Antarctic ice sheets and shelves and contributions to sea-level rise: 1992–2002. *Journal of Glaciology* 51: 509–527.

Zweig, D., and J. Bi. 2005. China's global hunt for energy. *Foreign Affairs* 84 (5): 25–38.

Name Index

Subject Index